THE FRONTIERS COLLECTION

THE FRONTIERS COLLECTION

Series Editors:
A.C. Elitzur M.P. Silverman J. Tuszynski R. Vaas H.D. Zeh

The books in this collection are devoted to challenging and open problems at the forefront of modern science, including related philosophical debates. In contrast to typical research monographs, however, they strive to present their topics in a manner accessible also to scientifically literate non-specialists wishing to gain insight into the deeper implications and fascinating questions involved. Taken as a whole, the series reflects the need for a fundamental and interdisciplinary approach to modern science. Furthermore, it is intended to encourage active scientists in all areas to ponder over important and perhaps controversial issues beyond their own speciality. Extending from quantum physics and relativity to entropy, consciousness and complex systems – the Frontiers Collection will inspire readers to push back the frontiers of their own knowledge.

Other Recent Titles

The Thermodynamic Machinery of Life
By M. Kurzynski

The Emerging Physics of Consciousness
Edited by J. A. Tuszynski

Weak Links
Stabilizers of Complex Systems from Proteins to Social Networks
By P. Csermely

Quantum Mechanics at the Crossroads
New Perspectives from History, Philosophy and Physics
Edited by J. Evans, A.S. Thorndike

Particle Metaphysics
A Critical Account of Subatomic Reality
By B. Falkenburg

The Physical Basis of the Direction of Time
By H.D. Zeh

Asymmetry: The Foundation of Information
By S.J. Muller

Mindful Universe
Quantum Mechanics and the Participating Observer
By H. Stapp

Decoherence and the Quantum-to-Classical Transition
By M. Schlosshauer

For a complete list of titles in The Frontiers Collection, see back of book

Alwyn C. Scott

THE NONLINEAR UNIVERSE

Chaos, Emergence, Life

With 86 Figures

 Springer

Alwyn C. Scott[†]
Tucson, USA

Series Editors:
Avshalom C. Elitzur

Bar-Ilan University,
Unit of Interdisciplinary Studies,
52900 Ramat-Gan, Israel
email: avshalom.elitzur@weizmann.ac.il

Mark P. Silverman

Department of Physics, Trinity College,
Hartford, CT 06106, USA
email: mark.silverman@trincoll.edu

Jack Tuszynski

University of Alberta,
Department of Physics, Edmonton, AB,
T6G 2J1, Canada
email: jtus@phys.ualberta.ca

Rüdiger Vaas

University of Gießen,
Center for Philosophy and Foundations of Science
35394 Gießen, Germany
email: Ruediger.Vaas@t-online.de

H. Dieter Zeh

University of Heidelberg,
Institute of Theoretical Physics,
Philosophenweg 19,
69120 Heidelberg, Germany
email: zeh@uni-heidelberg.de

Cover figure: Image courtesy of the Scientific Computing and Imaging Institute,
University of Utah (www.sci.utah.edu).

ISSN 1612-3018
ISBN 978-3-642-07057-0   e-ISBN 978-3-540-34153-6

Springer is a part of Springer Science+Business Media

springer.com

© Springer-Verlag Berlin Heidelberg 2007
Softcover reprint of the hardcover 1st edition 2007

Production: LE-TEX Jelonek, Schmidt & Vöckler GbR, Leipzig
Cover design: KünkelLopka, Werbeagentur GmbH, Heidelberg

A hydrodynamic soliton

Frontispiece: Recreation of Scott Russell's soliton on Scotland's Union Canal, 12 July 1995

Preface

It has been suggested that the big questions of science are answered – that science has entered a "twilight age" where all the important knowledge is known and only the details need mopping up. And yet, the unprecedented progress in science and technology in the twentieth century has raised questions that weren't conceived of a century ago. This book argues that, far from being nearly complete, the story of science has many more chapters, yet unwritten. With the perspective of the century's advance, it's as if we have climbed a mountain and can see just how much broader the story is.

Instead of asking how an apple falls from a tree, as Isaac Newton did in the 17th century, we can now ask: What is the fundamental nature of an apple (matter)? How does an apple (biological organism) form and grow? Whence came the breeze that blew it loose (meteorology)? What in a physical sense (synaptic firings) was the idea that Newton had, and how did it form?

A new approach to science that can answer such questions has sprung up in the past 30 years. This approach – known as *nonlinear science* – is more than a new field. Put simply, it is the recognition that throughout nature, the whole is greater than the sum of the parts. Unexpected things happen. Minute causes can explode into mighty effects. Metaphorically, a butterfly flaps its wings in the Gobi Desert, causing a tornado in Texas. Nonlinear science provides the tools to study these phenomena. It is a metascience, a tree trunk that supports and governs the organization of almost every other branch of inquiry. Like Copernicus putting the Sun at the center of the Solar System, nonlinear science is a revolution. And just as Newton's work offered a basis for scientific discoveries in the three centuries that followed, so nonlinear science will support research in the 21st century and beyond.

Yet for all of its usefulness, nonlinear science is not widely known. The public still thinks Newton's laws, and others of a similar nature, are sufficient to explain what causes a plane to crash or a cancer to grow. But this *reductionism*, as will be shown, is inadequate to deal with the more intricate questions of the 21st century.

As a physical scientist who graduated from the university in the middle of the twentieth century, I have been engaged in studies of nonlinear science over the past four decades. Such studies include applications to computers, living creatures, weather prediction, oceanography, planetary motions, the brain

– in fact, all fields of science from physics and chemistry to the biological and social sciences. Over the years I have become convinced that this new perspective is essential for every scientist working today.

Why? Consider this example. It's commonly assumed that the Earth orbits the Sun in a precisely regular way. Of course, we only know Earth's present position within a certain range of measurement error. Our intuition tells us that this small error stays small over time, implying that Earth's position was about the same in the past. However, nonlinear science shows that our intuition is wrong. Massive computations possible only within the past decade prove that the error in our knowledge of the Earth's distance from the Sun doubles about every four million years. This error is trivial over the few millennia of recorded human history, but over geological eons, it compounds into a giant uncertainty. This means *we don't know* where the Earth was in relation to the Sun half a billion years ago, when multicelled life began. Most people are unaware of this startling fact. Do scientists studying climate and evolution take it into account? Mostly no. Because of the parochial nature of the many branches of science, nonlinear science is often ignored even by those working at the cutting edges of their fields.

My aim in writing this book is to show general readers and academics how nonlinear science is applied to research, both in physical and biological sciences. Some of the outstanding questions have practical applications and are being studied. Can we observe gravitational waves and understand their dynamical nature? Is there – as Einstein believed – a nonlinear theory that incorporates quantum theory and explains elementary particles? Is Einstein's geometrical theory of gravity related to quantum theory, as string theorists believe? How did living organisms manage to emerge from the lifeless molecules of the hot chemical soup during Earth's Hadean eon, four billion years ago? To what new vistas are Internet developments in the storage, transmission and manipulation of information leading us? How does the human brain work? Where is Life headed?

Other problems are more philosophical, pushing the envelope of what is possible to know. Are predictions of all future events from present knowledge possible "in principle" – as reductionist science continues to believe – or are important aspects of dynamics *chaotic*, precluding this possibility? What is the relationship between quantum theory and chaos? Where does the above-noted "butterfly effect" leave the concept of causality? Indeed, what do we *mean* by causality? What is the fundamental nature of *emergence*, when qualitatively new entities come into being? Is it possible for there to be something "new under the sun"? Does emergence lead to new *things* or mere *epiphenomena*, which can be explained in other ways? How is the phenomenon of biological evolution to be viewed? Are chaos and emergence related? Can we comprehend Life?

Far from esoteric, all of these questions can be addressed by nonlinear science. In describing such intricate phenomena as planetary motion and the state of the weather, nonlinear science can't always give a precise answer, but it can tell us whether a precise answer exists. To paraphrase the old saying, nonlinear science gives us the insight to accept the things we cannot calculate, the ability to calculate the things we can and the wisdom to know the difference.

Written for general readers who would understand science and for university undergraduates who would become researchers in or teachers of science, the book begins with descriptions of the three fundamental facets of nonlinear science: chaos; the emergence of independent entities in energy-conserving systems; and the quite different emergence of independent entities in dissipative (nonconservative) systems. Like the legs of a milk stool, these three facets are interrelated, with more general systems, like rungs, linking them. Chapters 6 and 7 then serve as backbones, presenting applications to the physical and biological sciences.

The book concludes with one of the most pressing questions in modern science: the debate over the Newtonian notion that all effects can be reduced to simple causes (the whole is equal to the sum of its parts). Most scientists believe this, because they hold that everything we experience is based on physical matter. Although I am committed to physicalism, I show that there are many phenomena – including Life, Mind and Spirit – that cannot be described merely by the actions of atoms, molecules, genes, synapses, memes, or whatever. Thus from the mountain top of nonlinear science, we see that reductionism is invalid, and that we can study the many fascinating questions it fails to address.

Tucson, Arizona,   *Alwyn Scott*
December 2006

Alwyn C. Scott (1931–2007)

My earliest memories of my father are of his intense desire to understand the world. It permeated everything he did, read, wrote, and said. Life for Alwyn Scott was a thing to be passionately explored, and life's meaning a thing to be pursued with every waking moment. Time with my father was spent building plastic models of molecules, carrying out amateur and sometimes messy chemistry experiments with household products, and investigating the capillary attraction of the candle on our kitchen table. The greatest lesson he taught me was to approach everything with curiosity and wonder.

This hunger for understanding was apparent in Alwyn Scott from an early age. As a boy, he was fascinated by water waves and built his own Ham radio. As a doctoral candidate at MIT, he studied shock waves and nerve impulses, which eventually led him to dive into the vast study of consciousness. His contributions to nonlinear science made him an undisputed pioneer of the field. Yet with admirable humility, he always expressed an acute awareness of the complexity that confronted him in his search for answers to the questions of the universe. During our many after dinner discussions on science and philosophy, he was fond of quoting Socrates, saying "The only thing I know is that I know nothing." For these words to come from someone with such a reservoir of knowledge, not only about science, but art, politics, history, philosophy, and culture, was humbling and inspiring. To have such a man as my mentor has been invaluable.

I know that I am not alone in having been touched by Alwyn Scott's overwhelming intellectual drive. In his seventy-five years, he profoundly affected both the scientific and philosophical world views of those he worked with. It was impossible to walk into a room where Alwyn Scott was making a point without stopping and taking notice; his was clearly a mind to be reckoned with.

Completed shortly before his diagnosis of lung cancer in 2006, *The Nonlinear Universe: Chaos, Emergence, Life* embodies the same spirit of voracious curiosity that drove my father's life. In this, his final work, he once again strives to understand the dazzling complexity of the world around us, a world so many of us take for granted. Those of us who mourn the loss of such a great thinker and dear human being can find solace in the fact that his life's work and his power to inspire will live on in his writing.

New York City
August 2007

Lela Scott MacNeil

Contents

1 Introduction

I just want to know what Truth is!

Thomas Kuhn

In addition to providing us with the above epigraph [516],[1] Thomas Kuhn wrote a book on the history and philosophy of science entitled *The Structure of Scientific Revolutions* [515], which has sold over a million copies and remains in print after four and a half decades. Upon first reading this book in the early 1960s, I did not realize that the program of research that I was then embarking upon – theoretical and experimental studies of nonlinear wave motion – would soon become part of a Kuhnian revolution.

In *Structure* (as he called his book), Kuhn famously proposed that the history of science comprises two qualitatively different types of activity. First are the eras of "normal science", during which the widely-accepted models of collective understanding (he called them paradigms) are agreed upon, and the primary activity is "puzzle solving" by adepts of those paradigms [515].[2] Occasionally, these eras of normal science are "punctuated" by "revolutionary periods" of varying magnitude and theoretical importance, during which the previously accepted paradigms change rapidly, leading the scientific community to see the world in new ways. According to Kuhn, such rapid changes in scientific perspective occur in a "Gestalt-like manner", qualitatively similar to the psychological changes of perception where one sees a familiar image (like the Necker cube) in a new way, and also with the rapid development of a new species in the "punctuated evolution" of NeoDarwinism [625]. After a collective switch of perceptions, the "lexicon" that scientists use to describe reality changes – new words are coined for new concepts, and old words are assigned new meanings, often with implications for the wider culture.

An example of this phenomenon that was studied in detail by Kuhn in the 1950s is the Copernican Revolution [514], during which the European percep-

---

[1]See the reference section for these citations.

[2]The third edition of Kuhn's classic book is particularly valuable as it includes a final chapter responding to critics of the first edition.

tion of planetary motions changed from the geocentric (Ptolemaic) picture to the heliocentric view proposed in 1543 by Copernicus in his *De Revolutionibus Orbium Caelestium*. Although the heliocentric idea had been suggested by Aristarchus of Samos in the third century BCE and seems evident to us now – familiar as we are with the amazing photographs that space probes have sent back from remote corners of the solar system – heliocentrism was not obvious to astronomers of the sixteenth century. In addition to the fact that Earth does not *seem* to move, Ptolemy's formulation of Aristotle's cosmology predicted all observations of celestial motions recorded over many centuries by Chinese, Greek, Islamic and European astronomers, and a geocentric Universe is in accord with the Christian myths that had recently been so vividly described by Dante Alighieri in his classic *La Divina Commedia*.

The basic structure of the Ptolemaic Universe consisted of two primary spheres, the inner one being the stationary surface of the earth and the outer (stellar) sphere carrying the stars around us, once every day. Within the stellar sphere were seven lesser spheres – the *orbium* of Copernicus's title – inhabited by the seven moving bodies: Saturn, Jupiter, Mars, Sun, Mercury, Venus, and Moon, all of which were driven in their motions by the daily rotation of the outer (stellar) sphere. Beyond the stellar sphere there was nothing, as the Ptolemaic Universe was finite.

Because a daily rotation of the stars about a fixed Earth is the simplest way to think about some calculations, the geocentric formulation is still used to teach celestial navigation, and although we modern scientists snicker at the absurd notions of astrology – which claims that motions of the heavenly bodies can influence terrestrial events – this concept is not unreasonable in the context of Ptolemaic astronomy. Aristotle supposed that the motion of the stars causes Saturn's motion, which in turn causes Jupiter's motion, and so on, suggesting that some terrestrial phenomena could be partially influenced by the motions of the stars and planets, and in accord with this view, it is well established that the ocean tides are governed by motions of the Sun and Moon. That we now find such difficulty seeing the Universe through Ptolemaic eyes shows not how dimwitted our ancestors were but how markedly our collective view of reality has changed. And as the polemics over Kuhn's work have shown, it is difficult for some modern thinkers to accept that Ptolemaic truth seemed as valid as ours in its day [515].

Why then did Copernicus propose a heliocentric formulation of astronomy? He had long been concerned with a "problem of the planets" – when Mercury, Venus, Mars, Jupiter and Saturn appear brightest, their motions through the heavens cease and reverse directions for a while before resuming their more regular westward paths. Although the Ptolemaic astronomers could describe and predict these retrograde motions through a well-defined system of deferents, epicycles, ecliptics and equants, there is an ad hoc character of their explanations that can be avoided by assuming Earth to be a planet lying between Venus and Mars, rotating on its polar axis each day, and revolving around the Sun. Although the Ptolemaics had a credible explanation for this retrograde motion, Copernicus did not believe that they had the *correct* explanation.

Within a century of Copernicus's death, Johannes Kepler used the careful observations of his colleague Tycho Brahe – the best pre-telescopic data then available – to show that planets, including Earth, can be more simply and accurately assumed to follow elliptical orbits about the Sun. Furthermore, he observed that a line joining a planet to the Sun sweeps out equal areas in equal times, a result that became known as Kepler's law. Galileo Galilei's application of the newly-invented telescope to celestial observations then revealed the moons of Jupiter and the phases of Venus, adding further data in support of the heliocentric theory.

In addition to the religious implications of Earth no longer being located at the center of the Universe, the Copernican revolution also altered the scientific concept of motion. Aristotle's fundamental picture was that moving objects are impelled to move toward or away from the center of the Universe, whereas the corresponding Galilean picture was of isolated massive bodies ideally moving in an infinite Universe with constant speed along straight lines. Galileo assumed that such uniform motion would continue until and unless a mass is acted upon by gravity, impact, friction, or other mechanical forces. Thus the stage was set for Isaac Newton – born just a century after the death of Copernicus and within a year of Galileo's death – to propose a self-consistent dynamical model of the Universe in his *Principia Mathematica* (entitled to complement René Descartes' *Principia Philosophiae*). Being from Britain, Newton was truly standing "on the shoulders of giants" [407], as the work leading to his *Principia* comprised the efforts of an impressive international group, involving essential contributions from Greece (Aristotle and Ptolemy), Poland (Copernicus), Denmark (Brahe), Germany (Kepler), Italy (Galileo), and France (Descartes) – all member states of modern Europe.

My central claim in this book is that the concepts of nonlinear science comprise a Kuhnian revolution which will have profound implications for scientific research in the present century. As we shall see, research in nonlinear science underwent significant changes over the last three decades of the twentieth century, particularly during the 1970s. Before this decade, important ideas lay undiscovered or were not widely noted, and communications among researchers doing mathematically related work in different fields of science varied from poor to nonexistent. Nowadays, these conditions have changed dramatically. Several international conferences on nonlinear science are held every year, mixing participants from a variety of professional backgrounds to a degree that was not imagined in the 1960s. Nonlinear science centers have spread across the globe, bringing together diversely educated young researchers to collaborate on interdisciplinary activities, combining their skills in unexpected ways. Dozens of nonlinear science journals have been launched, and hosts of textbooks and monographs are now available for introductory and advanced courses in nonlinear science.

In addition to its many interesting and important applications in the physical sciences and technology, we shall see that nonlinear science offers new perspectives on biology and answers a deep question that arose in the context of Newton's mechanistic model of the Universe: What is the nature of Life?

1.1 What Is Nonlinear Science?

When asked this question at a cocktail party, I often paraphrase Aristotle, saying that nonlinear science is the study of those dynamic phenomena for which the whole differs from the sum of its parts [31] – or just claim it is the science of Life. In other words, particular effects cannot be assigned to particular causal components (as is so for linear systems) because all components interact with each other. If she does not disappear to refresh her drink, I proceed by pointing to dynamic phenomena in virtually every area of modern research that are currently being investigated under the aegis of nonlinear science, including the following:

- **Chaos.** Sensitive dependence on initial conditions or the *butterfly effect*, strange attractors, Julia and Mandelbrot sets, problematic aspects of weather prediction, executive toys, electronic circuits.
- **Turbulence.** Wakes of ships, aircraft and bullets; waterfalls; clear air turbulence; breaking waves; fibrillation dynamics of heart muscle.
- **Emergent Structures.** Chemical molecules, planets, tornadoes, rogue waves, tsunamis, lynch mobs, optical solitons, black holes, flocks of birds and schools of fish, cities, Jupiter's Great Red Spot, nerve impulses.
- **Filamentation.** Rivers, bolts of lightning, woodland paths, optical filaments, rain dripping down window panes.
- **Threshold Phenomena.** An electric wall switch, the trigger of a pistol, electronic flip-flop circuits, tipping points, the all-or-nothing behavior of a neuron.
- **Spontaneous Pattern Formation.** Fairy rings of mushrooms, the Gulf Stream, ecological domains, biological morphogenesis.
- **Phase Changes.** Freezing and boiling of liquids, the onset of superconductivity in low temperature metals, superfluidity in liquid helium, magnetization in ferromagnetic materials, polarization in ferroelectric materials.
- **Harmonic Generation.** Digital tuning of radio receivers, conversion of laser light from red to blue, symphonic music and overdriven amplifiers for rock bands.
- **Synchronization.** Coupling of pendulum clocks, mutual entrainment of electric power generators connected to a common grid, circadian rhythms, hibernation of bears, coordinated flashing of Asian fireflies.

- **Shock Waves.** Sonic booms of jet airplanes, the sound of a cannon, bow waves of a boat, sudden pile-ups in smoothly-flowing automobile traffic.
- **Hierarchical Systems.** Stock markets, the World Wide Web, economies, cities, living organisms, human cultures.
- **Psychological Phenomena.** Gestalt perceptions, anger, depression, startle reflex, love, hate, ideation.
- **Social Phenomena.** Lynch mobs, war hysteria, emergence of cultural patterns, development of natural languages.

All of these phenomena and more comprise the subject matter of nonlinear science, which is in some sense a metascience with roots reaching into widely diverse areas of modern research.[3]

In the United States, the first use of the term "nonlinear science" may have been in a 1977 letter written by Joseph Ford to his colleagues, which defined our subject and is included here as the epigraph to Chap. 5 [877].[4] This letter was historically important as it introduced Ford's *Nonlinear Science Abstracts*, an ambitious project that soon evolved into *Physica D: Nonlinear Phenomena* – the first journal devoted to nonlinear science. Since the middle of the twentieth century, of course, the adjective "nonlinear" has been employed to modify such nouns as: analysis, dynamics, mechanics, oscillations, problems, research, systems, theory, and waves – particularly in the Soviet Union [92, 746] – but Ford defined a broad and cohesive field of interrelated activities; thus it is his sense of the term "nonlinear science" that is used in this book.

A yet deeper characterization of nonlinear science recognizes that the definition of nonlinearity involves assumptions about the nature of causality. Interestingly, the concept of causality was carefully discussed by Aristotle some twenty-three centuries ago in his *Physics*, where it is asserted that [29]:

> We have to consider in how many senses *because* may answer the question *why*.

As a "rough classification of the causal determinants of things", Aristotle went on to suggest four types of cause [136].

- **Material Cause.** Material cause stems from the presence of some physical substance that is needed for a particular outcome. Following Aristotle's suggestion that bronze is an essential factor in the making of a bronze

---

[3] A fairly complete listing and description of such applications can be found in the recently published *Encyclopedia of Nonlinear Science*, which aims to make the facets of the field available to students at the undergraduate level [878].

[4] Joseph Ford (1927–1995) was both fun to be around and an inspiration to many in the early years of research in nonlinear science. Ever striving to understand the philosophical implications of chaos, Joe was often at odds with the physics community, but without his research and his encouragement of others, the revolution described in this chapter would have had an even more difficult birth.

statue, many other examples come to mind: atoms of iron are necessary to produce hemoglobin, obesity in the United States is materially caused by our overproduction of corn, water is essential for Life. At a particular level of description, a material cause may be considered as a time or space average over dynamic variables at lower levels of description, entering as a slowly varying *parameter* at a higher level of interest.

- **Formal Cause.** For some particular outcome to occur, the requisite materials must be arranged in an appropriate form. The blueprints of a house are necessary for its construction, the DNA sequence of a gene is required for synthesis of the corresponding protein, and a pianist needs the score to play a concerto. At a particular level of description, formal causes might arise from the more slowly varying values of dynamic variables at higher levels, which then enter as *boundary conditions* at the level of interest.

- **Efficient Cause.** For something to happen, there must be an *"agent* that produces the effect and starts the material on its way". Thus, a golf ball moves through the air along a certain trajectory because it was struck at a particular instant of time by the head of a properly swung club. Similarly, a radio wave is launched in response to the alternating current that is forced to flow through an antenna. Following Galileo, this is the limited sense in which physical scientists now use the term causality [136]; thus an efficient cause is usually represented by a *stimulation–response* relationship, which can be formulated as a differential equation with a *dependent variable* that responds to a *forcing term*.

- **Final Cause.** Events may come about because they are desired by some intentional organism. Thus a house is built – involving the assembly of materials, reading of plans, sawing of wood, and pounding of nails – because someone wishes to have shelter from the elements, and economic transactions are motivated by future expectations [448]. Purposive answers to the question "why?" seem problematic in the biological sciences, and they emerge as central issues in the social sciences because such phenomena don't conform to a general belief in reductionism [873,876]. As we will see in Chap. 8, final causes offer additional means for closed causal loops of dynamic activity which must be included in realistic models, leading to a class of physical systems that cannot be simulated [817].

In more modern (if not more precise) terms, Aristotle's material and formal causes are sometimes grouped together as *distal causes*, his efficient cause is called a *proximal cause*, and his final cause is either disparaged as a *teleological cause* or disregarded altogether. The present-day disdain of many scientists for final causes is a serious oversight, as an event often transpires because some living organism wills it so. The ignoring of such phenomena is rooted in Newtonian reductionism, which we will consider in the closing chapter of this book.

While these classifications may seem tidy, reality is more intricate, as Aristotle was aware [29]. Thus causes may be difficult to sort out in particular cases, with several of them often "coalescing as joint factors in the production of a single effect". Such interactions among the components of complex causes are a characteristic property of nonlinear phenomena, where distinctions among Aristotle's "joint factors" are not always easy to make. There is, for example, a subtle difference between formal and efficient causes that appears in the metaphor for Norbert Wiener's *cybernetics*: the steering mechanism of a ship [1036]. If the wheel is connected directly to the rudder (via cables), then the forces exerted by the helmsman's arms are the efficient cause of the ship's executing a change of direction. For larger vessels, however, control is established through a servomechanism in which changing the position of the wheel merely resets a pointer that indicates the desired position of the rudder. The forces that move the rudder are generated by a feedback control system (or servomechanism) that minimizes the difference between the actual and desired positions of the rudder. In this case, one might say that the position of the pointer is a formal cause of the ship's turning, with the servomotor of the control system being the efficient cause. And of course the overall direction of the ship is determined as a final cause by the intentions of the captain and his navigator.

Another example of the difference between formal and efficient causes is provided by the conditions needed to fire the neurons in our brains. If the synaptic weights and threshold are supposed to be constants, they can be viewed as formal causes of a firing event. On a longer time scale associated with learning, however, these parameters change; thus they can be considered collectively as a *weight vector* that is governed by a learning process and might be classified as efficient causes of neuron ignition [874]. Although the switchings of real neurons are far more intricate than this simple picture suggests, the point remains valid – neural activity is a nonlinear dynamic process, melding many causal factors into the overall outcome.

Finally, when a particular protein molecule is constructed within a living cell, sufficient quantities of appropriate amino acids must be available to the messenger RNA as material causes. The DNA code, determining which amino acids are to be arranged in what order, is a formal cause, and the chemical (electrostatic and valence) forces acting among the constituent atoms are efficient causes. Thus in the realms of the chemical and biological sciences, it is not surprising to find several different types of causes involved in a single nonlinear event – parameter values, boundary conditions, forcing functions, and intentions combining to influence the outcome of a particular dynamics. Can these ideas be extended to social phenomena?

Just as supercooled water, resting quietly in its fluid state, may experience the onset of a phase change during which it suddenly turns to ice, collective social phenomena can unexpectedly arise, sweeping away previous assumptions and introducing new perspectives. Examples of such "social phase changes" abound – the revolutions in eighteenth-century France and twentieth-century Russia, lynch mobs, the outbreak of war, England's collective heartbreak over the untimely death of Princess Diana, and the Copernican revolution [514], among many others. In the 1970s, I claim, something similar happened in the organization and practice of nonlinear science.

Although research in nonlinear dynamics goes back at least to Isaac Newton's successful treatment of the two-body problem of planetary motion, such activities were until recently scattered among various professional areas, with little awareness of the common mathematical and physical principles involved. Beginning around 1970, this situation changed. Those interested in nonlinear problems became increasingly aware that dynamic concepts first observed and understood in one field (population biology, for example, or flame-front propagation or nonlinear optics or planetary motion) could be useful in others (such as chemical dynamics or neuroscience or plasma stability or weather prediction). Thus research activities began to be driven more by an interest in generic types of nonlinear phenomena than by specific applications, and the concept of nonlinear science began to emerge.[5] Apart from particular applications, we shall see, there are three broad classes of nonlinear problems.

- **Low-Dimensional Chaos.** As discussed in the following chapter, an important discovery of nonlinear science is that one cannot – not even "in principle" – predict the behaviors of certain very simple dynamical systems. Due to a phenomenon now popularly known as the butterfly effect, systems with as few as three dependent variables can exhibit "sensitive dependence on initial conditions". Errors in such systems grow exponentially with time, which renders predictions of future behaviors mathematically impossible beyond a certain characteristic (Lyapunov) time.
- **Solitons.** In energy-conserving nonlinear fields, it is often observed that energy draws itself together into localized "lumps", becoming particle-like entities (new "things") that remain organized in the subsequent course of the dynamics. For an example see the frontispiece, where a hydrodynamic soliton has been generated on a Scottish canal by suddenly stopping a motorboat, whereupon a soliton emerges from the bow wave. Similar examples arise in optics, acoustics, electromagnetics, and theories of elementary particles, among other nonlinear dynamical systems.
- **Reaction-Diffusion Waves.** Since the middle of the nineteenth century, it has been known that localized waves of activity travel along nerve fibers, carrying signals along motor nerves to our muscles and from one neuron to another within our brains. As nerve fibers do not conserve energy, their dynamics are characterized by an interplay between the release of stored electrostatic energy and its consumption through dissipative processes

---

[5]This development has recently been noted by Rowena Ball in her introduction to *Nonlinear Dynamics: From Lasers to Butterflies* [50], where she points out that studies in nonlinear science are now often driven by new ideas generated within the field rather than merely responding to national needs like research on cancer, plasma confinement, or weapons technology.

(circulating ionic currents). To grasp this phenomenon, think of a candle where chemical energy is stored in the unburned wax and released by a moving flame at the same rate that it is dissipated by radiation of heat and light. Thus a candle models the nonlinear processes on nerve fibers, with the flame corresponding to a nerve impulse – exemplifying a second general type of emergence, distinctly different from that of energy conserving systems.

These three types of nonlinear phenomena – low-dimensional chaos, solitons, and reaction-diffusion fronts – are of central concern in this book. "Chaos" is a familiar word of Greek origin, describing, perhaps correctly, the original character of the Universe, but it is now also used in a new sense to imply "low-dimensional chaos" in nonlinear science. The term "soliton", on the other hand, was coined in 1965 by Norman Zabusky and Martin Kruskal to indicate the particle-like properties of the solitary-wave solution of energy-conserving wave systems [1074].[6] Describing processes in which energy (or some other essential quantity) is released by the ongoing dynamics, the adjective "reaction-diffusion" is widely but not universally used; thus such phenomena are also referred to as self-excited waves or self-organizing waves. Following a coinage by Rem Khokhlov in 1974, they are also called autowaves in the Russian literature [678].

1.2 An Explosion of Activity

Although the roots of these three components of modern nonlinear science go back at least to the nineteenth century, the frequency with which they appeared in scientific publications began to grow explosively 1970, as Fig. 1.1 shows. More precisely, the curves indicate an exponential rise (or Gestalt switching to use Kuhn's metaphor) from 1970 to 1990, with a doubling time of about three years. This was followed by an apparent leveling off (or saturation) around the beginning of the present century at a rate of more than 3000 papers per year, or about eight per day – evidently a lower estimate because some nonlinear-science papers don't use the terms "chaos", "soliton" or "reaction-diffusion" in their titles or abstracts. From the perspectives of other manifestations, these curves look much like the heat emitted from a freshly lit bonfire, the onset of applause in a theater, or the initial growth of a biological population.

How are we to understand these data? What causes the early rise of the curves? Why do they saturate? How do they get started? More generally, can

---

[6]This new term took a couple of decades to work its way into English dictionaries, and it was not uncommon for manuscripts being published in the 1970s to have "soliton" replaced with "solution" by overly zealous copy-editors of physics journals. The word entered the public mind from a *Star Trek* episode of the early 1990s, and to my great relief it is now in the official Scrabble dictionary.

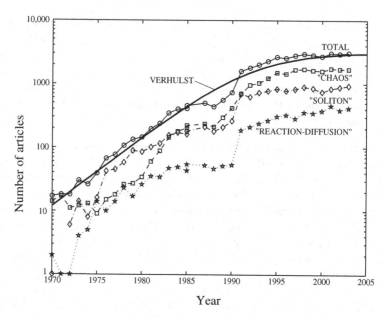

Fig. 1.1. The annual number of articles in scientific publications that have used the term CHAOS, SOLITON, and REACTION-DIFFUSION in their titles, abstracts, or key words, and the TOTAL of these three plots. (Data from the Science Citation Index Expanded.) The VERHULST curve is calculated from (1.1) to approximate the TOTAL curve with the parameters given in Table 1.1. (Note that the values of the CHAOS plot are misleadingly high before about 1975, as authors then used the term in its classical meaning)

the tools of nonlinear science help us understand a social phenomenon: the way that modern nonlinear science emerged and grew?

Before introducing a brief mathematical representation of the data in Fig. 1.1, let's consider qualitative descriptions of the above examples. To get the dynamics started, some *threshold* must be overcome: a lighted match for the bonfire, one person's burst of enthusiasm in the theater, and the presence of at least one male and one female of a biological species. Once triggered, these processes begin to grow as a result of *positive feedback* around *closed causal loops*. Thus the initial burning of the bonfire releases more heat which burns more fuel which releases even more heat, and so on, in an endless loop of causality. Similarly, the initial applause in a theater induces more enthusiasm, which elicits more applause, which induces yet more enthusiasm, etc. In the example of population growth, the early reproduction rate is proportional to the population size, which again leads to a positive feedback loop and growth at an exponentially increasing rate. Other examples, albeit with very different time scales, are the explosive increase of neutrons inside an atomic bomb and the growth of dandelions on a poorly tended lawn.

Table 1.1. Fitting parameters in (1.1)

Parameter	Symbol	Value	Units
Limiting rate	N_0	3100	papers/year
Initial rate	N_{1970}	12	papers/year
Growth exponent	λ	0.25	years^{-1}

To obtain an analytic description of this nonlinear phenomenon, note that the TOTAL curve in Fig. 1.1 can be closely fitted by the function [778, 993]

$$N(t) = \frac{N_0 N_{1970} e^{\lambda(t-1970)}}{N_0 + N_{1970}\left[e^{\lambda(t-1970)} - 1\right]}, \tag{1.1}$$

with the parameters given in Table 1.1. This function was first derived in 1845 by Pierre-François Verhulst (1804–1849) to represent the growth of biological populations [993]. He showed that (1.1) is an exact solution to the nonlinear ordinary differential equation (ODE) [993]

$$\frac{dN}{dt} = \lambda N \left(1 - \frac{N}{N_0}\right) \tag{1.2}$$

for the limited growth of a biological population. Equation (1.2) is now widely known as the logistic or Verhulst equation, and its solution is one of the early and exact results from nonlinear science. Impressively, Verhulst used (1.1) to predict the limiting population of his native Belgium to be 9 400 000, whereas the 1994 population was 10 118 000.[7]

Additional empirical evidence for an explosion of activity in nonlinear science is provided by the many introductory textbooks [2, 182, 205, 245, 248, 470, 477, 517, 521, 526, 527, 599, 711, 798, 875, 980, 1033] and advanced monographs [6, 35, 46, 50, 63, 148, 214, 299, 341, 360, 379, 383, 391, 480, 560, 701, 736, 827, 933, 962, 969, 1039, 1078] recently written in the area. Also the histogram in Fig. 1.2 shows an impressive growth in the number of journals committed to publishing research in nonlinear science over recent decades [877]. This is not the whole story, of course, because many of the traditional journals of physics, applied mathematics, theoretical biology, and engineering have carried papers on nonlinear science over the past three decades and continue

---

[7]To see how these two equations work, note that $e^{\lambda(t-1970)} = 1$ at $t = 1970$, so the right-hand side of (1.1) reduces to N_{1970} as it must. For $N_{1970} \ll N_0$ at $t = 1970$, both the second term on the right-hand side of (1.2) and the second term in the denominator of the right-hand side of (1.1) are small, indicating that the solution, $N(t)$, begins to grow exponentially as $N(t) \approx N_{1970} e^{\lambda(t-1970)}$. For $e^{\lambda(t-1970)} \gg 1$, on the other hand, the right-hand side of (1.1) is approximately equal to N_0, which makes the right-hand side of (1.2) zero, confirming that $N(t)$ is no longer increasing with time. In other words, the growth has *saturated* at $N(t) = N_0$.

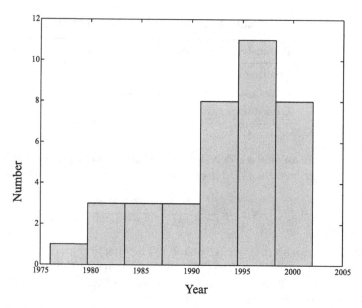

Fig. 1.2. A histogram of the number of new journals launched for publications in nonlinear science between 1980 and 2000

to do so, but the space in these new journals has contributed largely to the rising publication rate shown in Fig. 1.1, just as publication pressure has led to the new journals.

Another factor associated with the sudden growth of research activity in nonlinear science has been the worldwide emergence of interdisciplinary nonlinear science centers, of which the *Center for Studies of Nonlinear Dynamics* at the La Jolla Institute (founded in 1978), the *Santa Cruz Institute for Nonlinear Science*, which grew out of the *Santa Cruz Chaos Collective* in the early 1980s, the *Center for Nonlinear Studies* at the Los Alamos National Laboratory (founded in 1980), the *Institute for Nonlinear Science* at UCSD (founded in 1981), and the *Santa Fe Institute* (founded in 1984) were among the first. There are now dozens of such centers around the globe, dedicated to promoting interrelated studies of chaos and emergent phenomena from the fundamental perspectives of physics, chemistry, mathematics, engineering, biology, psychology, economics, and the social sciences and with emphasis upon particular areas of applied research.

Thus three types of data – publication rates of research papers, publications of books and monographs, and the launching of centers for the study of nonlinear science – all suggest that a Kuhnian revolution in this area took place in the latter decades of the twentieth century. But why did this explosion of activity begin in the 1970s? What *caused* it?

1.3 Causes of the Revolution

Following Aristotle, causes for the above-described explosion of nonlinear science activity can be formal and efficient, where formal causes set the conditions for an explosion to occur, and efficient causes lit the fuse. As is suggested by Edward Lorenz in his book *The Essence of Chaos* [576], a fundamental cause for the growth of nonlinear science research in the 1970s was the recent increase in computing power. Evidence of this dramatic increase is shown in Fig. 1.3, which plots the number of transistors on an Intel processor against time [462]. Between 1970 and 1990, the doubling time of this growth is about two years, which can be compared with Gordon Moore's 1965 estimate of a one year doubling time (referred to as Moore's law in the public press) [672].

Although many of today's "computers" are used for other tasks – word processing, record keeping, email, internet searches, and so on – the first electronic computers of the 1950s were designed to calculate solutions to partial differential equations (PDEs), which hitherto had been solved either analytically or with agonizing slowness on mechanical hand calculators. Importantly, it was on one of the very first digital computers – the vacuum-tube MANIAC at the Los Alamos National Laboratory in the early 1950s – that Enrico Fermi, John Pasta and Stan Ulam carried out the now famous FPU computations, which eventually led Zabusky and Kruskal to their numerical rediscovery of the soliton in the mid-1960s. And numerical studies of mathematical models for weather prediction on an early vacuum-tube computer (the Royal-McBee LPG-30 with a 16 KB memory) in the late 1950s led Lorenz to his unanticipated observation of low-dimensional chaos [575]. In the lexicon of nonlinear science, therefore, the steady increase in computing power shown in Fig. 1.3 is a progressive change in a *control parameter*, much like the progressive desiccation of a forest as it prepares to burn, the gradual temperature reduction in a beaker of supercooled water, or the rising level of discontent of a rebellious population.

Yet another formal cause of the explosion of activity shown in Fig. 1.1 has been the accumulation of seemingly unrelated research results on various nonlinear problems over previous decades and centuries, including nineteenth-century research in population growth, hydrodynamics, and planetary motions; mid-twentieth-century investigations of electron beam devices such as backward-wave oscillators and traveling-wave tubes, but also particle accelerators and plasma confinement machines, and nerve-impulse dynamics; and studies of the newly-invented laser and of various types of tunnel diodes in the early 1960s.

In Aristotelian terms, efficient causes of a Gestalt-like switching behavior are pressures that build up and eventually overcome the collective resistance to changes in attitude by practitioners of normal science. According to Kuhn, there are several reasons for resistance to such change. First, of course, most proposals of new theoretical perspectives turn out to be wrong (internally inconsistent, at variance with empirical observations, or both). Second, a

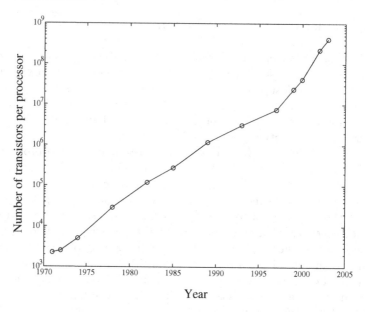

Fig. 1.3. The number of transistors on an Intel processor vs. time (data from [462])

new idea may not be wrong but fail to make new predictions, relegating the choice between competing theories to considerations of analytic convenience or taste. For a century, as we have seen, this was so for the Copernican theory of planetary motions vis-à-vis the Ptolemaic geocentric formulation [515], and physicists will recall the competition between the seemingly different quantum theories of Werner Heisenberg and Erwin Schrödinger, which are theoretically equivalent but differ in appearance and ease of application. Third, available texts and university courses limit their content to standard formulations, making it less likely for students to learn of new ideas. Finally, there is the possibility that a new theory is indeed a better representation of reality (more consistent, less ad hoc, in better agreement with experimental data, etc.), but influential leaders have vested interests in the established paradigm, motivating them to construct firm, if irrational, defenses of their traditional positions. An extreme example of such resistance was Galileo's trial and conviction by the Church of Rome as an heretic for the heliocentric views published in his *Dialogue on the Two Principal Systems of the World* [514]. The objections raised by nineteenth-century religious leaders and also by Harvard's Louis Agassiz against Charles Darwin's theory of natural selection as presented in his 1859 book *The Origin of Species* [209] provide other examples, where, although differently motivated, both religious and scientific leaders favored the previously dominant paradigm of divine creation of immutable biological species [130].[8]

Although studies of nonlinear problems have long been carried on in diverse areas of science, these efforts were largely balkanized, with little awareness of the mathematical principles relating them. Just as progressively supercooled water becomes more and more inclined to freeze (a sudden transition that can be triggered by gently shaking the liquid or dropping in a small crystal of ice), a drying forest becomes more flammable (ready to burn at the careless drop of a match or the strike of a lightning bolt) and a suppressed population becomes more restive (ready to revolt), the scattered results of nonlinear research – particularly during the nineteenth century and the first half of the twentieth century – seems to have reached an unstable level by 1970, leaving the exponential growth shown in Fig. 1.1 ready to emerge. What shook the beaker, lit the match or fired the first shot? How did the exponential growth get started?

A trigger event is also an efficient cause, several of which occurred to ignite the explosion of activity shown in Fig. 1.1. In the summer of 1966, Zabusky and Kruskal organized a NATO-supported International School of Nonlinear Mathematics and Physics at the Max Planck Institute for Physics and Astrophysics in Munich [1069]. Following general surveys by Heisenberg on nonlinear problems in physics [418] and by Ulam on nonlinear problems in mathematics, there were focused talks by Nicholas Bloembergen on nonlinear optics [93], Clifford Truesdale on nonlinear field theories in mechanics, John Wheeler on cosmology, Philip Saffman on homogeneous turbulence, Ilya Prigogine on nonequilibrium statistical mechanics, and Roald Sagdeev on nonlinear processes in plasmas, among those of several other scientists.

A three-week workshop on Nonlinear Wave Motion was organized in July of 1972 by Alan Newell, Mark Ablowitz and Harvey Segur at Clarkson College of Technology (now Clarkson University) in Potsdam, New York [700]. At this meeting, which was attended by about 60 budding nonlinear scientists of varied backgrounds from several different countries, Pasta described the hitherto perplexing FPU problem [305], and both Kruskal and Peter Lax explained the general structure of the *inverse scattering transform* (IST), among many other presentations. This IST method had recently been formulated by Kruskal and his colleagues for constructing multi-soliton solutions of the Korteweg–de Vries (KdV) equation, which describes the dynamics of shallow water waves in a one-dimensional channel (see frontispiece) [346,545]. Making connection with real-world phenomena, Joe Hammack showed that a tsunami can be viewed as a hydrodynamic soliton of the KdV equation [401], and Kruskal, among others, discussed the sine-Gordon (SG) equation, which was known to have an exact two-soliton solution and an infinite number of independent conservation laws. Importantly, the SG equation had previously

---

[8]Interestingly, some of Agassiz's arguments are still used to maintain the anti-evolutionary position in benighted subcultures of North America.

arisen in the context of dislocation dynamics in crystals [331], nonlinear optics [523], Bloch-wall dynamics in magnetic materials [294], and elementary particle theory [758].

One of the most important contributions to Newell's meeting arrived unexpectedly through the mail. A current copy of *Zhurnal Eksperimental'noi i Teoreticheskoi Fiziki* appeared, containing an article by Russian scientists Vladimir Zakharov and Alexey Shabat in which an IST formulation was developed for multi-soliton solutions of the nonlinear Schrödinger (NLS) equation [1079], a model that had independently arisen in studies of nonlinear optics both in Russia [731] and in the United states [486] during the mid-1960s. When Hermann Flaschka began to translate this unanticipated contribution at a crowded evening session, a hush fell over the room, and as it became clear that yet another nonlinear dynamical system of practical interest shared the IST properties, excitement among the participants grew. According to my notes from the subsequent discussions, Fred Tappert pointed out that the NLS equation appears in many different applications – including deep water waves [76] and plasma waves [944] in addition to nonlinear optics – because it is *generic*, arising whenever a wave packet experiences nonlinearity. Robert Miura then spoke about another equation that is closely related to KdV – thus called the modified KdV equation or MKdV. Finally, Lax explained how his operator formulation for KdV had been ingeniously applied by Zakharov and Shabat to the NLS equation. Thus many left Newell's workshop expecting that the special IST properties would generalize to an important class of integrable nonlinear wave equations in which localized lumps of energy are found to emerge. This exciting expectation was soon fulfilled in an important paper by Ablowitz, Newell, Segur, and David Kaup, which brought the KdV, MKdV, SG and NLS under the aegis of a unified theoretical picture that included several other nonlinear PDEs [4].

In the following years there were many such workshops, bringing together researchers from diverse academic backgrounds and areas of science. These new relationships and the enlarged perspectives they offered helped to fix nonlinear paradigms in the minds of recently converted acolytes. Communications between Soviet and Western nonlinear scientists were greatly improved by an exciting conference held at the Institute of Theoretical Physics in Kiev in September of 1979 with participants comprising many nonlinear researchers from both sides of the Cold War barrier. During these meetings, many scientific collaborations were begun and friendships formed which significantly influenced the future course of research in nonlinear science.

Also in 1972, a talk entitled "Predictability: Does the flap of a butterfly's wings in Brazil set off a tornado in Texas?" was presented by Lorenz at the annual meeting of the American Association for the Advancement of Science in Washington, D.C. [576]. In other words, Lorenz asked: Is the weather significantly influenced by very small causes? Although similar questions had been raised before, his metaphor for sensitive dependence of a nonlinear system

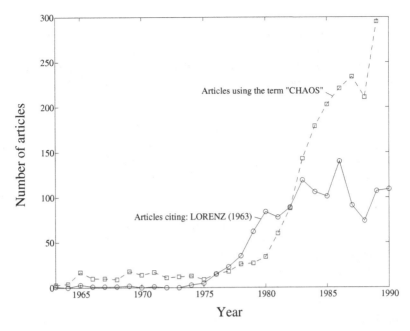

Fig. 1.4. Plots of the annual number of citations to Lorenz's 1963 paper [575] and the annual number of papers using the term "chaos". (Data from Science Citation Index Expanded)

on its initial conditions caught the imagination of the scientific world – not to mention the general public – and studies of chaos in nonlinear dynamical systems of low dimension began to take off.

Lorenz's 1972 talk was based on an article that he had written for meteorologists a decade earlier [575]. This paper presented results of numerical studies on a highly simplified model of the weather system (only three dynamic variables) which showed that solutions evolving from closely spaced initial conditions would rapidly diverge into very different non-periodic trajectories. Although the implications of Lorenz's result for weather prediction are now accepted – computer power must grow exponentially in order to achieve a linear increase in prediction time – his paper was largely ignored, accumulating only ten citations (by meteorologists) between 1963 and 1973, less than one per year and an order of magnitude less than the number of papers using the word "chaos" in its traditional sense. After 1975, however, this situation changed dramatically, as is shown by Fig. 1.4. From 1975 to the end of 2003, Lorenz's paper accumulated a total of about 3300 citations for an average of 114 per year, placing it among the most highly cited publications on nonlinear science.

As the importance of Lorenz's studies was recognized by the wider scientific community, the term "chaos" came to be used in a new sense, implying sensitive dependence of solution trajectories on their initial conditions and the unanticipated non-periodic behavior of low-dimensional systems that Lorenz had reported a decade earlier. During the rapid rise of interest in deterministic chaos after 1975, amusingly, references to Lorenz's early paper even exceeded the number of papers using the term "chaos" for several years, reflecting the fact that physical scientists were not yet comfortable with the new definition of this ancient term.

Yet another trigger for the explosive growth of nonlinear science research in the early 1970s was the interest of US applied mathematicians in reaction-diffusion problems. In 1952, the importance of these systems had been established in Britain through experimental studies of nerve impulse propagation by Alan Hodgkin and Andrew Huxley[9] [440] and contemporary theoretical work on the problem of biological morphogenesis by Alan Turing [973]. Following these leads, reaction-diffusion systems had been of interest to electrical engineers in several countries, including Denmark, Japan, Russia, and the US, throughout the 1960s. Such problems were taken up by applied mathematicians in the early 1970s after the publication of a highly-visible paper by Henry McKean at New York University [639], which soon led to others [164, 296, 499, 807]. Around this time, Charles Conley – one of the most brilliant and innovative applied mathematicians at the University of Wisconsin – began using reaction diffusion as an example of his geometrical approach to dynamics [188], and he encouraged his students to study reaction diffusion [156, 157, 472].[10] Thus by the mid-1970s, research on reaction-diffusion systems was deemed a respectable mathematical activity, leading to many of the publications counted in the lowest curve of Fig. 1.1.

Based on these ideas and events, this book is organized as follows. The next three chapters (Chaps. 2–4) give general descriptions of the three components of nonlinear science (chaos, solitons, and reaction-diffusion phenomena), followed by a chapter (Chap. 5) showing how the components are interrelated and describing general approaches to nonlinear science. Filling out the details of these general chapters are two (Chaps. 6 and 7) on applications of nonlinear concepts to physical problems and to the life sciences. A closing chapter (Chap. 8) then presents a critical evaluation of reductionism, developing a precise definition of the term "complex system" and suggesting how the realms of nonlinear science may be expanded to include the phenomena of Life.

---

[9]In 1963 Hodgkin and Huxley were awarded the Nobel Prize in medicine for this research.

[10]As I was then working on reaction-diffusion processes in the UW electrical engineering department, Conley drew me into the interdisciplinary applied mathematics community that was gathering at the Mathematics Research Center.

2 Chaos

If we knew exactly the laws of nature and the situation of the universe at the initial moment, we could predict exactly the situation of that same universe at a succeeding moment. but even if it were the case that the natural laws had no longer any secret for us, we could still only know the initial situation *approximately*. If that enabled us to predict the succeeding situation with *the same approximation*, that is all we require, and we should say that the phenomenon had been predicted, that it is governed by laws. But it is not always so; it may happen that small differences in the initial conditions produce very great ones in the final phenomena. A small error in the former will produce an enormous error in the latter. Prediction becomes impossible, and we have the fortuitous phenomenon.

Henri Poincaré

Oddly, the story of low-dimensional chaos begins with Isaac Newton's formulation of mathematical laws governing the motions of the planets and moons of the Solar System in his *Principia Mathematica* which was published in 1687. Based on calculus, which Newton also invented, these laws of motion describe how a massive body responds to forces (friction and gravitational attraction) acting on it. I write "oddly" because the motions of the planets in their elliptic orbits about the Sun – with lines from each planet to the Sun sweeping out equal areas in equal times, in accord with Kepler's law – seem to be among the most regular phenomena in our experience. The Sun rises every morning at times that were accurately predicted centuries ago, and the positions of the planets against the stellar background have been retrodicted for many centuries into the past, in agreement with earlier observations.

In addition to launching the field of celestial mechanics, Newton formulated the two-body problem (the Sun and one planet, say), showing that planetary orbits are elliptical, in accord with the heliocentric theory of Copernicus and Galileo, and that planetary motions should obey Kepler's law. Today the adjective "Newtonian" implies regularity and predictability – so how can the Solar System's mechanics have any connection with low-dimensional chaos?

2.1 The Three-Body Problem

Following Newton's geometrical solution of the two-body problem, Johann Bernoulli showed in 1710 that two masses moving through space under a mutual gravitational attraction must trace out some conic section (an ellipse, parabola, or straight line formed by the intersection of a cone with a plane), and Leonhard Euler published a complete analytic solution in 1744, seventeen years after Newton's death.[1] With the two-body problem completely solved in terms of computable functions, therefore, it was natural for eighteenth-century physical scientists to turn their collective attention to the three-body problem, and more generally to the N-body problem with $N > 2$, and this they did, fully expecting these problems also to be solvable.

Applied to the motions of the Sun, Earth, and Moon, the three-body problem is of great practical interest as it offers a basic theory of our ocean tides, which a maritime power like Britain could not ignore. Briefly, the three-body (or N-body) problem can be stated as follows [234]:

> Given a system of three (or arbitrarily many) mass points that attract each other according to an inverse-square law and assuming that these points never collide, find a representation of each point as a series in a variable that is some known function of time and for which all the series converge uniformly (i.e., for all times in the past and future).

To understand the language here, note first that an inverse-square law says that two masses m_1 and m_2 separated by a distance r mutually attract with a force equal to

$$G\frac{m_1 m_2}{r^2},$$

where G is a universal constant for gravitational attraction. Because heavenly bodies *do* occasionally collide, second, the assumption that this does not occur is a necessary idealization to find exactly formulated solutions. Finally, a convergent series comprising a possibly infinite number of computable terms is the most general means that mathematicians have for defining smooth functions.

Beyond its central importance for planetary motions, interestingly, the N-body problem also arises in atomic physics. As the electrostatic forces between atomic particles with negative and positive charges $-q_1$ and $+q_2$ mutually attract with a force equal to

$$\varepsilon_0 \frac{q_1 q_2}{r^2},$$

---

[1] An excellent brief history of the early years of celestial mechanics is given by Florin Diacu and Philip Holmes in their recent book *Celestial Encounters: The Origins of Chaos and Stability* [234].

(where r is their distance of separation and ε_0 is the dielectric permittivity of the vacuum), electrons and atomic nuclei also obey inverse-square laws, and the N-body problem appears in attempts to find classical solutions to atomic structures. As we will see in a later chapter, such classical solutions to problems of atomic dynamics are important, as they provide bases for constructing corresponding quantum-mechanical solutions [844]. Thus the hydrogen atom is a two-body problem for the motions of one negatively charged electron and one positively charged proton, and it has been completely solved both classically and in the context of quantum mechanics. The helium atom (comprising two negatively charged electrons and a positively charged nucleus), on the other hand, is a three-body problem.

Separated by some twenty-three orders of magnitude in spatial and temporal scales, the problems of planetary and atomic motions are nonetheless closely related from a mathematical perspective, and eventually, in the words of Edmund T. Whittaker, the three-body problem became "the most celebrated of all dynamical problems" [1034].

2.2 Poincaré's Instructive Mistake

Considering both its practical importance and the fact that the corresponding two-body problem had been solved, the three-body problem was studied by the best mathematical minds of the eighteenth and nineteenth centuries, even exciting interest in the salons of Paris [234]. Although this work had led to many important results in the theory of differential equations – including perturbation theory, stability, and the definitions of functions – the three-body problem remained unsolved as the nineteenth century drew to a close and Sweden's King Oscar II approached his sixtieth birthday on 21 January 1889.

To celebrate this event, Sweden's renowned mathematician, Gösta Mittag-Leffler, suggested that his sovereign offer a prize to the mathematician who could solve the three-body problem of mechanics, as stated above. The closing date for submissions was 1 June 1888; the Award Committee comprised Mittag-Leffler, his friend and former teacher, Karl Weierstrass, and Charles Hermite; and the winner would receive a cash prize of 2 500 Swedish Crowns, with the prize-winning essay to be published in Sweden's *Acta Mathematica* of which Mittag-Leffler was then editor-in-chief.[2]

In 1885 at the age of 31, Henri Poincaré learned of King Oscar's Prize and decided to take up the challenge. On 17 May 1888 – two weeks before the closing date – he submitted a 158-page essay using novel geometrical methods

---

[2] Although the oft-repeated gossip that there is no Nobel Prize in mathematics because Mittag-Leffler had a love affair with Alfred Nobel's wife is clearly incorrect (Nobel was never married), ill-will between the two Swedish scientists may nonetheless provide an explanation for this strange lapse [970].

and claiming among other results to have found a stability proof guaranteeing existence of a solution to the three-body problem. For this essay, Poincaré was awarded the prize on 21 January 1889, but in the course of editing it for publication in *Acta Mathematica*, unfortunately, an oversight in Poincaré's stability proof was uncovered, leading him to request on 30 November 1889 that printing be stopped, pending arrival of a correction. In January of 1890, Poincaré submitted a revised paper of 270 pages, in which it was proven that the three-body problem *cannot* be solved in general because for some initial conditions solution variables can wander about in the space of their allowed values and velocities (called *phase space*) in an aimless and unpredictable manner.[3] For small uncertainties in the initial conditions, in other words, there would be no error bounds on the subsequent solution trajectory in phase space. The total cost to Poincaré for recalling the previously distributed issues of *Acta Mathematica*, resetting the revised manuscript, and reprinting it was over 3 500 Swedish Crowns, consuming his entire prize and then some [59,234].

How could Poincaré – one of the brightest and best mathematicians of the late nineteenth and early twentieth centuries – have made such an embarrassing and costly blunder? The answer, it seems to me, is that mathematicians, like all of us, tend to prove what they know to be true, and scientists of that era collectively believed that the solutions of simple nonlinear ordinary differential equations (ODEs) are predictable, at least numerically. If the initial conditions of a solution trajectory are known with sufficient accuracy, the corresponding errors in the subsequent behavior would be bounded. In other words, the series defining the solution trajectory would converge "uniformly, for all times in the past and future". In Kuhnian terms, this tenet was one of the universally accepted "paradigms" of nineteenth and early twentieth-century science. His eventual realization that the universal belief was incorrect led Poincaré to include the epigraph to this chapter in his *Science and Method*, which appeared in 1903 [764].

Although the fact that the solution to a simple nonlinear ODE system (the three-body problem, for example) can wander in an unpredictable manner through a bounded region of phase space without ever repeating itself was proven by Poincaré in the corrected version of his prize essay and clearly stated in a widely-published book in 1903, most scientists remained blissfully unaware of it. Why was this striking result not shouted from the rooftops of early twentieth-century science? Just as the pre-Copernican astronomers were not ready to accept a heliocentric planetary system, perhaps, our scientific grandparents didn't want to know how complicated reality can be. Scientists wanted to know what *could* be done, not what couldn't.

Of course, this criticism cannot be made of all scientists during the first half of the twentieth century; the better ones knew of Poincaré's revolutionary result. Albert Einstein, for example, was aware that nonlinear ODE solutions can wander unpredictably, and in 1917 he raised this as an objection to the

---

[3]For a general discussion of phase space, see Appendix A.

Bohr–Sommerfeld formulation of quantum theory which required that the lengths of closed (i.e., periodic) trajectories be an integer number of young Louis de Broglie's matter wavelengths [275]. Also the American mathematician George David Birkhoff discussed such solutions in his classic *Dynamical Systems*, which was published in 1926 [91], but he didn't *like* them. Instead of entering the theoretical forest where they led, he called them "irregular solutions" to be ignored because there would never be tabulated functions describing them.[4] Although proven to exist, these irregular solutions were not part of the lexicon of science during the first half of the twentieth century, and most students of physical science were not taught about them. Not until the late 1950s, at long last, did the earlier studies by Poincaré and Birkhoff on non-periodic solution trajectories of conservative dynamical systems start to influence those studying the behaviors of plasmas and particle accelerators and observing what appeared to be "stochastic motions" [559, 560], as an awakening began. A seminal contribution to this awakening was a paper by Boris Chirikov that provided a criterion for the onset of chaotic motion in plasma-confinement systems, which is now known as the Chirikov resonance-overlap criterion [174].

2.3 The Lorenz Attractor

An important and independent contribution to the development of modern chaos theory was Edward Lorenz's above mentioned paper, which appeared in 1963 [177,575]. In a recent book written for the general reader [576], Lorenz provides valuable insight into the circumstances surrounding the creation of this seminal work. Whereas his background and interests were in dynamical weather forecasting, Lorenz found himself in charge of a project devoted to linear statistical forecasting, exploring how the newly available digital computer could be put to meteorological use. As a key issue was to know how well such numerical tools can predict complex weather patterns, he sought a set of simple nonlinear differential equations that would mimic meteorological variations, thus providing a test example for linear statistical studies. It was necessary to find a system with non-periodic behavior, because the future of a periodic solution can be exactly predicted without limit from knowledge of the trajectory over a single cycle, and the model system had to be simple because the computers of that era weren't very powerful. After several failed attempts and some successes with more complicated models, Lorenz arrived in 1961 at the following system of ordinary differential equations (ODEs):

---

[4] Amusingly, Thomas Kuhn found Birkhoff to be a very poor teacher, unable to get simple concepts across to his students [516].

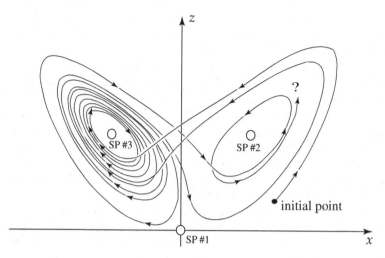

Fig. 2.1. Sketch of Lorenz's strange attractor in the phase space of (2.1) with $\sigma = 10$, $r = 28$ and $b = 8/3$. The view is looking down from the y-axis onto the (x,z)-plane. There are three singular points. SP#1 is at $(x, y, z) = (0, 0, 0)$, and SP#2,3 are at $(\pm 8.48, \pm 8.48, 27)$. The axes are not to scale

$$\frac{\mathrm{d}x}{\mathrm{d}t} = \sigma(y - x) \ ,$$

$$\frac{\mathrm{d}y}{\mathrm{d}t} = rx - y - xz \ , \qquad\qquad (2.1)$$

$$\frac{\mathrm{d}z}{\mathrm{d}t} = xy - bz \ ,$$

which now bears his name. In these equations, σ is the Prandtl number (ratio of fluid viscosity to thermal conductivity), r represents a temperature difference driving the system, and b is a geometrical factor.

With the parameter values indicated in Fig. 2.1, Lorenz found a region of phase space occupied by a solution trajectory having two wings, with the phase point jumping back and forth between nearly periodic oscillations on the wings at seemingly random intervals. Somewhat accidentally, he further discovered that this behavior requires neighboring trajectories to diverge exponentially over some regions of the phase space, a property that is described as *sensitive dependence on initial conditions* (SDIC) and can be detected by looking for one or more positive *Lyapunov exponents* (or logarithms of local rates of divergence) [1054]. In other words, if λ is a Lyapunov exponent, corresponding errors go as $e^{\lambda t}$, whence positive Lyapunov exponents indicate errors that increase with time. If time increases by $\tau \equiv 1/\lambda$ – called the *Lyapunov time* – errors increase by a factor of e. Thus even for Lorenz's simple model, a linear increase in the time over which predictions can be made requires an exponential increase in computation. To double the prediction time requires four times the computational effort and to increase prediction time by a factor of 10 requires a computer that is 100 times as large. This is the reason for the question mark in Fig. 2.1.

Although he avoided the term until the early 1980s (preferring Birkhoff's *irregularity*), Lorenz eventually defined chaos to imply such seemingly random behavior of a determinate dynamical system [576].[5]

2.4 Other Irregular Curves

The mid-1960s saw several other publications in the general theory of chaos that added to the pressure that was eventually released by the research explosion of Fig. 1.1. One of these was the horseshoe map proposed by Stephen Smale in 1967 to study SDIC in a general dynamical system [910, 1067]. In a sign of the poor interdisciplinary communications that prevailed in nonlinear science before the mid-1970s, however, the 132 references cited in Smale's paper did not include Lorenz's seminal work. Smale used a clever topological construction to show that the irregular (or non-periodic) solutions previously noted by Birkhoff and Poincaré are generic – to be anticipated and welcomed in a wider view of dynamics rather than shunned as impediments to finding analytic solutions. To this end, Smale proposed a method for distorting a two-dimensional *Poincaré surface* or *section* (see Fig. 2.4a), which intersects a higher-dimensional phase space, with points on the section indicating a passage of the solution trajectory. Stretching the section introduces SDIC, and folding it back upon itself (by bending it into the shape of a horseshoe) keeps the trajectories from escaping to infinity; thus this construction is now universally called Smale's horseshoe [1067].

Also in 1967, another seminal contribution appeared with the provocative title "How long is the coast of Britain?" by Benoit Mandelbrot [598]. Again there was no mention of Lorenz's paper – perhaps more understandably as Mandelbrot considers the structure of a static spatial curve (a coastline rather than a dynamic trajectory) that becomes infinite even though the area enclosed remains finite. A simple example of this seemingly paradoxical geometrical phenomenon is provided by the fractal "snowflake" proposed in 1904 by Niels Fabian Helge von Koch (a former student of Mittag-Leffler) and shown in Fig. 2.2a [502].[6] As n increases by one in this succession of figures, each line segment is divided into three parts which are then increased to four parts, increasing each segment of the perimeter p by a factor of 4/3. Thus the total the perimeter increases with n as

---

[5] For a film by Steve Strogatz showing several mechanical and electronic analogs of Lorenz attractors, see http://dspace.library.cornell.edu/bitstream/1813/97/3/Strogatz+Demos.mov

[6] Another example is the hierarchical set of closed Lineland universes, sketched in Fig. 6.18.

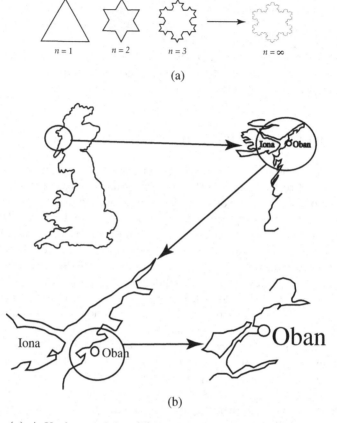

$n = 1$ $n = 2$ $n = 3$ $n = \infty$

(a)

(b)

Fig. 2.2. (a) A Koch snowflake. (b) Portions of Britain's West Coast seen on various scales

$$p = \left(\frac{4}{3}\right)^{n-1} \, ,$$

and as $n \to \infty$ the perimeter approaches infinity, whereas the enclosed area evidently remains finite. Mandelbrot showed the scientific community that such self-similar curves are often observed in nature, for example in geographical coastlines, and he went on to define a *fractional dimension* – lying between 1 for a smooth line and 2 for a smooth surface to characterize such curves.

To compute the dimension of a smooth line, it can be divided into N nonoverlapping segments, each of which is similar to the original line but reduced by the ratio $r(N) = 1/N$. Then the formula

$$D \equiv -\frac{\log N}{\log r(N)} = 1 \tag{2.2}$$

recovers the well-known dimension of a smooth line, and an appropriate generalization of this concept is readily defined for smooth surfaces.

In applying (2.2) to the Koch snowflake, observe that at each stage of the construction edges are divided into 4 parts which are reduced in size by a factor of 3. Thus $N = 4$ and $r(N) = 1/3$ so

$$D = -\frac{\log 4}{\log(1/3)} = \frac{\log 4}{\log 3} \doteq 1.261\,9 \,.$$

To show that this concept of a fractional dimension (lying between 1 for a smooth line and 2 for a smooth surface) is not a bizarre concept dreamed up by impractical mathematicians but a phenomenon often found in nature, Mandelbrot applied (2.2) to the rugged West Coast of Britain, obtaining $D \approx 1.25$, which is among the largest fractal dimension observed for oceanic coastlines. Evidently, any measurement of the length of this coastline depends on the size of the ruler being used.

Thus we see that realistic curves can be continuous but nowhere differentiable – which seems strange only because mathematicians of the past have focused our collective attention on smooth curves and surfaces. Making a connection with dynamics, such curves are exemplified by the trajectories of colloidal particles in water under Brownian motion, as was first observed by Scottish botanist Robert Brown in 1827. Mandelbrot's original paper and a lovely book [599] set the stage for research on fractal structures which are now studied under the aegis of chaos theory [519].

In a widely-cited reference on fluid turbulence published in 1971, David Ruelle and Floris Takens introduced the term "strange attractor" for an irregular solution trajectory [828] – a problem that Jerry Gollub and Harry Swinney studied experimentally [365] still without noticing Lorenz's 1963 paper. Lorenz's paper first came to the attention of the international physics community in 1974 in connection with studies by John McLaughlin and Paul Martin on computational fluid dynamics [642, 643], followed by Hermann Haken's deep analysis of nonlinear laser dynamics [388, 389], and in the seminal paper by Tien Yien Li and James Yorke entitled "Period three implies chaos" [557], which may be the first time that the term "chaos" appeared in print with its modern technical meaning.

The pace of chaos research picked up in the mid-1970s, with: (i) a paper by Otto Rössler that introduced a chaotic dynamical system having about the same form as Lorenz's (2.1) (but with only a single product nonlinearity) as a model for nonlinear chemical reactions [824], (ii) a two-variable discrete map proposed by Michel Hénon as a simplified model for a Poincaré section of Lorenz's system [424], and (iii) Robert May's discrete version of the logistic (or Verhulst) equation, which models the seasonal growth of a biological population [621]. Thus it suddenly became evident that – like fractal curves – chaotic dynamical systems of low dimension are not at all rare, but instead occur widely throughout applied science. They were present all along, but scientists weren't interested in looking for them.

Table 2.1. Behaviors of (2.3)

k	α_k	$\alpha_{k-1} < \alpha < \alpha_k$
0	1	$x_n \to 0$
1	3	$x_n \to (1 - 1/\alpha)$
2	$1 + \sqrt{6}$	$x_n = x_{n+2}$
3	3.544 090 359 6 ...	$x_n = x_{n+4}$
4	3.564 407 266 1 ...	$x_n = x_{n+8}$
5	3.568 759 419 5 ...	$x_n = x_{n+16}$
6	3.569 691 609 8 ...	$x_n = x_{n+32}$
etc.	etc.	etc.

Often called the logistic map, May's system has the deceptively simple form

$$x_{n+1} = \alpha x_n (1 - x_n) , \tag{2.3}$$

where x_n is a normalized population, n is a discrete time variable (indicating successive years in a biological application), and the growth parameter α corresponds to $\lambda + 1$ in (1.2). The basic idea is that seeds that ripen in year n germinate and reproduce in year $n + 1$.

Although (1.1) provides a complete solution to the ODE defined in (1.2), the situation for the corresponding difference-differential equation (DDE) defined in (2.3) is vastly more intricate and interesting, as the following numerical facts indicate:

- For $0 \le \alpha < 1$, equation (2.3) implies a steady-state population of zero, corresponding to $\lambda < 0$ in the ODE case of (1.2).
- For $1 < \alpha < 3$, there is a steady-state population of $1 - 1/\alpha$, much as in the ODE case, but this can also be viewed as a periodic solution with period 1.
- At $\alpha = 3$, this steady-state (or period 1) solution *bifurcates* into two solutions, each of period 2; thus $x_n = x_{n+2}$.
- At $\alpha = 1 + \sqrt{6}$, these two solutions bifurcate again into four solutions, each of period 4; thus $x_n = x_{n+4}$.
- At certain larger values of α (as given in Table 2.1), solutions bifurcate again into eight with period 8, sixteen with period 16, and so on.

But don't take my word for it; all of these facts can be readily checked on your pocket calculator.

Interestingly, there is a critical value of α

$$\alpha_\infty = 3.569\,945\,672\ldots$$

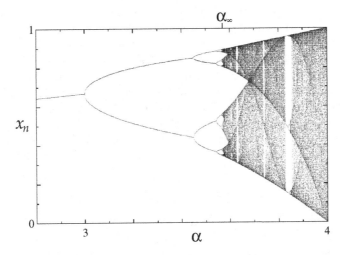

Fig. 2.3. A bifurcation diagram for solutions of (2.3). (Courtesy of Mikhail Rabinovich and Nikolai Rulkov)

beyond which the solution becomes irregular or chaotic. A bifurcation diagram for this succession of solutions is shown in Fig. 2.3, which displays one of several "routes to chaos" that are exhibited by nonlinear dynamical systems [361].

In 1975, a young physicist named Mitchell Feigenbaum observed that the limiting ratio

$$\lim_{k \to \infty} \left(\frac{\alpha_{k+1} - \alpha_k}{\alpha_{k+2} - \alpha_{k+1}} \right) \quad \longrightarrow \quad 4.669\,201\,609\,102\,990\,671\,853\ldots$$

(now called the *Feigenbaum number*) is the same for all such discrete maps with quadratic maxima, a property he termed *universality* [124, 302]. It is a reflection on the controversial nature of chaos research that persisted well into the mid-1970s that Feigenbaum struggled to get this important result into print. When he presented a paper on his work at an international nonlinear science conference that was held on Lake Como in Italy during the summer of 1977, corresponding manuscripts had been rejected by several editors over the past two years, and he was becoming discouraged [303]. Upon learning of this difficulty, Joel Lebowitz – in an example of both enlightened editorship and the growth of awareness of nonlinear science – accepted the work in his *Journal of Statistical Physics* without refereeing. Five years later, Feigenbaum was famous for having written this seminal paper [302].

For several reasons, the logistic map (2.3) is an interesting mathematical object. First, it is motivated by the real-world problem of seasonal population growth, which is observed for many biological species. In this problem, the number of seeds produced in one year depends on conditions during that year, but these seeds germinate in the following year. Second, the logistic map displays a surprisingly wide variety of dynamic behaviors under variations of the single parameter α, as Table 2.1 and Fig. 2.3 show. Finally, it is computationally simple. For these three reasons, David Campbell has referred to May's logistic map as the hydrogen atom of chaotic modeling.

In contrast to a hydrogen atom, however, all of the above-mentioned chaotic systems – Lorenz, Rössler, Hénon, and May's logistic map – are *non-conservative* or *dissipative* in the sense that an initial volume of phase space grows or shrinks with increasing time, implying that some essential quantity (Q) is being supplied and consumed in the course of the dynamics rather than being conserved.

2.5 The KAM Theorem

For systems that conserve energy and so can be derived from a Hamiltonian functional (H), on the other hand, initially defined volumes of phase space are conserved under the dynamics [559, 560, 586]. In such cases, there is a body of work collectively called the KAM theorem which qualitatively describes the onset of chaos as perturbations of an originally integrable system are increased. This perspective was developed over the 1950s and early 1960s through work by the Russian mathematician, Andrei Kolmogorov, his student-cum-colleague Vladimir Arnol'd, and Jürgen Moser from NYU's Courant Institute. (An informative account of how KAM developed is given by Diacu and Holmes [236].)

Assuming an unperturbed system to be integrable, its solutions lie on tori (see Fig. 2.4a) and can be described in terms of *action–angle variables* – for example, fixed angular momenta and corresponding angles of rotation, which increase linearly with time. When the ratios of two angular frequencies – one (ω_1) around the hole of the torus and the other (ω_2) in the transverse direction – are given by an irrational number (not a ratio of integers), the corresponding torus is said to be non-resonant. As an energy-preserving perturbation is slowly turned on, most solution trajectories continue to lie on nested, non-resonant toroidal surfaces in the phase space, which is then governed by coupled dynamics of the action–angle variables. If the perturbation (controlled by a parameter ε) destroys integrability (i.e., the perturbed system is non-degenerate), the main KAM results can be informally stated as follows [229]:

> *KAM theorem*: If an unperturbed Hamiltonian system is not degenerate, then for a sufficiently small ε most non-resonant invariant tori do not vanish but are only slightly deformed, so that in the phase space of the perturbed system there are invariant tori densely filled with conditionally periodic phase curves winding around them, with

the number of independent frequencies equal to the number of degrees of freedom. These invariant tori form a majority in the sense that the measure of the complement to their union is small when the perturbation is small.

Thus the non-resonant tori persist under perturbations, but are separated in phase space by irregular (chaotic) regions that stem from resonant tori.

We can understand both the nature of this result and the qualitative reason behind it by recalling that points on the real line are either rational numbers (expressible as the ratio of two integers, like $1/2$, $2/3$, $3/4$, $3/5$ and so on) or irrational numbers (not so expressible, like π, e, $\sqrt{2}$, $\sqrt{3}$, and so on). Although the rational numbers are dense on the real line, surprisingly, their measure is negligible compared with that of the irrational numbers. Thus as the persistent tori stem from irrational frequency ratios, their measure continues to dominate the phase space for weak to moderate values of perturbation.

In studies of planetary motion, KAM theory has applications to the weak perturbation of integrable two-body dynamics by a small third body – for example, the motion of the Earth around the Sun under perturbations from the motion of the Moon – which is one of the classical problems of nonlinear science [236]. To perform numerical studies related to KAM, Hénon and Carl Heiles proposed the relatively simple system

$$H = \frac{1}{2}(\dot{x}^2 + \dot{y}^2) + \frac{1}{2}(\omega_1^2 x^2 + \omega_2^2 y^2) + \varepsilon\left(x^2 y - \frac{1}{3}y^3\right) \tag{2.4}$$

in 1964 [425], which comprises two harmonic oscillators for x and y with frequencies ω_1 and ω_2 respectively. In a three-body application to planetary motion, ω_1 might correspond to the rotation frequency of a planet and its moon around the Sun and ω_2 might be the rotation frequency of the moon about the planet. In the Hénon–Heiles model, these two oscillations are coupled by a nonlinear term: $\varepsilon(x^2 y - y^3/3)$. With $\varepsilon = 0$ the two oscillators are uncoupled; thus by increasing ε from zero at a fixed energy, the effects of the nonlinear coupling are turned on and magnified.

As one of the simplest non-integrable, energy-conserving systems known to be chaotic, this Hénon–Heiles Hamiltonian has become a widely-used model for demonstrating the KAM theorem. Numerical studies are conveniently carried out by defining a Poincaré section (usually a plane) in the four-dimensional phase space $(\dot{x}, \dot{y}, x, y)$ and marking this surface with a dot whenever the solution trajectory goes through it in a specified direction. For small values of the perturbation (ε), solution trajectories lie on tori and the corresponding Poincaré dots lie on closed curves – see Fig. 2.4a. At moderate values of ε – see Fig. 2.4b – small regions of irregular (or chaotic) motion appear between the closed curves and these regions grow with further increases in ε [386].

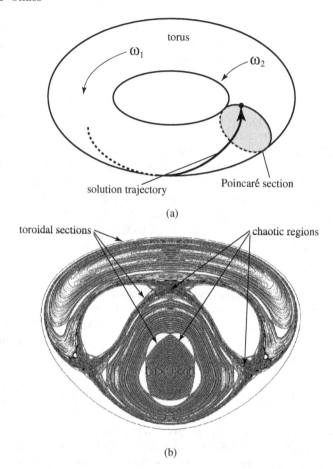

Fig. 2.4. (a) Sketch of a solution trajectory on a torus, intersecting a Poincaré section in a point. (b) A Poincaré section for the Hénon–Heiles Hamiltonian of (2.4) at $x = 0$, for a moderate value of the perturbation parameter ε. Notice how the nested tori are separated by chaotic regions

To complement such numerical studies, an analytic method for detecting chaos in Hamiltonian systems that are periodically perturbed was developed by Alexey Mel'nikov in 1963 [653]. Based on ideas introduced by Poincaré in the revised version of his prize-winning essay, Mel'nikov's method can detect the existence of tangled homoclinic orbits which approach singular points of the phase space as $t \to \pm\infty$ and intersect an unbounded number of times. For sufficiently large perturbation, this construction thus demonstrates the existence of homoclinic chaos [147].

An important paper by Grayson Walker and Joe Ford appeared in 1969, which drew these many of developments together and brought them to the attention of the physics community [1011]. Citing the works of both Poincaré and Birkhoff, these authors described the KAM theorem, used numerical studies to show in detail how KAM applied to both the Hénon–Heiles model and the FPU Hamiltonian [305], and discussed the profound implications for statistical mechanics, in which macroscopic laws of thermodynamics are derived from the Newtonian dynamics of microscopic behavior. Viewed warily by physicists at the time of its publication, this paper has subsequently been highly cited and is now accepted as a prophetic contribution to nonlinear science. In related work at about the same time, Ilya Prigogine and his colleagues used models proposed by Henry McKean and by Mark Kac to show how the recognition of molecular chaos leads to sharper derivations of Boltzmann's H-theorem for increasing entropy in a gas [422].

In 1979, Chirikov considered the nonlinear map

$$\theta_{n+1} - 2\theta_n + \theta_{n-1} = K \sin \theta_n \, , \tag{2.5}$$

which bears the same relation to the ODE for a nonlinear pendulum

$$\frac{\mathrm{d}^2\theta}{\mathrm{d}t^2} = k \sin \theta$$

as (2.3) does to the Verhulst ODE of (1.2). Thus as $K \to 0$, (2.5) approaches an integrable limit, conserves mechanical energy, and has a family of periodic (Jacobi elliptic) functions as solutions [529, 875]. If n is viewed as a spatial variable, (2.5) is closely related to the system proposed in 1939 by Yakov Frenkel and Tatiana Kontorova to model the dynamics of crystal dislocations, which will be discussed in a later chapter [312, 331].

Defining $I_{n+1} \equiv \theta_{n+1} - \theta_n$, (2.5) can also be written as the two-variable system

$$\begin{bmatrix} I_{n+1} \\ \theta_{n+1} \end{bmatrix} = \begin{bmatrix} I_n + K \cos \theta_n \\ I_n + \theta_n + K \cos \theta_n \end{bmatrix} \, , \tag{2.6}$$

which was introduced in the late-1960s in a report by John Bryan Taylor to model the motion of a charged particle in a magnetic field [950]. Published during the years when such nonlinear problems were of scant interest to the physics community, Taylor's work has been largely ignored, but it is mentioned in a related study by John Greene in 1979 [374], which appeared at the same time as the widely-cited paper by Chirikov [175]; thus (2.6) is now called either the Taylor–Chirikov map or the standard map [652].

As (2.6) can be derived from a Lagrangian formulation, the KAM theorem applies [229, 652]. The implications of this can be appreciated by noting from (2.5) that the solutions will be regular for $K \ll 1$, which in turn implies that plots of θ_{n+1} against θ_n will be families of closed curves. As K is increased from zero but is still sufficiently small, many of these closed curves persist, interspersed with irregular (chaotic) regions, and as K is further increased the chaotic regions grow at the expense of the closed curves, a prediction that agrees with numerical studies [652].

Fig. 2.5. Leonardo da Vinci's sketch turbulence generated by the flow of water into a pool. (The Royal Collection ©2005 Her Majesty Queen Elizabeth II)

From Figs. 1.1 and 1.2, it is seen that the burst of chaos research between 1976 and 1979 led directly into the the current era of nonlinear science, in which more than five papers are published on chaos each day. In addition to many experimental studies, this current work comprises analyses of additional theoretical models for chaos (both dissipative and conservative) [779], discussions of various routes to chaos [361], and considerations of the relationship between chaos and the complex phenomenon of *turbulence* – a phenomenon that was studied by Leonardo da Vinci in complex fluid flows (see Fig. 2.5) [580, 707]. We will visit turbulence again in Chap. 6.

2.6 More Early Discoveries of Low-Dimensional Chaos

Although he is often given credit for it, Lorenz was not the first to announce the discovery of deterministic chaos. In 1961, Yoshisuke Ueda, an electrical engineering graduate student working with Chiro Hayashi in Kyoto, observed that periodically driving a nonlinear oscillator leads to chaotic behavior [409], but this publication was overlooked in part because it was a technical report written in Japanese, so not accessible to western scientists. A deeper reason, however, is that the Japanese scientific community was not then ready to accept the concept of low-dimensional chaos; thus Ueda's results were not agressively promoted, to his present-day chagrin [980]. In 1959, as previously noted, Chirikov had discussed the onset of "stochastic motions" in plasma-confinement machines, and in 1958, Tsuneji Rikitake had observed paradoxically complex behavior of a two-disk dynamo model [806].

The effects of periodically driving a simple oscillator had been investigated by Mary Cartwright and John Littlewood in 1945 [161], but they regarded their non-periodic trajectories as "bad", whereas Birkhoff had merely labeled them "irregular" [91]. As in the Hénon–Heiles model of (2.4), the basic idea in this early study was to take an oscillator – either an electronic circuit, a mechanical system, or a numerical model – and then drive it with a commensurate frequency. Suppose, in other words, that the oscillator frequency is ω_1 and the (commensurate) driving frequency is ω_2. Then to obtain irregular behavior, the ratio

$$\frac{\omega_1}{\omega_2}$$

should be a rational number (a number equal to the ratio of two integers). In the Hénon–Heiles model, ω_1 and ω_2 would be the frequencies of the originally uncoupled oscillators.

Nor was recognition of the difficulties with numerical weather prediction original with Lorenz; in the mid-1950s, Norbert Wiener had expressed his concerns about this problem [1037], and in 1898 physicist William Suddards Franklin had written [323]:

Long range detailed weather prediction is therefore impossible, and the only detailed prediction which is possible is the inference of the ultimate trend and character of the storm from observations of its early stages; and the accuracy of this prediction is subject to the condition that the flight of a grasshopper in Montana may turn a storm aside from Philadelphia to New York!

Thus even Lorenz's butterfly metaphor had its precedent!

With retrospective awareness of these many precursors, we again ask why Poincaré's admonition – given in the epigraph to this chapter – was so long ignored by the main-stream scientific community? Why was it necessary for Lorenz to discover SDIC numerically, rather than learning about irregular solutions in his undergraduate or graduate mathematics lectures?

The answer, I claim, is that the nonlinear-science revolution had not yet happened in the late 1950s when Lorenz began his meteorological studies. Just as it was difficult for Nicolaus Copernicus and Galileo Galilei to convince their geocentric colleagues that our Earth revolves around the Sun [514], it seems that scientists in the first seven decades of the twentieth century were unable to accept the idea that long-range predictions can be impossible in a well-defined ODE system. Thus we recognize the explosive growth of interest in the subject of chaos shown in Fig. 1.1 as an example of a Kuhnian paradigm shift. Prior to 1970, most scientists erroneously agreed among themselves that prediction is possible in principle for any determinate ODE system and it is possible in practice – given adequate computing power – for any sufficiently simple ODE system. The mathematical textbooks of the time (as I recall from my student days) supported this perspective by omitting to mention the contrary views of Poincaré and Birkhoff, among other knowledgeable mathematicians and physicists. Since 1990, on the other hand, most physical scientists have come to accept low-dimensional chaotic phenomena as a typical feature of reality and dozens of textbooks are currently being used in university courses on nonlinear science to teach the basic ideas to undergraduate students. In addition, the qualitative nature of chaotic phenomena has often been explained to the general public, and executive toys that demonstrate irregular trajectories are readily available in boutiques.

Looking back on the decade-long neglect of his 1963 paper prior to its later enthusiastic acclaim, Lorenz remarks [576]:

> One may argue that the absence of an early outburst [of interest in chaos] was not *caused* by a prevailing lack of interest; it *was* the lack of interest. To some extent this is true, yet it may have been caused by the priorities of the leaders in the field. [...] Certainly Poincaré and Birkhoff and most other leaders did not suggest that the problems of the future would lie in chaos theory.

Before the 1970s, it appears, the importance of of low-dimensional chaos was not yet accepted – even by the most clear-sighted of the scientific community.

2.7 Is There Chaos in the Solar System?

Finally, let us return to consideration of our Solar System with which this chapter began, noting an apparent inconsistency. It was the regularity of planetary motions that: (i) excited the imaginations of ancient astronomers, (ii) suggested the periodic components (deferents, epicycles, ecliptics and equants) of the Ptolemaic system, (iii) became better explained through the heliocentric theory proposed by Copernicus and promoted by Galileo, and (iv) was finally formulated as mathematical laws of celestial dynamics by Newton. Yet our survey of subsequent studies in nonlinear dynamics indicates that the Solar System is described by nonlinear ODEs which exhibit irregular (chaotic) behavior, at least in some instances. Are the Solar System dynamics regular or chaotic? This question can be answered at several different levels of observation, including the following.

Planetary Moons

Discovered in 1848 by George Bond and William Lassel, Hyperion is believed from its many craters to be the oldest of Saturn's thirty known moons. It is only about 165 miles across with an irregular shape, and during the Voyager-2 flyby of August 1981, it was discovered that the orientation of Hyperion tumbles chaotically, which is probably caused by perturbations from the motions of Saturn's other moons. Similar chaotic tumbling motions have been observed for the Martian moons: Phobos and Deimos [68,261,691,1052]. With a mass about 40% greater than that of the planet Pluto and an eccentric orbit, Triton, now Neptune's largest moon, may once have been in a planetary orbit, suggesting instability in the structure of the Solar System [10,675]. Our Earth, on the other hand, has only a single Moon, the axial rotation of which has become synchronized – through motions of the tides – to its rotation about Earth; thus we see only one side of it.

Halley's Comet

Widely viewed as a paragon of celestial regularity after Edmond Halley (1656–1742) correctly predicted in 1705 that its 1682 visit would be repeated in 1758, the comet that bears his name has been observed 29 times with an average interval of about seventy-six years since the first recorded sighting by Chinese astronomers in 240 BCE [176,261,691]. As shown in Fig. 2.6, however, the intervals between successive returns are variable, ranging randomly over 4.8 years from a smallest interval in 1920 to a greatest interval in 530. This variability suggests low-dimensional chaos, which has recently been convincingly demonstrated by Boris Chirikov and Vitold Vecheslavov [176]. To support their claim, these authors have shown that the dynamics of Halley's Comet can be fairly well modeled as a three-body problem comprising the Sun, Halley's Comet and Jupiter, and more closely as a four-body problem involving additional gravitational perturbations from Saturn. Interestingly, they find the Lyapunov time (over which nearby trajectories diverge by a factor of e) to be about 29 returns; thus the earliest recorded observations of this phenomenon lie at the edge of retrodictability.

The Planets

Using the powerful computers that have been developed over the past few decades, Jacques Laskar has integrated the nonlinear equations of the Solar System far into the future and the past, leading to the discovery that the motions of the planets are chaotic with Lyapunov times of about 5 million years (Myrs) [261,537,538]. For example, an error in the position of Earth will grow by a factor of

$$e^{20} \approx 4.8 \times 10^8$$

in 100 Myrs (0.1 Gyr), making planetary motions unpredictable on that time scale. Thus there are difficulties in retrodicting the Earth's orbit beyond a few tens of millions of years, and climate reconstruction over geological

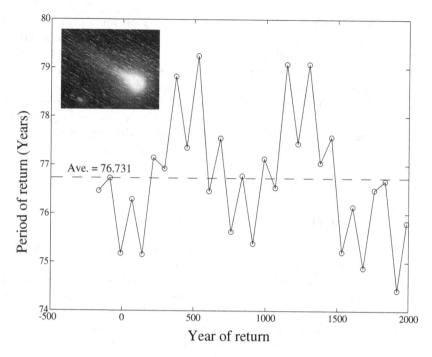

Fig. 2.6. Variations in the intervals between returns of Halley's Comet. Data from [176]. *Inset image* taken on 14 March 1986, from the Giotto spacecraft, courtesy of the European Space Agency

time scales becomes problematic. Such computer studies of the Solar System's chaotic history are now an interesting and exciting branch of nonlinear research, opening possibilities for understanding the patterns of the Solar System planets and those of other stars [463, 938].

Numerical patterns of the planetary orbits have been speculated upon since the eighteenth century when Johann Titius and Johann Bode noted a regular relationship among their mean distances from the Sun as indicated in the left-hand column of Table 2.2. Defining the 93 million miles between Earth and Sun as one AU (astronomical unit), the Titius–Bode (TB) rule is

$$a = 0.4 + 0.3 \times 2^n , \tag{2.7}$$

where a is the mean distance from a planet to the Sun in AU and n is the planetary number which is taken to be 0 for Venus and $-\infty$ for Mercury. As Saturn was the most distant planet known when this rule was first put forward in 1772, it offered fair agreement with the available data and accurately predicted the position of Uranus which was subsequently discovered in 1781. Although Neptune (discovered in 1846) has a mean distance from the Sun that is significantly at variance with the Titius–Bode rule, Pluto (discovered in 1930) was at about the position expected for $n = 7$, leading apologists for

Table 2.2. Mean planetary distances from the Sun (a in AU) and corresponding values from the Titius–Bode rule of (2.7), Laskar's results from (2.8) [539], and the macroscopic quantum model of Nottale et al. from (2.9) [722]

Planet	Measured	Titius–Bode	Laskar	Nottale et al.
Mercury	0.387	0.4 ($-\infty$)	0.52	0.40
Venus	0.723	0.7 (0)	0.74	0.72
Earth	1	1 (1)	1	1.125
Mars	1.524	1.60 (2)	1.30	1.62
Asteroid Belt	2.77	2.80 (3)	–	2.21
Jupiter	5.203	5.20 (4)	10.05	4.5
Saturn	9.539	10.00 (5)	15.84	10.12
Uranus	19.18	19.60 (6)	22.94	18.0
Neptune	30.06	38.80 (7)	31.36	28.12
Pluto	39.44	77.20 (8)	–	40.5

TB to suggest that Neptune might be some sort of exception. As it lacks a theoretical basis, however, most astronomers have dismissed the TB rule as idle numerology.[7]

Interestingly, Laskar has recently used a combination of analytical and numerical techniques to derive and compute statistically probable patterns of planetary orbit sizes and masses [539]. Starting with a large number (10 000) of small planets (planetesimals) that have random orbits, he *allows* collisions, thus breaking out of the confines of the classical N-body problem because colliding planetesimals stick together, introducing an "arrow of time" that is lacking in Newtonian dynamics. For a variety of initial conditions and several thousand simulations, he finds that our Sun's planets settle into two stable groups: an "inner" group corresponding to Mercury, Venus, Earth, and Mars, and an "outer" group comprising Jupiter, Saturn, Uranus, Neptune and Pluto. Importantly, the planets of the outer group (particularly Jupiter and Saturn) carry most of the Solar System's angular momentum. In both the inner and outer groups, successive planets in Laskar's model have mean distances from the Sun that increase with the squares of the planetary numbers, in accord with empirical observations for our Solar System. This same relation holds for the recently discovered planets of ν-Andromedae [143], suggesting some general phenomenon in which

$$\sqrt{a} = \sqrt{a_0} + kn , \qquad (2.8)$$

where a is a mean planetary distance from the Sun, n is a planetary index number, and k is the slope of a plot of $\sqrt{a}$ vs. n. Laskar has found a probable

---

[7]Since 24 August 2006, Pluto is considered to be a dwarf planet [441, 738].

value of k to be 0.14 for the inner planets, whereas the observed value is 0.20 (see Table 2.2). For the outer planets, his probable value is 0.81 and the observed value is 1.06 [539].

Both the Titius–Bode rule and Laskar's numerical computations suggest some underlying order, as did the regularity of the hydrogen-atom spectrum prior to Niels Bohr's proposal of an elementary version of quantum theory in 1913 [102] and its full realization by Erwin Schrödinger in 1926 [850] and Max Born in 1927 [113]. As Born then pointed out, a key feature of this new quantum theory (QT) was that it does not give the exact trajectory of an electron orbiting about an atomic nucleus; instead QT merely provides a means for computing the *probability* of finding an electron in a particular region of space-time (see Appendix B). Considering that the life of the Solar System is about 5 Gyr and the Lyapunov time for planetary motions is about 5 Myr, it follows that only *probabilities* for planetary trajectories can be computed over about 99% of their histories, leading to a probabilistic situation qualitatively similar to that in standard quantum mechanics [844].

At the beginning of this chapter, it was noted that the classical N-body problem of planetary science is formally identical to the classical theory of atomic dynamics. Recently, Laurent Nottale, Gérard Schumacher and Jean Gay have suggested that this similarity may extend to quantum realms as well [722]. Their conjecture, in other words, is that the trajectories of both electrons in atoms and of planets of the Solar System on a time scale longer than 0.1 Gyr undergo something like Brownian motion, which is continuous but nowhere differentiable like the perimeter of Koch's snowflake shown in Fig. 2.2 or the coast of Britain. Thus these authors have proposed a theory for computing the probabilities of planetary orbits that is formally identical to the QT proposed by Schrödinger and Born for electronic orbits in atoms, albeit with vastly different parameters. Just as the radii of hydrogen-atom orbits are proportional to n^2, where n is the index of the orbit and the proportionality constant is the Bohr radius (5.29×10^{-11} meters), so Nottale et al. claim that the most probable planetary orbits are given by

$$a = \alpha_i n^2 \quad \text{or} \quad a = \alpha_o n^2 \,, \tag{2.9}$$

where n is an orbital index and α_i (α_o) is a proportionality constant for the inner (outer) planets.

As the proportionality constant they find for the inner planets is $\alpha_i = 0.045$ AU and the index of Mercury is 3, two orbits may lie inside that of Mercury, with a planet at the smaller orbit (at 0.045 AU) probably evaporated by the Sun's heat and a planet at the larger orbit (at 0.18 AU) possibly waiting to be discovered. Taking the inner planets to collectively comprise the $n = 1$ orbit of the outer system, the proportionality constant for the outer planets is increased by a factor of 5 (the index of Earth, which is in the middle of the inner planets) to $\alpha_o = 5\alpha_i = 1.125$. Considering that it only estimates the probabilities of planetary orbits and ignores the accretions of planetesimals through their many collisions, one should not expect this "macroscopic quantum theory" to provide an exact description of planetary dynamics, yet variations on this theme by Nottale et al. [722] and others [431, 797] lead to impressive agreements with the positions of the inner planets, components of the Asteroid Belt, and even the moons of Jupiter and Saturn.

Are these credible scientific ideas or mere numerology? Observations of other planetary systems and further numerical and theoretical research in this area may provide answers.

3 Solitons

> I was observing the motion of a boat which was rapidly drawn along a narrow channel by a pair of horses, when the boat suddenly stopped – not so the mass of water which it had put in motion; it rolled forward with great velocity, assuming the form of a large solitary elevation which continued its course along the channel apparently without change of form or diminution of speed ...
>
> John Scott Russell

When I became a graduate student in science just a half century before this book is scheduled to be published, my peers, our professors and I shared two beliefs about the qualitative nature of nonlinear dynamics, both of which turned out to be dramatically wrong. The first of these was that simple systems of nonlinear ordinary differential equations (ODEs) were well understood and scarcely worth the effort of serious study. As we have seen in the previous chapter, this belief overlooked the "fortuitous phenomena" uncovered by Henri Poincaré and the "irregular orbits" of George Birkhoff, which would become of central interest in the outburst of chaos studies during the 1970s. The second erroneous belief of my graduate student years was that systems of nonlinear partial differential equations (PDEs) were so complicated that little or no analytical progress was to be expected, and only developments of the digital computer could lead to progress. As with chaos studies, the roots of an alternative perspective on nonlinear PDEs also go back to the nineteenth century, and the scientific establishment would soon discover that interesting and important dynamic entities often emerge from such systems. This and the following chapter present an overview of emergence in energy-conserving and dissipative PDE systems.

3.1 Russell's Solitary Waves

Hydrodynamic solitary waves (or *solitons* as they came to be called in the 1970s) had been coursing up the fjords and firths of Europe since the dawn of time, but they were not scientifically studied until 1834. In August of that year, a young Scottish engineer named John Scott Russell was engaged in an urgent project. The future of Britain's horse-drawn canal boats was threatened by competition from the railroads, and Russell was conducting a series of experiments on Scotland's Union Canal to measure the dependence of a boat's speed on its propelling force. His aim was two-fold: to understand a curious anomaly in this relationship in which the required force had recently been observed to *decrease* with increasing speed at around eight miles per hour [208], and to establish design considerations for conversion from horsepower to steam.[1]

As chance would have it during Russell's experiments, a rope parted in his apparatus and he first observed the solitary wave shown in the frontispiece and vividly described in the epigraph to this chapter [833]. Characteristically, Russell did not ignore this unexpected event; instead, he "followed it on horseback, and overtook it still rolling on at a rate of some eight or nine miles and hour, preserving its original figure some thirty feet long and a foot to a foot and a half in height" until the wave became lost in the windings of the channel. As is exhaustively recorded in his now classic *Report on Waves* to the British Association for the Advancement of Science (BAAS) [833], he continued to study this unanticipated phenomenon in wave tanks (see Fig. 3.1), canals, rivers, and the Firth of Forth over the following decade, showing that the above-mentioned force–speed anomaly stems from interaction of the boat with a solitary wave, thereby putting prow design on a scientific basis and establishing himself as one of the great naval architects of the nineteenth century [289].

In the course of these studies, Russell found his "Great Wave of Translation" to be an independent dynamic entity that moves with constant shape and a speed given by

$$v = \sqrt{g(d + h)}\,, \tag{3.1}$$

where d is the depth of the water, h is the height of the wave, and g is the acceleration of gravity. Furthermore, he demonstrated that a sufficiently large initial mass of water will spontaneously resolve itself into two or more independent solitary waves (each moving at its own speed), and that solitary waves cross each other "without change of any kind".

Importantly, (3.1) indicates that a solitary wave travels *faster* than the small amplitude waves (of speed $\sqrt{gd}$) out of which it emerges, suggesting that it is a fundamentally different entity.

Although one might expect the fruits of Russell's decade-long investigation to be readily accepted by the scientific community, this did not happen.

---

[1]Dug between 1818 and 1822, the Union Canal linked Glasgow and Edinburgh, offering serene passenger transport between the two cities in addition to providing a means for carrying coal in and garbage out. Presently, it is preserved by canal enthusiasts as a delightful walking path.

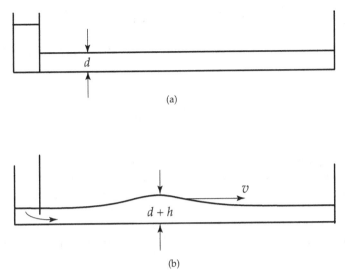

Fig. 3.1. A hydrodynamic soliton created in a wave tank by John Scott Russell in the 1830s. (**a**) A column of water is accumulated at the left-hand end of the tank. (**b**) Release of this water by lifting the sliding panel generates a solitary wave that travels to the right with velocity v. (Redrawn from [833])

Already in 1841, George Airy, the Astronomer Royal, had inferred from his theoretical investigations that a solitary wave could not be obtained from the hydrodynamic equations and so concluded that Russell's interpretations of his observations must be in error [14], an opinion that was soon supported by independent theoretical studies of young George Stokes [931]. Contributing to this dispute may have been disdain felt by mathematically oriented wave theorists at Cambridge University for papers published by the recently organized BAAS, which was dominated by amateurs and engineers. In retrospect, the fact that a solitary wave travels faster than corresponding low-amplitude waves is important in these polemics, as this means that a solitary wave cannot be constructed from low-amplitude waves, reinforcing the idea that it is an independent dynamic entity, lying beyond the purview of the linearly-based analyses of Airy and Stokes.[2]

Empirical confirmation of Russell's canal observations was eventually presented by Henri Bazin in France in 1865 [64], and in 1872 Joseph Boussinesq published a long and detailed analysis showing that the solitary wave was indeed a possible solution of the hydrodynamic equations [118], a conclusion that was supported by Lord Rayleigh, Adhémar Saint-Venant and John McCowan [632,791,836]. Finally in 1895, Diederik Kortweg and Gustav de Vries

---

[2]Very readable accounts of these differences of opinion in the context of nineteenth-century wave science have recently been published by Robin Bullough [134], Alex Craik [197], Alexandre Filippov [306], and Oliver Darrigol [208].

derived the following nonlinear PDE for Russell's shallow water waves:[3]

$$\frac{\partial u}{\partial t} + c\frac{\partial u}{\partial x} + \gamma u\frac{\partial u}{\partial x} + \varepsilon\frac{\partial^3 u}{\partial x^3} = 0 , \tag{3.2}$$

where

$$c = \sqrt{gd} \tag{3.3}$$

is the speed of low ampliude (linear) waves, $\varepsilon = c(h^2/6 - T/2\rho g)$ is a dispersion parameter, $\gamma = 3c/2h$ is a nonlinear parameter and T and ρ are respectively the surface tension and density of water [510]. Using Carl Jacobi's recently developed elliptic functions, they obtained a nonlinear periodic solution of (3.2) (which they called cnoidal for the Jacobi cn function), and in the limit of infinite wavelength this solution becomes Russell's solitary wave, with just the dynamic properties that he had recorded and published five decades earlier.

As the amplitude of the solitary wave is decreased to zero, the last two terms on the left-hand side become small, and the wave speed reduces to $\sqrt{gd}$ in accord with (3.1). At larger values of h, the dispersion introduced by the ε term is balanced by the nonlinearity of the γ term for an exact solution with the "smooth and well defined" shape that was observed experimentally by Russell. By this time, however, Russell was resting in his grave and interest in his solitary wave had waned, as we see from three facts. Based on the views of Stokes, first, the importance of the solitary wave was discounted in the *Encyclopedia Britannica* of 1886 [306]. Second, Horace Lamb's opus on hydrodynamics allots a mere 3 of 730 pages to the solitary wave [522]. Finally, there were only about two dozen citations of reference [833] from its publication in 1845 to the beginning of the nonlinear science explosion in 1970 [306].[4]

3.2 The Inverse Scattering Method

This problem rested on a back burner of science until the early 1950s when Enrico Fermi, John Pasta, and Stan Ulam began using the newly-assembled MANIAC computer at Los Alamos to conduct numerical studies of thermalization on a one-dimensional mass–spring system, finding unexpectedly

---

[3]Although this KdV equation lies buried in Boussinesq's tome [118], it is the specific focus of the Korteweg–de Vries paper [510].

[4]In fairness to the nineteenth-century British scientific community, however, it should be noted that Russell's case for the importance of the solitary wave was not aided by the publication of several misguided claims in a posthumous book that was prepared from his unpublished manuscripts in 1885 [834].

frequent recurrences of the initial state.[5] Attempts to understand and explain this anomaly led Norman Zabusky and Martin Kruskal to approximate the FPU mass–spring chain by KdV, discovering in 1965 that the initial conditions would fission into a spectrum of solitary waves which then reassembled themselves into approximately the original state after an unexpectedly short interval [1074]. To emphasize the particle-like properties of these dynamically independent solitary waves – which confirmed the intuition of Russell in contrast to the linearly-based calculations of Airy and Stokes – they were called solitons.

In 1967, Kruskal and his colleagues (Clifford Gardner, John Greene and Robert Miura) considered KdV written in the normalized form

$$\frac{\partial u}{\partial t} - 6u\frac{\partial u}{\partial x} + \frac{\partial^3 u}{\partial x^3} = 0 , \tag{3.4}$$

with a solitary wave solution

$$u_s(x,t) = -\frac{v}{2}\operatorname{sech}^2\left[\frac{\sqrt{v}}{2}(x - vt - x_0)\right] , \tag{3.5}$$

and discovered an unexpected result. If a general solution of (3.4) is introduced as a time-varying potential in the Schrödinger equation

$$\frac{d^2\psi}{dx^2} + \left[\lambda + u(x,t)\right]\psi = 0 , \tag{3.6}$$

the eigenvalues of (3.6) are independent of time [346], with each eigenvalue corresponding to a particular soliton in the total solution. Furthermore, the time evolution of ψ is given by

$$\frac{d\psi}{dt} = \left(-4\frac{\partial^3 \psi}{\partial x^3} + 3u\frac{\partial \psi}{\partial x} + 3u_x\psi\right) . \tag{3.7}$$

Together, (3.6) and (3.7) imply that the Schrödinger equation scattering data (eigenvalues and reflection coefficient) computed for the initial column of water in Fig. 3.1a can be used to compute the time evolution $u(x,t)$ as follows [702, 723]:

- Given an initial disturbance $u(x,0)$, calculate the Schrödinger equation scattering data SD(0).
- As $\psi \to 0$ for $x \to \pm\infty$, use the asymptotic form of (3.7)

$$\frac{d\psi}{dt} = -4\frac{\partial^3 \psi}{\partial x^3}$$

to find the scattering data at a later time SD(t).

---

[5]Nick Metropolis told me that this discovery was accidental. Because on a short time scale the initial energy seemed to proceed toward a thermalized state (confirming the motivating intuition), early computations were terminated before the energy returned to its original configuration. One afternoon, they got to talking and unintentionally left MANIAC running, leading to their unexpected observation.

- Determine $u(x,t)$ from SD(t), by using an inverse scattering calculation.

As a diagram, this inverse scattering transform (IST) can be represented as:

$$\begin{array}{ccc} \boxed{u(x,0)} & \longrightarrow & \boxed{u(x,t)} \\ \downarrow & & \uparrow \\ \boxed{\text{SD}(0)} & \longrightarrow & \boxed{\text{SD}(t)} \end{array}$$

where the upper arrow corresponds to numerical integration and the lower three arrows indicate stages in the analytic IST computation. Thus a nonlinear numerical computation was replaced by a sequence of linear analytical calculations in solving KdV, a problem that had been of central interest to hydrodynamics for more than seven decades. During the 1970s, there were several experimental studies establishing the relationship between KdV solitons and shallow water waves [108, 401, 402, 940, 1073].

Importantly, this approach is not limited to KdV. Writing (3.6) and (3.7) respectively as $L\psi = 0$ and $\psi_t = M\psi$, Peter Lax generalized IST to the class of operators for which

$$L_t = ML - LM \equiv [M, L] \,,$$

where L and M are called a Lax pair [545]. Lax's formulation is now used to obtain exact solutions for a broad class of nonlinear wave systems, many of which have applications in applied science as will be discussed in later chapters of this book. For certain initial conditions on KdV, these exact solutions comprise N-solitons and correspond to members of a family of reflectionless potentials which had been known for the Schrödinger equation since the 1950s [482].

Interestingly, the IST method of analysis can be viewed as a nonlinear generalization of Fourier-transform (FT) analysis in the sense that IST reduces to FT as the nonlinearity (or wave amplitude) is reduced to zero [875]. In appreciating this reduction, however, it is important to emphasize that solitons *disappear* at low nonlinearity, showing again that they are *new dynamic entities* that emerge from nonlinearity. It was this feature of the problem that was missed by the *Encyclopedia Britannica* of 1886.

Thus an important aspect of the soliton concept is to focus attention on fully nonlinear solutions (expressed as hyperbolic, Jacobi elliptic or theta functions or numerically) rather than starting with Fourier expansions and calculating interactions among component modes. Although the FT approach is practical in some (quasilinear) cases when solitons have not appeared, a fully nonlinear analysis is usually easier to implement and empirically more illuminating. This difference of perspective was at the core of Russell's disagreement with Airy and Stokes, and it has arisen in other applications to be discussed below (fluxons on long Josephson junctions and slinky modes on reverse-field-pinch plasma confinement machines).

An example of an exact 2-soliton solution to (3.4) is

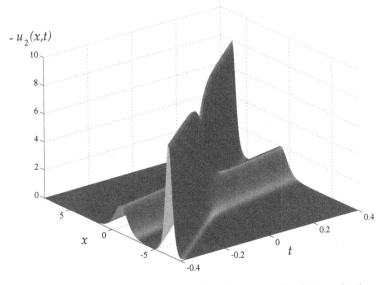

Fig. 3.2. A plot of a 2-soliton KdV solution $-u_2(x, t)$ from (3.8)

$$u_2(x, t) = -12 \frac{3 + 4\cosh(2x - 8t) + \cosh(4x - 64t)}{[3\cosh(x - 28t) + \cosh(3x - 36t)]^2} , \qquad (3.8)$$

which was published by Zabusky in 1968 based on the GGKM work [346, 1070]. Figure 3.2 shows a plot of this function, confirming – after an interval of a century and a half – Russell's observation that two solitary waves pass through each without change of shape. But (3.8) is not merely the sum of two solitons. From a careful examination it can be seen that the trajectory of the faster soliton is displaced slightly ahead and that of the slower is pushed backward by the interaction – a feature of the interaction phenomenon that Russell missed.

In defining the term soliton, there is an interesting difference of attitude between mathematicians and physicists. The former tend to reserve this term for solitary-wave solutions of PDEs (like KdV) for which an IST can be formulated, leading to exact N-soliton solutions. Some physicists, on the other hand, prefer to call any solution of an energy-conserving (Hamiltonian) system that exhibits particle-like properties a soliton, without worrying about the existence of exact N-soliton solutions. Thus the physicists' soliton might be a hydrodynamic vortex (tornado or hurricane) or an elementary particle that changes its character under collisions with other particles. This question arose at a soliton conference in Gothenburg, Sweden in the summer of 1978. Speaking as a particle physicist, Tsung-Dao Lee argued for the broader usage, and when Alan Bishop (a condensed-matter physicist with many connections to the applied mathematics community) objected, Lee suggested that the mathematicians' special solitary waves should be called aristocratic solitons

– to which Bishop replied that those without the special IST properties should then be called plebian solitons.

3.3 The Nonlinear Schrödinger Equation

This was the international meeting where Alexander S. Davydov announced his theory of solitons on the alpha-helical regions of natural proteins.[6] In a continuum approximation, these solitons are governed by the nonlinear Schrödinger (NLS) equation, which can be written in normalized form as

$$\mathrm{i}\frac{\partial u}{\partial t} + \frac{\partial^2 u}{\partial x^2} + 2|u|^2 u = 0 \,, \tag{3.9}$$

where $u(x,t)$ is the complex amplitude of an oscillating field. A solitary wave solution of this equation is

$$u_s(x,t) = a \exp\left[\mathrm{i}\frac{v_e}{2}x + \mathrm{i}\left(a^2 - \frac{v_e^2}{4}\right)t\right] \mathrm{sech}\left[a(x - v_e t - x_0)\right] \,, \tag{3.10}$$

which has an envelope moving at envelope velocity v_e and a carrier wave moving with velocity $v_c = (v_e/2 - 2a^2/v_e)$; thus the envelope velocity and the wave amplitude are independent parameters. This is evidently more complicated than a KdV soliton, but just as in (3.2), the second term of (3.9) introduces dispersion which is exactly balanced by the nonlinearity of the third term, resulting in a stable solitary wave with particle-like properties. As was shown by Zakharov and Shabat in 1972, the initial value problem for (3.9) can be solved by an IST method, leading to N-soliton formulas corresponding to (3.8) [1079]. The NLS equation has been of interest to applied scientists since the mid-1960s as a model for deep water waves [76], nonlinear optics [478], nonlinear acoustics [398,699,945], and plasma waves [944]. Since the early 1970s, it has been used as the fundamental description of pulses on an optical fiber [405], among other technical applications [597].

Although both the KdV equation and the NLS equation were first considered in the context of particular engineering applications, their importance is more general, as they are both *canonical equations*. KdV, for example, arises whenever one studies a low-frequency, nonlinear wave system with lowest-order representations of dispersion and nonlinearity. Similarly, NLS is a lowest-order description of a high-frequency nonlinear wave system; thus these two equations are generic, arising in a wide variety of applications.

---

[6]Organized by plasma physicist Hans Wilhelmsson, this meeting was typical many interdisciplinary nonlinear science conferences that were held in the late 1970s.

3.4 The Sine-Gordon Equation

A physically motivated soliton model is the sine-Gordon (SG) equation

$$\frac{\partial^2 u}{\partial x^2} - \frac{1}{c^2}\frac{\partial^2 u}{\partial t^2} = \sin u \,, \tag{3.11}$$

which was first formulated by Frenkel and Kontorova in 1939 in connection with their studies of crystal dislocations [331].[7] Investigations of the SG system were carried on in the postwar years by Alfred Seeger and his collaborators, who were aware of the special properties of non-destructive solitary-wave collisions [433, 896]. In this application, $u(x,t)$ is a real variable describing the dynamics of a row of atoms in a spring–mass approximation (where the $\sin u$ term on the right-hand side of the equation accounts for the periodic forces induced by a static neighboring row) and c is the speed of a sound wave along a row. Another application of (3.11) is as a model for the simple mechanical system shown in Fig. 3.3, where the first term accounts for the spring (elastic) forces between adjacent pendula and the other two terms describe the dynamics of the coupled nonlinear pendula [859].

The solution of (3.11) shown in Fig. 3.3 corresponds to one of the two functions

$$u_s(x,t) = 4\arctan\left[\exp\left(\pm\frac{x - vt - x_0}{\sqrt{1 - v^2/c^2}}\right)\right]\,, \tag{3.12}$$

where the $\pm$ signs account for the fact that the solution can wind in either the clockwise or counterclockwise direction.

These *kinks* are now known to be solitons in the strict (mathematicians') sense because (3.11) was shown to possess an IST formulation by two groups of scientists in the early 1970s. One group, comprising Mark Ablowitz, David Kaup, Alan Newell, and Harvey Segur at Clarkson College in Potsdam, New York, published soon after their successful soliton workshop in the summer of 1972, and the other, consisting of Leon Takhtajan and Ludvig Faddeev, published at about the same time from the Steklov Institute at the University of Leningrad (now St. Petersburg) [943]. The fact that such an important discovery was made almost simultaneously at two such distant locations shows the momentum in nonlinear science research that was developing in the early 1970s and boded well for future collaborations among Russian and American scientists.

The IST formulation allows many exact analytic solutions to be computed, and as is demonstrated in Fig. 3.4, SG solitons experience varying amounts

---

[7]Although it should be called the Frenkel–Kontorova equation, the curious name of (3.11) is a pun on the well-known Klein–Gordon equation ($u_{xx} - u_{tt} = u$), which was independently formulated in 1926 by Oskar Klein and Walter Gordon (among several others) as a linear relativistic wave equation with a rest mass [512]. The pun is due to Martin Kruskal [826].

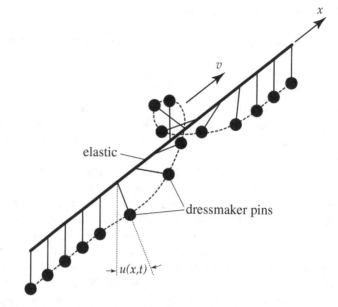

Fig. 3.3. A single soliton (kink) on a mechanical model of (3.11)

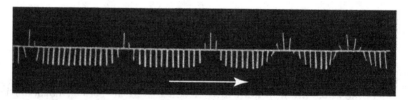

Fig. 3.4. Stroboscopic photograph showing Lorentz contraction of a kink soliton moving on a mechanical model of (3.11) [859]. (Dissipative effects cause the kink to slow down as it moves to the right, whereupon Lorentz contraction lessens and its size increases)

of *Lorentz contraction* as they move with different speeds. This stems from the analytical fact that (3.11) is invariant under the *Lorentz transformation*:

$$x \longrightarrow x' = \frac{x - vt}{\sqrt{1 - v^2/c^2}} \,, \qquad t \longrightarrow t' = \frac{t - vx/c^2}{\sqrt{1 - v^2/c^2}} \,, \tag{3.13}$$

which leaves the form of (3.11) unchanged – a property that is shared by both Maxwell's electromagnetic field equations and elementary particles of matter.

The Lorentz transformation (LT) of (3.13) provides a basis for the special theory of relativity (SRT) [272, 765]. To see this, imagine that the space and time axes in (3.11) are infinitely long and comprise a universe – the Lineland of Edwin Abbott's Flatland, wherein the kinks of (3.12) are cognizant entities

[1]. As the laws of physics (e.g., the SG equation) remain the same under LTs, these kinks might conclude that their universe is not based on space and time but on *space-time*, under which the expression

$$s^2 = x^2 - c^2 t^2 = x'^2 - c^2 t'^2 \tag{3.14}$$

also remains the same under LTs. In other words, primed and unprimed observers cannot agree on measurements of distance and time, but they do find the same values for s. Why? Because in terms of a complex time ($\tau \equiv it$), (3.14) is a form of the Pythagorean theorem, which states that the distance between two points on the $(x, c\tau)$-plane remains unchanged under rotations of the axes.

Evaluation of the energy (E) and momentum (p) for a kink gives

$$E = \frac{8c^2}{\sqrt{1 - v^2/c^2}} , \qquad p = \frac{8v}{\sqrt{1 - v^2/c^2}} ,$$

so

$$E^2 = p^2 c^2 + m_0^2 c^4 , \tag{3.15}$$

where $m_0 = 8$ is the rest mass of a kink. Also

$$E = mc^2 , \tag{3.16}$$

where $m = 8/\sqrt{1 - v^2/c^2}$ is the mass of a moving kink. In Appendix B, (3.15) is used to construct a quantum theory for a SG kink, whereas (3.16) is the famous mass–energy equation upon which the atomic bomb is based.

Although seemingly specific, SG models diverse physical phenomena [135], including the dynamics of domain walls in ferromagnetic and ferroelectric materials [247], the propagation of splay waves on biological (lipid) membranes [304], self-induced transparency of short optical pulses [523], the propagation of quantum units of magnetic flux (called fluxons) on long Josephson (superconducting) transmission lines [892], slinky modes in reversed-field pinch plasma confinement machines [258], and as a one dimensional model for elementary particles [65, 293, 758, 813]. If the kink is associated with a positron and an antikink with an electron, kink–antikink annihilation on the mechanical system of Fig. 3.3 provides a simple model for electron–positron decay into cosmic (electromagnetic) radiation.

Unaware of this previous work [433] but motivated by the Korteweg–de Vries results [306], John Perring and Tony Skyrme undertook numerical studies of SG as a classical model for fermions and published the following exact kink–kink and kink–antikink solutions in 1962 [758]:

$$u_{\mathrm{kk}}(x, t) = 4\arctan\left[\frac{v\sinh(x/\sqrt{1 - v^2})}{\cosh(vt/\sqrt{1 - v^2})}\right] , \tag{3.17}$$

$u_{\text{sbr}}(x,t)$

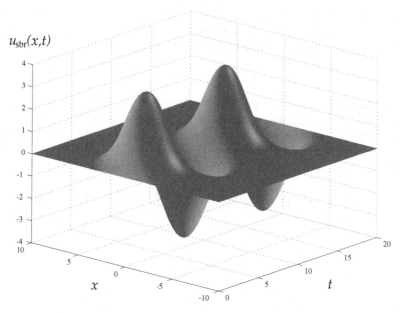

Fig. 3.5. A stationary breather solution of SG plotted from (3.20) with $\omega = \pi/5$

$$u_{\text{kak}}(x,t) = 4\arctan\left[\frac{\sinh(vt/\sqrt{1-v^2})}{v\cosh(x/\sqrt{1-v^2})}\right], \qquad (3.18)$$

where $c = 1$ for notational convenience. As further evidence of the poor communications among nonlinear scientists in the 1960s, Kruskal and Zabusky did not know about this work when they wrote their seminal paper on the KdV soliton [1071].

The kink–antikink solution of (3.18) takes an interesting form if the velocity parameter (v) is allowed to be imaginary. For example, setting $v = i\omega/\sqrt{1-\omega^2}$ with $\omega < 1$, this equation becomes the *stationary breather* [3] or *bion* [166]

$$u_{\text{sbr}}(x,t) = 4\arctan\left[\frac{\sqrt{(1-\omega^2)}}{\omega}\frac{\sin\omega t}{\cosh\sqrt{(1-\omega^2)}x}\right], \qquad (3.19)$$

an example of which is plotted in Fig. 3.5 for $\omega = \pi/5$. Under the elementary particle analogy, this solution provides a classical, one-dimensional model of *positronium*: the hydrogen- like entity comprising a positron and an electron rotating about each other [231].

Taking advantage of the fact that SG is invariant under a Lorentz transformation, this stationary breather can be boosted to a breather moving at speed v_{e}, which is described by the formula

$$u_{\mathrm{mbr}}(x,t) = 4\arctan\left\{\frac{\sqrt{(1-\omega^2)}}{\omega}\sin\left[\frac{\omega(t-v_e x)}{\sqrt{1-v_e^2}}\right]\mathrm{sech}\left[\frac{\sqrt{1-\omega^2}(x-v_e t)}{\sqrt{1-v_e^2}}\right]\right\}.$$

$$(3.20)$$

In the low-amplitude limit, (3.20) approaches the NLS soliton in (3.10), where both expressions are localized dynamic entities that exhibit both wave and particle properties – a combination of properties that has seemed mysterious to quantum theorists for many decades. This reduction is also a specific example of the above-mentioned generic nature of NLS.

Although these examples of exact analytic results are among the many that emerge from the IST cornucopia, it is important to consider how robust they are under various *structural perturbations*. In other words, do they persist under small changes in the PDE? Five types of SG structural perturbations are particularly important [641].

1. A constant term can be added to the right-hand side (RHS) of (3.11). This perturbation preserves the energy conserving (Hamiltonian) property and acts as a force, pushing a kink in one direction and an antikink in the other without destroying their localized natures [641]. In the context of Josephson junction technonogy, this forcing phenomenon can be used as the basis of a millimeter-wave oscillator, which has good frequency stability because a fluxon bounces back and forth between the ends of a finite system [194, 875].

2. If a time-periodic term is added to the RHS of (3.11), the resulting inhomogeneously driven SG system may exhibit low-dimensional chaos [270]. This phenomenon is related to the sinusoidally driven oscillator [409, 980] and to the three-body problem of planetary dynamics [234].

3. The RHS function of SG can be changed from a sinusoid to a general periodic function $f(u) = f(u + u_0)$. This preserves the Hamiltonian property and Lorentz invariance (LI), but the IST formulation is destroyed. Thus there are no more N-soliton formulas or breathers, but kinks and antikinks survive. As discussed under item 1 above, these kink-like objects tend to move in one direction or another, depending on the spatial average of $f(u)$ [641].

4. The RHS function can be changed to a polynomial function with at least three zeros – say $u(1 - u^2)$, which is called the phi-four equation because the corresponding potential function is quartic [151]. Again, the Hamiltonian property and LI survive, and there may be kink or antikink solutions, undergoing transitions between zeros of $f(u)$.

5. Dissipative perturbations can be added, such as LHS terms proportional to $\partial u/\partial t$ or to $-\partial^3 u/\partial x^2 \partial t$. In this case, the Hamiltonian (energy conserving) property is lost but an energy balance can be established between energy input to a kink (or antikink) in the above three cases and the effects of the dissipation. This balance leads to a steady propagation speed that depends on the parameters of the equation [641, 875].

6. Because (3.11) provides only an approximate description of the mechanical system of coupled pendula shown in Fig. 3.3, a corresponding difference-differential equation (DDE) might better be used, as in the crystal dislocation studies of Frenkel and Kontorova [331]. Under such a perturbation, the Hamiltonian property is preserved but LI and the IST formulation are lost, although the qualitative phenomenon of Lorentz contraction is observed under weak discretization, as is seen in Fig. 3.4. More generally, a time-independent DDE system corresponds to the standard map of (2.5), which can exhibit spatial chaos [652].

3.5 Nonlinear Lattices

An important participant in the 1978 Gothenburg meeting was Morikazu Toda from Japan, who had independently developed soliton theory for a particular form of the mass–spring lattice during the 1960s [968, 969]. His system is shown in Fig. 3.6, comprising equal unit masses connected by nonlinear springs of potential energy $U(r)$ where r is the change in distance between adjacent masses from its resting value at which the spring energy is a minimum. Newton's second law then leads to the following set of DDEs:

$$\frac{d^2 r_n}{dt^2} = \frac{dU(r_{n+1})}{dr_{n+1}} - 2\frac{dU(r_n)}{dr_n} + \frac{dU(r_{n-1})}{dr_{n-1}}, \tag{3.21}$$

where n is an index running along the chain. For the special spring potential

$$U(r) = \frac{a}{b}\left(e^{-br} + br - 1\right), \tag{3.22}$$

this Toda lattice (TL) system is exactly integrable through an IST method [311], and a TL soliton has the form

$$r_{n,s} = -\frac{1}{b}\log\left[1 + \sinh^2\kappa\,\text{sech}^2\left(\kappa n \pm t\sqrt{ab}\sinh\kappa\right)\right], \tag{3.23}$$

where the $\pm$ signs allow waves to go in either direction. These are compression waves on a discrete structure, one of which is shown on the lower part of Fig. 3.6. Interestingly, this compression wave travels at velocity

$$v = \sqrt{ab}\left(\frac{\sinh\kappa}{\kappa}\right) > \sqrt{ab} \tag{3.24}$$

lattice points per second, which is faster than the speed ($\sqrt{ab}$) of small amplitude waves on the lattice as for KdV solitons in (3.1). This again suggests that the solitary wave has escaped from the confines of a linear representation.

If the nonlinear spring potential is altered from the special Toda form of (3.22), the property of exact integrality is lost, but stable supersonic compression waves (as shown in Fig. 3.6) survive under a wide class of realistic

mass points at rest

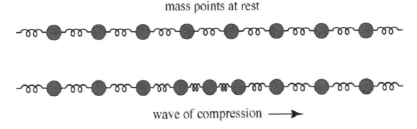

wave of compression ———▶

Fig. 3.6. The *upper figure* shows a Toda lattice at rest. The *lower figure* shows the Toda-lattice soliton given in (3.23)

interatomic potentials in solids, liquids and gases [340]. The fact that these solitary waves can be both stable and supersonic is in accord with observations by Captain William Edward Parry, a celebrated British explorer of the early nineteenth century, who sought to find a Northwest Passage from the Atlantic to Pacific Oceans across Northern Canada. In an appendix to the published version of Parry's journal [742], the Reverend G. Fisher records that:

> On one day and one day only, February 9, 1822, the officer's word of command "fire" was several times heard distinctly both by Captain Parry and myself about one beat of the chronometer [nearly half a second] after the report of the gun.

A dozen years before Russell's discovery of the soliton, in other words, the sound of a cannon firing was observed to travel faster than the command to fire it! This unanticipated acoustical phenomenon was noted by Russell and related to his observation that water waves of large amplitude travel faster than those of low amplitude, leading him to correctly estimate the height of the Earth's atmosphere from (3.3) [834].[8]

Thus we have encountered four nonlinear wave systems (KdV, NLS, SG, and TL), all of which arose in applications prior to 1970 and all of which share the property of being exactly integrable under an IST method. In addition, the solitons of these systems survive under many structural perturbations, leading to a rather large class of nonlinear PDE systems upon which stable solitary waves can propagate, lending support to Russell's speculation that solitary waves are a general physical phenomenon [834].

At this point, the reader may ask how the striking IST results could have escaped the notice of the mathematics community. In fact they didn't [433, 525]. Along with the nineteenth-century developments in physics and engineering there was a line of research in differential geometry, going back to Ferdinand Minding who studied families of surfaces of constant negative

---

[8]Equation (3.3) was first derived for shallow water waves by Joseph Lagrange in 1786 following ideas suggested by Newton in his *Principia* [197].

Gaussian curvature (pseudospherical surfaces) in 1839 [659], and to Edmond Bour [117] and others [109, 292] who discovered that such surfaces can be described by the PDE

$$\frac{\partial^2 u}{\partial \xi \partial \tau} = \sin u \,, \tag{3.25}$$

which is identical to SG under the independent variable transformation: $\xi = (x + t)/2$ and $\tau = (x - t)/2$. Building upon work by Luigi Bianchi [85] and Sophus Lie [561], Albert Bäcklund showed in 1883 that an infinite number of solution surfaces to (3.25) can be generated through successive applications of the transformation [44]

$$
\begin{aligned}
\frac{\partial u_1}{\partial \xi} &= 2a \sin \left(\frac{u_1 + u_0}{2} \right) + \frac{\partial u_0}{\partial \xi} \,, \\
\frac{\partial u_1}{\partial \tau} &= \frac{2}{a} \sin \left(\frac{u_1 - u_0}{2} \right) - \frac{\partial u_0}{\partial \tau} \,.
\end{aligned}
\tag{3.26}
$$

In this *Bäcklund transformation* (BT), $u_0(\xi, \tau)$ is a known solution of (3.25) and $u_1(\xi, \tau)$ is a new solution that can be obtained by integrating a first-order ODE system [811]. In 1892, Bianchi showed that two successive BTs (with different values of the parameter a) commute, which leads directly to algebraic constructions of (3.17) and (3.18) [86]. A decade earlier, Gaston Darboux had published a related transformation [207] that can be applied in a similar way to the linear equations of an IST formulation for KdV [39]. Although several other PDE systems were considered from these perspectives prior to 1914 – some of which were rediscovered during the explosion of soliton studies during the 1970s [433] – this promising line of activity died out after 1920, possibly because many of the young researchers were killed in the First World War [524]. In 1936 a summary of this early work was published by Rudolf Steuerwald, which would have been of great interest to Frenkel and Kontorova, but they didn't know about it [927].

3.6 Some General Comments

Now it is known that in all four of the above-mentioned cases (KdV, NLS, SG, and TL) the IST formulation is equivalent to the existence of a corresponding BT and from these transformations, N-soliton formulas and countably infinite sets of conservation laws can be derived, just as for the corresponding ISTs. N-soliton formulas for these four systems (among others) can also be constructed directly using a method devised by Hirota in 1971 [437].

As previously noted, both the KdV equation and the NLS equation are generic, arising when wave dispersion and nonlinearity are accounted for to lowest order. The other two cases (SG and TL) are models of specific physical structures (see Figs. 3.3 and 3.6) with solitary wave solutions that are robust under changes in the particular analytic functions – the $\sin u$ for SG and $U(r)$ for TL – on which the IST formulation depends. Importantly, all of these solitons or solitary waves can be viewed as Newtonian particles, which are acted on by forces derived from perturbing influences. In the face of such interesting empirical and theoretical results, the almost total neglect of the solitary wave concept by five generations of the scientific community between 1845 and 1970 is a sad tale of lost opportunities.

Recently, as all know, a devastating tsunami was generated in the Indian Ocean by an earthquake off the coast of Sumatra. Figure 3.7a by Kenji Satake shows a computer simulation of this wave, providing a vivid example of Russell's Great Wave of Translation. Thus tsunami energy is transmitted without dispersion over long distances at several hundred kilometres per hour in accord with (3.1), and from its long, smooth shape, a tsunami is difficult to detect on the deep ocean. Upon approaching a continental shelf, tsunamis slow down, gain amplitude and fission into several independent solitary waves, as is expected from Russell's tank experiments.

Internal ocean waves also provide examples of oceanic solitons that are well described by (3.2) with parameters given by $c = \sqrt{gh_1\Delta\rho/\rho}$, $\varepsilon = ch_1h_2/6$, and $\gamma = -3c/2h_1$, where $\Delta\rho$ is the density difference of the two layers, h_1 is the depth of the upper layer and $h_2 \gg h_1$ is the depth of the lower layer [727, 734]. Figure 3.7b shows a satellite photograph of such internal wave solitons in the Andaman Sea, which are visible from above due to the surface rip phenomenon. Whereas tsunamis are episodically initiated by earthquakes, internal waves appear regularly every 12 hours and 26 minutes, with greater intensities during the Full and New Moons; thus they are evidently driven by tidal flows that respond to motions of our Moon and Sun.

With tsunamis such a familiar and frightening phenomenon, why were Russell's observations and conclusions ignored by the scientific community for so many decades? One answer is that until recently scientists were conditioned – through their collective experiences with electromagnetism, acoustics, and quantum theory – to think about dynamics in terms of linear normal modes, which obey the principle of superposition. Since the discovery of normal modes by Johann and Daniel Bernoulli in the eighteenth century [306], there has evolved a general belief that any nonlinear solution can be resolved into its normal mode components, and then (if necessary) interactions among these modes can be subsequently computed. Based on this belief, physics and engineering students were taught a variety of methods (Fourier transforms, Laplace transforms, Green functions) which are appropriate for analyses of linear systems, but they learned almost nothing about the tools of nonlinear analyses: phase-space analysis, traveling-wave studies, chaotic trajectories, Jacobi elliptic functions, etc. When its component modes interact (according to the quasilinear view), the shape of a large amplitude wave would seem to become distorted, as was asserted by Airy and Stokes, incorrectly suggesting

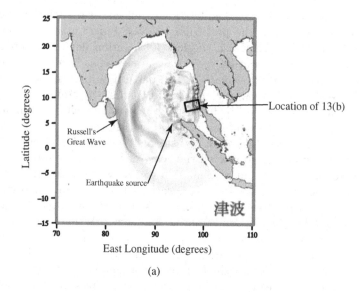

(a)

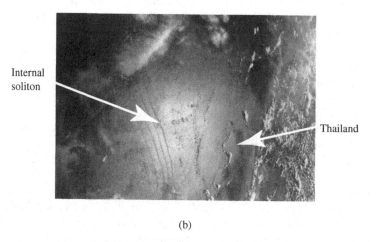

(b)

Fig. 3.7. (a) A computer simulation of the deadly tsunami (an example of Russell's Great Wave) in the Indian Ocean on 26 December 2004, 1 hour and 50 minutes after it was launched by an earthquake source (green area) near Sumatra. See http://staff.aist.go.jp/kenji.satake/animation.html for a colored and animated version of this figure. (Courtesy of Kenji Satake, National Institute of Advanced Industrial Science and Technology, Japan.) (b) Apollo–Soyuz photograph of the Andaman sea, showing surface evidence of internal waves [727]. The location of (b) is shown as a *small rectangle* in (a). (Courtesy of Alfred Osborne)

that Russell's empirical observations of a solitary wave were theoretically impossible. From this normal-mode perspective, nonlinearity came to be widely regarded as an analytic nuisance – an inconvenience to be avoided rather than a feature of realistic dynamics to be embraced for its unique properties.

At a deeper level, Russell's difficulty in obtaining recognition for the importance of his solitary-wave concept is an example of what Thomas Kuhn has described as the reflexive resistance of the scientific community to revolutionary ideas [515]. Truly new ideas, Kuhn pointed out, are often first recognized by those who are not fully indoctrinated into the prevailing paradigms and rejected by those who are. Approaching hydrodynamic wave studies as an engineer, Russell brought eyes that were unclouded by details of analytic manipulations as he became thoroughly familiar with the empirical facts. Both Airy and Stokes, on the other hand, were mathematicians, invested in the quasilinear results of their theoretical publications, the intricacies of which provided opportunities for them to insist that solitary waves would necessarily experience distortion, rendering traveling-wave solutions impossible. As outstanding scientists who made many important contributions to nineteenth-century research, Airy and Stokes had considerable influence and probably inhibited young British scientists from pursuing Russell's revolutionary concept of the solitary wave as an independent dynamic entity. Thus although Russell's observations were independently confirmed by Bazin [64] and the theoretical validity of his work was subsequently established in papers by Boussinesq [118], Rayleigh [791], Saint-Venant [836], and Korteweg and de Vries [510], these results didn't get into the textbooks. Only after the solitary-wave paradigm was forced upon those engaged in numerical computations during the 1960s did his work begin to be taken seriously by the scientific community.

As the solitary-wave phenomena described in this chapter are energy conserving (Hamiltonian), however, we are left with a question: Can solitary waves occur in *dissipative* PDE systems where energy is released and destroyed?

4 Nerve Pulses
and Reaction-Diffusion Systems

There is no better, there is no more open door by which you can enter into the study of natural philosophy than by considering the physical phenomena of a candle.

Michael Faraday

Five years after Russell's now classic *Report on Waves* appeared in Great Britain in 1845 [833], a twenty-nine year old German physicist named Hermann Helmholtz published measurements of the propagation speed of electrical pulses along the sciatic nerve of frog, a bundle of fibers that carries signals from the spinal cord to leg muscles [420, 864]. Influenced by comments of Isaac Newton suggesting that the speed of propagation of nerve activity is very high (like that of light [706]), most scientists then considered the speed of a nerve pulse too large to measure. Helmholtz was further advised not to undertake this project by his father, a philosopher who believe that a muscular response was identical to its cause, theoretically precluding any time delay between the two pheomena. Hermann Helmholtz, characteristically, decided to ask Nature. Using a clever experimental device to measure the relevant time delay as the pulse passed two points on the nerve, he found the surprisingly low propagation speed of 27 meters per second, which is close to presently measured speeds [874]. In 1868, Julius Bernstein used an even more ingenious measurement apparatus to record the localized shape of a nerve pulse [80]. Following these experiments, therefore, neuroscientists were pressed to explain why the speed of a pulse is so low – a question they would puzzle over for a century.

4.1 Nerve-Pulse Velocity

Several answers to this pulse velocity question appeared as new information accumulated, corresponding to Thomas Kuhn's prescientific phase of theoretical activity in which a widely-accepted paradigm has not yet been established [515]. Helmholtz suspected that some material motion must be involved, but 1902 Bernstein astutely proposed his membrane hypothesis which assumed that a breakdown of the nerve's surface resistance to ionic current plays a key role in pulse conduction [81]. At about the same time, Robert Luther demonstrated a propagating chemical reaction to a meeting of the German Society for Applied Physical Chemistry, pointing out that the wave speed was of order

$$v \sim \sqrt{D/\tau} \,, \tag{4.1}$$

where τ is the reaction time for an energy releasing process and D is the diffusion constant of a chemical species [583]. Because both diffusion constants and reaction times can vary widely for physical systems, he suggested that such a *reaction-diffusion process* might provide a credible explanation for the modest speed of a nerve pulse.[1] Then in 1914, young Edgar Douglas Adrian observed and stated the *all-or-nothing* principle of nerve-pulse excitation [9] – which implies the existence of a *threshold* for excitation [886] – and went on to pioneer applications of the newly invented vacuum-tube amplifier to studies of nerve-pulse dynamics, observing a *refractory period* of diminished excitability (higher threshold) following the passage of a pulse along a nerve.

In the mid-1920s, Ralph Lillie developed a passive-iron-wire model of pulse conduction, thus showing that nerve-pulse propagation need not be part of a living system. In this physical model, a length of iron wire initially rests in a bath of weak nitric acid and is prevented from dissolving by the rapid formation of a thin surface oxide layer. Upon being disturbed (by scratching, say, or from a pulse of voltage), the oxide layer collapses locally, allowing a pulse of ionic current to flow. In accord with Bernstein's membrane hypothesis, this disturbance propagates along the wire, after which the oxide layer reconstitutes itself to reestablish the resting state [563]. Thus Lillie's iron-wire model exhibits both a threshold for excitation and a refractory period.

Starting from Lillie's model, Nicholas Rashevsky published the first detailed analytic study of nerve-pulse propagation in 1933, arriving at conditions under which the propagation velocity is proportional to the square root of the fiber diameter [787]. In the mid-1930s, John Zachary ("J.Z.") Young recognized that two long and relatively large cylindrical structures running along the back of the common squid (*Loligo vulgaris*) are in fact nerves [1066]. The diameters of these *giant axons* are about a half millimeter, allowing electrophysiologists of the day to directly measure both the time course of the transmembrane voltage and the membrane permeability (or electrical conductivity), as Kenneth Cole did in the classic oscilloscope photograph of a nerve pulse shown in Fig. 4.1 [187]. In this image, the solid line indicates the transmembrane voltage, which is measured inside the axon with a glass

---

[1] Although Walther Nernst was both present at the meeting and clearly interested in these ideas, Luther's prescient observation was neglected for several decades.

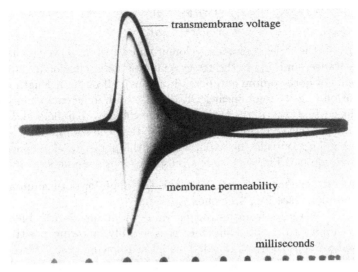

Fig. 4.1. An early cathode-ray oscilloscope photograph of a nerve pulse. The dots on the lower edge indicate time in milliseconds, increasing to the right. The solid line shows the time course of the transmembrane voltage (V), becoming increasingly positive inside the axon. The wide band shows the transmembrane permeability which is large (small membrane resistance) during the pulse. (Courtesy of Kenneth Cole)

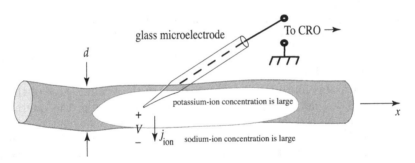

Fig. 4.2. Diagram showing a squid axon and the measurement of internal voltage using a glass microelectrode (not to scale)

microelectrode that is filled with a saline solution (see Fig. 4.2). The wide band in Fig. 4.1 records the transmembrane permeability (or electrical conductivity), simultaneously measured by an ac bridge.

An electromagnetic analysis of nerve-axon dynamics that is based upon Maxwell's equations plus Ohm's law shows that the transmembrane voltage is governed by the nonlinear diffusion equation [864]

$$\frac{1}{rc}\frac{\partial^2 V}{\partial x^2} - \frac{\partial V}{\partial t} = \frac{j_{ion}}{c},\qquad(4.2)$$

where r is the series resistance to longitudinal current flow, c is the membrane capacitance, and j_{ion} is the transmembrane ionic current (comprising components of both sodium and potassium ions), all per unit length of the axon. Equation (4.2) is a nonlinear PDE because both ionic components of j_{ion} are nonlinear functions of the transmembrane voltage. To understand the dynamics of squid nerve membranes, note first that the sodium-ion concentration remains large outside the axon, whereas the potassium-ion concentration is large inside (see Fig. 4.2); thus a pulse develops as follows [874]:

- At rest, the membrane resistance is very high for both sodium and potassium ions (see Fig. 4.1); thus $j_{ion} \doteq 0$.
- As V increases from its resting value by about 25 mV, membrane permeability to sodium ions increases sharply, allowing positively-charged sodium ions to flow inward. This inward-flowing charge further increases V in a positive feedback loop that drives V to a maximum value of about 100 mV, which is the peak value of V shown on Fig. 4.1.
- At this peak value of V, the sodium-ion current ceases to flow because the tendency of ions to diffuse inward is balanced by the opposing tendency of them to be conducted outward.
- On a time scale of about one millisecond, sodium permeability falls back to zero, and potasium-ion permeability rises.
- Increased potasium-ion permeability allows positively-charged potassium ions to flow outward, carrying V back to its resting value.
- After the resting state is reestablished (the refractory period), the nerve is ready to carry another pulse.

Briefly, the story is that sodium ions flow into the axon along the leading edge of a squid nerve pulse, causing the transmembrane voltage (V) to rise, and then potassium ions flow outward along the trailing edge of the pulse, causing V to fall. Under the passage of a single pulse, the internal ionic concentrations remain essentially constant, although axonal fatigue caused by changes in these concentrations does occur when large numbers of pulses are conducted in rapid succession. In this case, the original resting concentrations are restored through the actions of *ionic pumps* which are located in the membrane [436].

In 1952, Alan Hodgkin and Andrew Huxley (HH) quantified the nonlinear dependencies of the transmembrane ionic currents and carried through a traveling-wave analysis of (4.2), finding values for pulse speed and time courses of V and of transmembrane conductivity that are in good agreement with Cole's measurements which are shown in Fig. 4.1 [440]. Done on a mechanical hand calculator in the last of the precomputer days, this was a computational tour de force.

Although there is a superficial resemblance between the nerve pulse of Fig. 4.1 and the hydrodynamic soliton shown in Fig. 3.1b, they should be viewed as qualitatively different dynamic entities for several reasons. First, the KdV equation (3.2) models an energy-conserving (Hamiltonian) system, whereas the reaction-diffusion systems modeled by (4.2) conserve nothing [875]. (Physicists, who are trained to treat energy-conserving wave systems, sometimes overlook this essential difference.) Second, the speed of a hydrodynamic solitary wave in a uniform channel depends on the initial conditions (the size and shape of the heap of water on the left-hand side of Fig. 3.1a), whereas the speed of a nerve pulse on a uniform axon depends on the parameters of the system. Third, energy-preserving solitary waves obey Newton's second law ($F = ma$) under perturbations, accelerating and decelerating in response to applied forces. A nerve pulse, on the other hand, doesn't remember its previous speed and responds directly to the local properties of the medium.[2] Finally, a soliton or energy-conserving solitary wave expresses a dynamic balance between the opposing effects of nonlinearity and dispersion, whereas a nerve pulse establishes a dynamic balance between the dissipation of energy and the rate of its release through a nonlinear threshold mechanism [875]. It is important to be aware of these essential differences, as physicists have incorrectly called nerve pulses "solitons" under their loose definition [306].

Interestingly, the shape of the leading edge of the pulse shown in Fig. 4.1 is mathematically similar to the growth curve of Fig. 1.1 – an initial exponential rise followed by a leveling off (or saturation). Because the leading edge sets the pace of propagation, the speed of a nerve pulse can be analytically estimated by representing the initial sodium-ion current as the cubic polynomial

$$j_{\text{sod}} \doteq g\frac{V(V - V_1)(V - V_2)}{V_2(V_2 - V_1)} , \tag{4.3}$$

where g is the conductance per unit length of the membrane at the peak of the pulse (the maximum value of the band in Fig. 4.1), V_2 is the peak value of voltage, and V_1 is a threshold voltage above which sodium current begins to flow into the axon. A mathematically identical system was studied by Yakov Zeldovich and David Frank-Kamenetskii as a model of flame-front propagation in 1938 – the same year that Cole took the classic photograph of Fig. 4.1.

This ZF equation [680, 1083]

$$D\frac{\partial^2 u}{\partial x^2} - \frac{\partial u}{\partial t} = \frac{u(u - a)(u - 1)}{\tau} , \tag{4.4}$$

is a simple reaction-diffusion system in which D is a thermal diffusion constant and τ is a turn-on time for exponential growth of the propagating

---

[2]The difference between a soliton (or energy-conserving solitary wave) and a nerve pulse is like that between a Galilean and an Aristotelian concept of a moving particle.

Table 4.1. Parameters for (4.7) corresponding to a standard Hodgkin–Huxley axon

Parameter	Value	Units
Diameter (d)	0.476	mm
Temperature (T)	18.5	°C
g	.0108	mhos/cm
r	2.0×10^4	ohms/cm
c	1.5×10^{-7}	F/cm

conflagration. While still in their mid-twenties, these two Russian scientists[3] obtained an exact solution for (4.4) as the traveling wave [1083]

$$u_{\text{tw}}(x - vt) = \frac{1}{1 + \exp\left[(x - vt)/\sqrt{2D\tau}\right]},\qquad (4.5)$$

with speed

$$v = (1 - 2a)\sqrt{\frac{D}{2\tau}}.\qquad (4.6)$$

If this result had been known to Cole, he could have immediately expressed the speed of a squid nerve pulse as

$$v = \sqrt{\frac{g}{rc^2}}\,\frac{V_2 - 2V_1}{\sqrt{2V_2(V_2 - V_1)}},\qquad (4.7)$$

where the parameter values for the standard axon studied by HH are given in Table 4.1 [874].

Taking V_1 (the voltage at which sodium current begins to flow) as 25 mV and V_2 (the maximum pulse voltage) as 100 mV, (4.7) gives a pulse velocity of 20.0 m/s, which compares favorably with the HH computed value of 21.2 m/s at 18.5°C [440].

Furthermore, g and c are both proportional to d, whereas r is proportional to $1/d^2$; thus (in a neurological version of Froude's law[4])

---

[3]Largely self-taught, Yakov Borisovich Zeldovich (1914–1987) began his scientific career as a laboratory assistant in 1931 and eventually rose to become a Soviet Academician who made many important contributions to combustion theory, relativistic astrophysics, and cosmic evolution. One wonders if the fact that Yakov Borisovich was not university trained helped him to appreciate the importance of this problem. David Albertovich Frank-Kamenetskii (1910–1970) was an internationally known expert in chemical kinetics, after whom a well-known parameter for spontaneous combustion is named [320]. His son believes that he was aware of the connection between (4.4) and nerve pulse propagation during the 1960s [322].

[4]Based on his own wave-tank measurements and in accord with earlier results obtained by John Scott Russell, William Froude proposed in 1862 that the speed of a ship is roughly proportional to the square root of its length [289].

$$v \propto \sqrt{g/rc^2} \propto \sqrt{d} \, , \tag{4.8}$$

which was one of the cases studied in 1933 by Rashevsky [787,788].[5] The pulse speed on a *myelinated nerve* (in which the pulse jumps from one active node to the next) is proportional to the fiber diameter, which is more satisfactory for the functioning of motor axons [874].

Thus we see that much can be learned about nerve-pulse propagation from analytical studies of simple reaction-diffusion systems. From a general perspective, (4.7) can be viewed as the product of a parameter factor

$$\sqrt{g/rc^2} \tag{4.9}$$

times a structure factor

$$\frac{V_2 - 2V_1}{\sqrt{2V_2(V_2 - V_1)}} \, . \tag{4.10}$$

The structure factor was found in 1938 by ZF [1083], and the parameter factor had been proposed by Luther in 1906 [see (4.1) and [583]], where the diffusion constant in (4.2) is evidently $D = 1/rc$ and the response time for the onset of sodium-ion current flow is $\tau = c/g$.

Clearly, it would have been useful for neuroscientists of the 1930s and 1940s to know what was happening in contemporary studies of flame-front propagation. In retrospect, this lack of scientific communication is particularly distressing because in the middle of the nineteenth century Michael Faraday had presented a series of Christmas Lectures to young people on "The Chemical History of a Candle", opening with the epigraph to this chapter [300].[6] From Faraday's lectures – which can be profitably read by nonlinear scientists a century and a half later – one sees that the candle is an apt metaphor for nerve-pulse propagation which Helmholtz vainly sought, no doubt while seated at a table under its very light. In the flame of a candle, chemical energy is released from the wax through vaporization, corresponding for a nerve pulse to the release of electrostatic energy from the fiber membrane that is triggered by a rise in transmembrane voltage.

4.2 Simple Nerve Models

Although the lack of communications between neuroscience and combustion chemistry which led to the decades-long neglect of the ZF analysis of (4.4) was

---

[5]This result is in accord with my measurements on about two-dozen axons of *Loligo vulgaris* at 18.5°C which gave

$$v = 20.3\sqrt{d/0.476} \text{ m/s} ,$$

to an accuracy of about five percent [874].

[6]An engraving of Faraday presenting his Christmas Lectures was on a recent British 20 pound note.

bad enough, a more egregious oversight occurred between two branches of biological science. In 1937 the famed English geneticist and statistician Ronald Fisher proposed a means for wave-like advance of advantageous genes through a two-dimensional population [308], and in the very same year traveling-wave propagation on this system was independently discussed in a long article by the Russian mathematician Andrey Kolmogorov and his colleagues, who changed the RHS term in (4.4) to $u(u-1)/\tau$, finding a minimum wave speed of $2\sqrt{D/\tau}$ [505]. Evidently, this model can also be obtained by adding a diffusion term $\partial^2 u/\partial x^2$ to the Verhulst equation discussed in Chap. 1, but in spite of the rather obvious model and the famous names involved, this work was also overlooked by the neuroscience community for several decades [864]. Prior to the 1970s, it seems, scientists working in neuroscience did not realize that they could learn from mathematical analyses in seemingly unrelated fields.

In the early 1960s, nonlinear wave propagation on reaction-diffusion systems became of interest to the electrical engineering community for three reasons. First was the proposal by Hewitt Crane of the *neuristor* as a nonlinear electronic line that can carry a signal without attenuation, as does a nerve axon [199, 884]. Various interconnections of neuristors, Crane showed, offer a means for realizing any Boolean logic circuit and thus a computer of arbitrary power. Second, the invention of several solid-state diodes with nonlinear differential conductance [848] brought the design of novel neuristor structures within the bounds of technical feasibility [861]. Finally, electronics has always been an important aspect of electrophysiology, so it was natural for electrical engineers to become involved with the scientific and modeling aspects of nerve conduction [697, 857]. During the 1960s, neuristor research was carried on in the electrical engineering communities of several countries, including Denmark, Japan, Russia and the United States [861].

Although Rashevsky had introduced and underscored the importance of nerve conduction studies in the 1930s as a component of his program in mathematical biophysics [787, 788] and Ludwig von Bertalanffy had emphasized the significance of open systems for theoretical biology in 1950 [1001], it was not until the early 1970s that US applied mathematicians became aware of the theoretical and practical importance of reaction-diffusion processes. Their collective interest was triggered by a highly-cited paper of New York University's Henry McKean which studied an augmentation of (4.4) to the third-order system [639]

$$D\tau\frac{\partial^2 u}{\partial x^2} - \tau\frac{\partial u}{\partial t} = u(u-a)(u-1) + r\,,$$
$$\tau\frac{\partial r}{\partial t} = \varepsilon(u+c-br)\,,$$

$$(4.11)$$

where r is a *recovery variable* which forces u back to its resting value of zero on a time scale of order $\tau/\varepsilon b$. In other words, r roughly models the outward flow of potassium ion current from an axon under the above description of

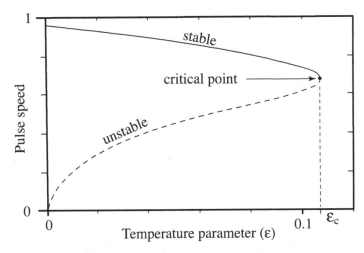

Fig. 4.3. Plot of pulse speed vs. temperature parameter for a typical FN nerve model [874]

the HH dynamics. In the limit $\varepsilon \to 0$, this potassium time constant (of $\tau/\varepsilon b$) becomes infinite and the third-order (FN) system of (4.11) reduces to the second-order (ZF) system of (4.4). As the potassium ion current responds more quickly at higher temperature for a real nerve, ε is often referred to as the temperature parameter in an FN model.

Called the FitzHugh–Nagumo (FN) equation, (4.11) was introduced by Jin-ichi Nagumo and his University of Tokyo colleagues in 1962 [697, 857], recognizing interactions with Richard FitzHugh at the US National Institute of Health [310]. During the 1960s, the FN system was extensively studied in the engineering literature, with which McKean seemed unacquainted in 1970 [861, 874]. (Applied mathematicians in the US may have been influenced in this tardy response by the US physics community, among whom reaction-diffusion processes were not of interest until the 1980s, perhaps because physicists were in the habit of dealing with energy-conserving fields.) Further interest in reaction-diffusion systems among the US applied mathematics community was generated during the 1970s by Charles Conley at the University of Wisconsin, who encouraged his research students to use generalizations of the FN system as an example of his "method of isolating blocks" for establishing the existence of pulse-like solutions in the singular limit $\varepsilon \to 0$ [156, 157, 188, 472]. From its timing, clearly, this sudden mathematical interest in reaction-diffusion processes was driven by the general explosion of research activity in nonlinear dynamics that occurred in the seventies.

A nerve pulse differs qualitatively from a soliton, and to see this difference, consider Fig. 4.3, which shows pulse speed plotted against the temperature parameter (ε) for a typical FN model, noting several features [874]:

- The speed of a nerve pulse depends entirely on the parameters of the system.
- There is a maximum value of the temperature parameter (ε_c) beyond which traveling-wave solutions don't exist. This corresponds to the empirical fact that real nerves do not function above a certain critical temperature.
- At values of the temperature parameter below (ε_c), there are two traveling-wave solutions, with different speeds. The solution of higher speed and amplitude is dynamically stable, whereas the lower speed solution is unstable. Thus the higher speed solution corresponds to a normal nerve pulse, and the lower speed solution is related to the threshold condition for launching a pulse.

4.3 Reaction Diffusion in Higher Dimensions

Prior to the nonlinear science paradigm shift, there were other insightful scientists whose work was ignored, prominent among whom was Boris Belousov, a chemist and biophysicist working at the Soviet Ministry of Health. In the 1940s, Belousov sought an inorganic equivalent to the Krebs cycle (which is a key part of the energy chain in biological organisms), and in 1951 he discovered a chemical reaction (involving citric acid, bromate ions and a catalyst) that oscillates periodically between an oxidized and a reduced state as indicated by changes in its color. His initial attempt to publish this important work failed because a benighted editor deemed his "supposedly discovered discovery" in violation of the Second Law of Thermodynamics and suggested that Belousov was merely observing artifacts of inhomogeneity (poor mixing) [1048]. After six years of careful work, Belousov submitted a revised manuscript with much additional empirical evidence to another chemical journal, again meeting skepticism. Angered that this second journal would only permit him to publish an abbreviated letter to the editor, deleting the empirical evidence, Belousov withdrew his manuscript with a vow never to publish it.

These refusals to publish Belousov's research are difficult to understand for several reasons. First, as mentioned above, Luther had demonstrated an oscillating chemical reaction to an interested audience at an important meeting of German physical chemists in 1906 [583]. Second, Alfred Lotka published a theoretical analysis of propagating chemical reactions in 1920 [578] which had received prompt experimental confirmation by William Bray [122], with both papers appearing in the *Journal of the American Chemical Society*. Third, Rashevsky and his students had studied nonlinear reaction-diffusion systems in the 1940s in the context of cellular growth [790].[7] Finally, Alan

---

[7]It was through reading Rashevsky's books in the 1950s that I became interested in mathematical biophysics.

Turing developed a mathematical theory of pattern formation in 1952, which was based on the concept of reaction-diffusion dynamics [973], extending the classic work of D'Arcy Wentworth Thompson on biological morphogenesis [961].[8] These were not obscure publications, and the reason that the editors of two different Russian chemistry journals failed to welcome Belousov's manuscript must have been that his ideas were not yet included in the lexicon of chemical science. Although encouraged by Simon Shnol' from the Institute of Biophysics in Pushchino to continue his work on oscillating chemical reactions, Belousov quit science and died a bitter man in 1970 – just as the explosion of interest in nonlinear science was dawning. In 1980, ironically, Belousov was posthumously awarded the Soviet Union's prestigious Lenin Prize for his research.

In 1961, Shnol' suggested Belousov's work to Anatol Zhabotinsky, a research student who slightly modified the chemistry (changing citric to malonic acid and adding ferroin sulfate as a color indicator) to obtain the Belousov–Zhabotinsky (BZ) reaction [252]. Thus in 1964, Zhabotinsky was able to confirm Belousov's results [1084], and in 1970 Albert Zaikin and Zhabotinsky published the photographs of self-oscillating *ring waves* in this reaction that are shown in Fig. 4.4a [1075].

It was during the mid-1960s that Art Winfree first saw self-exciting *spiral waves* on an acid-iron-wire screen (Lillie's model [563], comprising a section of iron screen in a bath of weak nitric acid) while visiting Nagumo's laboratory at the University of Tokyo. This phenomenon piqued his curiosity, and it was while trying to prove that spiral waves cannot exist in a continuous reaction-diffusion medium that he discovered them in the two-dimensional BZ reaction as is shown in Fig. 4.4b.[9]

Examples of two-dimensional reaction-diffusion waves in nature include prairie fires, lichens [875], social amoeba (slime molds), and fairy rings of mushrooms [48], among others [790]. Interestingly, these phenomena can all be modeled by a two-dimensional version of the FN system of (4.11) with

$$D\frac{\partial^2 u}{\partial x^2} \longrightarrow D\left(\frac{\partial^2 u}{\partial x^2} + \frac{\partial^2 u}{\partial y^2}\right) = D\left(\frac{\partial^2 u}{\partial r^2} + \frac{1}{r}\frac{\partial u}{\partial r} + \frac{1}{r^2}\frac{\partial^2 u}{\partial \theta^2}\right),$$

where x and y are Cartesian coordinates and r and θ are cylindrical coordinates.

As is evident from Fig. 4.4, circular and spiral waves are characterized by two qualitative features:

- The formation of cusps where wave fronts meet.

---

[8]Turing was a well-known English mathematician who had done seminal work on the theory of computers, which he used to crack the German military code *Enigma* during World War II.

[9]For an interesting collection of films demonstrating a variety of oscillating chemical reactions, see http://www.williams.edu/Chemistry/epeacock/.

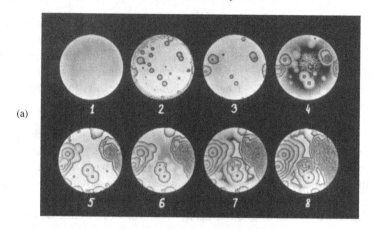

(a)

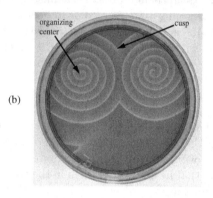

(b)

Fig. 4.4. (a) Self-oscillatory ring waves in a two-dimensional BZ system. (The circular diameter is 100 mm and the successive photographs are at intervals of 1 min.) (Courtesy of A.M. Zhabotinsky.) (b) Two-dimensional BZ waves are spiraling outward from organizing centers and intersecting to form cusps. (Courtesy of A.T. Winfree)

- Smooth locally circular curves.

The former property is typical of wave processes having refractory zones behind the moving front because two-dimensional waves cannot then cross. The latter property arises because the velocity of a front decreases as the local value of the curvature increases. To see this, note that neglecting θ dependence leads to a PDE of the form

$$D\frac{\partial^2 u}{\partial r^2} + D\frac{1}{r}\frac{\partial u}{\partial r} - \frac{\partial u}{\partial t} = \frac{u(u-a)(u-1)}{\tau} \, ,$$

which is identical to (4.4) except for the second term $(D/r)(\partial u/\partial r)$. Under the assumption that $r \gg D/v_0$ (which validates the neglect of θ dependence) the velocity v of a circularly symmetric front is given by [875]

Fig. 4.5. Dynamic elimination of a local bump

$$v = v_0 - D/R , \qquad (4.12)$$

where v_0 is the zero-curvature velocity of a line wave and R is the local radius of curvature of the wave front.

As shown in Fig. 4.5, (4.12) forces the local geometry to be circular because arcs with larger radii of curvature (R) propagate faster than those with smaller R. In other words, a small irregularity will be overtaken by a broader wave and thereby evolve into a smooth and circular front.

In three space dimensions, there are many self-exciting configurations because an organizing center can be extended into a *vortex line*, and this line can be tied into a variety of linked rings and knots. The simplest of these is the straight line of a *scroll wave*, from which the propagating surface unrolls like a piece of ancient Chinese art.

From analyses corresponding to those used for the two-dimensional case, the propagation velocity of a three-dimensional reaction-diffusion front will be

$$v = v_0 - D \left(\frac{1}{R_1} + \frac{1}{R_2} \right) , \qquad (4.13)$$

where v_0 is the speed of a plane wave and R_1 and R_2 are maximum and minimum values of the local radii of curvature.

An interesting example is the three-dimensional *scroll ring* displayed in Fig. 4.6, which evolves outward at a speed given by (4.13) from a circle of radius R. The global dynamics of a scroll ring can be studied in the following way. The outer surface at $R + \rho$ propagates outward at speed

$$v_{\text{outer}} = v_0 - D \left(\frac{1}{\rho} + \frac{1}{R + \rho} \right) ,$$

where ρ is the distance from the vortex ring to the propagating surface. Likewise, an inner surface at radius $R - \rho$ propagates inward at speed

$$v_{\text{inner}} = -v_0 + D \left(\frac{1}{\rho} - \frac{1}{R - \rho} \right) ,$$

because the curvature around the inside of the ring is negative, while that across the ring is positive (see Fig. 4.6). Thus the average of v_{outer} and v_{inner} is

$$\langle v \rangle = \frac{v_{\text{outer}} + v_{\text{inner}}}{2} = -\frac{D}{2} \left(\frac{1}{R + \rho} + \frac{1}{R - \rho} \right) = -\frac{DR}{R^2 - \rho^2} .$$

Fig. 4.6. A computer plot of a scroll ring, generated from a three-dimensional version of the FN equations. (Courtesy of A.T. Winfree)

As $\rho/R \to 0$, the average velocity approaches that of the vortex ring of radius R. In other words

$$\langle v \rangle \longrightarrow \frac{\mathrm{d}R}{\mathrm{d}t} = -\frac{D}{R} \, .$$

Upon integration, this differential equation implies that the radius of a ring shrinks slowly with time as

$$R = \sqrt{R_0^2 - 2D(t - t_0)} \, , \tag{4.14}$$

which is in accord with experiments carried out by Winfree and his colleagues on active chemical preparations [464].

I have presented this calculation to demonstrate that much can be done with simple analyses of reaction-diffusion waves in three-dimensional space, but the scroll ring is more than a mathematical curiosity. The human heart is – among other things – a reaction-diffusion medium for electrochemical activity that can be disrupted by the spontaneous formation of an internal scroll ring. If this "fibrillating" behavior is not immediately suppressed, the owner of the heart may succumb to "sudden cardiac death" [1049, 1050].

Generalization of the above ideas leads to a nonlinear geometrical optics for reaction-diffusion systems, which has been developed by Oleg Mornev [678]. From this perspective, a nonlinear version of Snell's law of refraction can be derived, in addition to conditions for blocking of a wave front at a boundary where the diffusion constant undergoes an abrupt change. This blocking condition, in turn, is useful in finding analytic conditions for nerve-pulse propagation to fail at branching regions of axons and dendrites [874].

Although (4.13) holds for stable three-dimensional wave fronts in relatively simple reaction-diffusion processes, such fronts are not always stable, especially when gravitational or convective forces come into play. In his above mentioned lectures for young people, Faraday noted that flames often flicker in a spatially chaotic pattern [300, 474], a phenomenon that was studied by Lord Rayleigh [792] and later by Geoffrey Taylor [949] and is now called the Rayleigh–Taylor instability [617].

After the speed of nerve-pulse conduction was first measured by Helmholtz in 1850, it was more than a century before the phenomenon was unambiguously explained as a *nonlinear* reaction-diffusion process, because research was impeded by at least three factors. First, we have seen again the extreme balkanization, the almost total lack of communication between areas of scientific inquiry that appear to be different but are nonetheless related by their underlying mathematical structure. Second, researchers in neuroscience were led astray by the qualitative character of *linear* diffusion, which is not at all wavelike. Thus a solution of the linear diffusion equation

$$\frac{\partial u}{\partial t} = D \frac{\partial^2 u}{\partial x^2}$$

necessarily decreases (increases) with time where the second spatial derivative is negative (positive), causing a pulse-like initial condition to spread out and disappear – like a puff of smoke – rather than propagate. A corresponding property, I suspect, was subconsciously anticipated for reaction-diffusion systems, for how else can one explain that a scientist with the analytical ability of Helmholtz did not formulate (4.4) and find the traveling-wave solution of (4.5)? Finally, the fact that US applied mathematicians ignored reaction-diffusion studies for two decades after the Hodgkin–Huxley and Turing papers were published [440, 973] suggests that before 1970 they were not comfortable with the concept of localized traveling waves emerging as solutions to a diffusive nonlinear PDE.

5 The Unity of Nonlinear Science

> Nonlinear science [comprises] mathematical studies from completely integrable to completely stochastic systems, studies on nonlinear systems described by stochastic governing equations (differential or otherwise) as well as deterministic nonlinear systems yielding stochastic behavior, and nonlinear studies throughout the physical and social sciences from astronomy to zoology.
>
> Joseph Ford

Scientists tend to concentrate on the work at hand, often ignoring the relations between our own research and that of others. Thus the aim of this chapter is to step back and give the reader a broader picture of what goes on under the banners of nonlinear science and related activities. We will see how the ideas of the previous chapters are interrelated, survey some general approaches to nonlinear science, and look at relations among these different approaches.

5.1 The Provinces of Nonlinearity

Like Caesar's Gaul, nonlinear science has been divided into three parts, yet there are many connections among

- low-dimensional chaos,
- energy-conserving solitary waves, and
- reaction-diffusion (RD)

systems which encourage us to consider them together as a unified realm. A fundamental reason behind this unity is that all three provinces involve closed causal loops of positive feedback, which influence their dynamics in essential ways. In chaotic systems, positive feedback loops lead directly to the positive Lyapunov exponents (exponential divergence of nearby solution trajectories) which are a basic requirement for Poincaré's "fortuitious phenomena". In solitonic and RD systems, on the other hand, positive feedback loops are

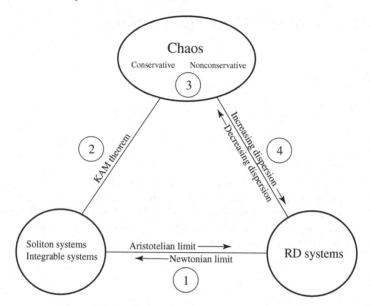

Fig. 5.1. Interrelations among low-dimensional chaotic systems, solitonic systems, and reaction-diffusion (RD) systems

essential for the emergence of localized regions of energy (a hydrodynamic or optical soliton) or dynamic activity (the flame of a candle or a nerve impulse). As is suggested by Fig. 5.1, the models of one province transform smoothly into those of another under continuous parameter variations, but let's look at some of the details.[1]

5.1.1 Solitons and Reaction Diffusion

One example of this interprovincial unity can be appreciated by comparing the sine-Gordon (SG) equation

$$c^2 \frac{\partial^2 u}{\partial x^2} - \frac{\partial^2 u}{\partial t^2} = \sin u \, ,$$

(where c is the limiting velocity of a kink) with the Zeldovich–Frank-Kamenetskii (ZF) equation

$$D \frac{\partial^2 u}{\partial x^2} - \frac{\partial u}{\partial t} = u(u - a)(u - 1) \qquad \text{with } 0 < a < 1 \, ,$$

where D is a diffusion constant. In both of these equations, a fundamental linear PDE is modified by adding a term to the right-hand side (RHS) that

---

[1]A variety of interesting films demonstrating features of Fig. 5.1 nonlinear dynamics can be found at http://www.pojman.com/NLCD-movies/NLCD-movies.html

depends nonlinearly on the dependent variable. Because the linear wave operator on the LHS of the SG equation (the wave equation) differs from the linear diffusion operator on the LHS of the ZF equation (the diffusion equation), one would expect these two equations to have very different dynamics, as indeed they do (see Chaps. 3 and 4).

The connection between these two systems becomes evident when we compare two related equations. First is a perturbed version of the SG equation (PSG) which has been augmented by the addition of terms that represent energy input and dissipation to[2]

$$c^2 \frac{\partial^2 u}{\partial x^2} - \frac{\partial^2 u}{\partial t^2} - \sin u = \gamma + \alpha \frac{\partial u}{\partial t} - \beta \frac{\partial^3 u}{\partial x^2 \partial t} \; . \tag{5.1}$$

Second is the FitzHugh–Nagumo (FN) system which was introduced in the previous chapter and is repeated here with a normalized notation [697, 874]:

$$D \frac{\partial^2 u}{\partial x^2} - \frac{\partial u}{\partial t} = F(u) + r \; , $$

$$\frac{\partial r}{\partial t} = \varepsilon(u + k - br) \; , \tag{5.2}$$

where $F(u) = u(u - a)(u - 1)$ or some other function with multiple zeros. Although these two systems seem to be quite different, note that (5.2) is equivalent to the third-order PDE

$$(D\varepsilon b) \frac{\partial^2 u}{\partial x^2} - \frac{\partial^2 u}{\partial t^2} - \varepsilon[u + bF(u)] = \varepsilon k + [\varepsilon b + F'(u)] \frac{\partial u}{\partial t} - \frac{\partial^3 u}{\partial x^2 \partial t} \; , $$

which has the same general form as (5.1) for $\varepsilon b \gg F'(u)$.

From this comparison of the PSG equation and the FN system, it follows that each can be smoothly changed into the other under continuous parameter variations. In other words, making the loss terms large in a PSG equation leads to an RD equation that is similar to the FN system, and increasing the parameter ε in a FN system leads to a lossy solitary wave system. This transition is indicated by the horizontal line at the bottom of Fig. 5.1 which is labeled 1 and indicates

$$\text{PSG} \longleftrightarrow \text{FN} \; .$$

Moving to the right along this line makes a PDE model more lossy, eventually leading to an RD system. Moving to the left makes an RD system less lossy, eventually arriving at a lossless solitary wave (or soliton) system.

---

[2]In the context of a long Josephson junction (see Sect. 6.6.7 and [641, 889]), γ is a forcing term that is proportional to a bias current, the α-term accounts for dissipation from normal electrons crossing the junction, and the β-term accounts for dissipation from normal electrons flowing parallel to the junction [858].

In the real world, of course, there are often dissipative effects (electrical resistance in electronic and ionic systems, viscosity in hydrodynamics, etc.) that modify the dynamical behavior of solitons and energy-conserving solitary waves, just as air resistance influences on the motion of a golf ball or a bullet and water resistance influences the dynamics of a torpedo. This phenomenon was at the heart of the famous disagreement between supporters of Newtonian and Aristotelian dynamics, which most believe was settled in favor of the Newtonians by Galileo's dropping of a heavy and a light stone from the famous leaning campanile of his home town, Pisa, but the matter is not so simple. If the stones are dropped in a viscous liquid (molasses, say, or water), the heavy stone moves faster than the lighter one (in accord with Aristotle's concepts), and if the stones are dropped in a medium of low viscosity (a perfect vacuum, say, or air), then they both move at about the same speed (in accord with Newton's laws of motion).

Thus the former case is called the Aristotelian limit and the latter one the Newtonian limit as is indicated on Fig. 5.1. In general, solitons (and lossless solitary waves) can be viewed as particles moving in the Newtonian limit, whereas nerve impulses (and other localized solutions of RD systems) move as particles do in the Aristotelian limit [679]. Newtonian particles have momentum, so if all external forces are zero the particle will continue to move at a constant speed – a key feature of Newton's concept of the Universe. Aristotelian particles, on the other hand, respond to the local conditions of energy input and loss without memory of their recent past – an essential aspect of Aristotle's physics [29].

The relationship between the PSG equation and the FN equation was presented in some detail as an explicit example, but the above concept is not restricted to these two systems. Consider, for example, the KdV equation defined in (3.2) or (3.4). If a small positive term proportional to $\partial^2 u/\partial x^2$ (representing visconsity of the water) is added to the RHS of these equations, the effect is dissipative; thus a KdV soliton will slow down and lose amplitude under the perturbation [875]. If this term is made large, the resulting equation becomes an RD system (the Aristotelian limit).

Such ranges of properties also arise in studies of equations related to the nonlinear Schrödinger (NLS) equation (see Sect. 3.3 and [597]), including the Swift–Hohenberg (SH) equation [549]

$$\frac{\partial u}{\partial t} = au + b\nabla^2 u + c|u|^2 u + d(1 + \nabla^2)^2 u , \qquad (5.3)$$

where a, b, c and d are complex parameters and

$$\nabla^2 \equiv \frac{\partial^2}{\partial x^2} + \frac{\partial^2}{\partial y^2}$$

is a two-dimensional differential operator on the spatial variables (x and y). This equation is of practical interest for those who would understand and control transverse instabilities (filamentation) in semiconductor and solid-state lasers (see Sect. 6.7.2 and [549, 837]). With $d = 0$ and a, b, c imaginary, (5.3) reduces to a two-dimensional version of the NLS equation with plane-wave soliton solutions given in (3.10). As we will see in Sect. 6.8.5, this two-dimensional NLS system is currently of great interest in modeling the dynamics of rogue waves on the surface of the oceans. With $d = 0$ and a, b, c real, on the other hand, SH reduces to the RD equation studied in 1937 by Ronald Fisher and by Andrei Kolmogorov and his colleagues as a model for genetic diffusion in two space dimensions (see Sect. 4.3 and [308, 505]).

Although this discussion is descriptive rather than exhaustive, the base line in Fig. 5.1 represents a variety of systems that are soliton-like (conserving energy and having constant momentum) in the Newtonian limit and nerve-impulse-like (consuming energy and moving with a speed that responds to the local conditions) in the Aristotelian limit. In many applications, realistic models will lie somewhere between these two limits, requiring the analyst to be familiar with the characteristic properties of both solitons and nerve impulses; thus nonlinear science is often concerned with the terrain between these two provinces.

5.1.2 The KAM Theorem

The left-hand edge of Fig. 5.1 (which is labeled 2) interpolates between Hamiltonian (energy-conserving) chaos and integrable systems, including soliton systems, and these sit on opposite ends of a gray scale that is governed by the KAM (for Kolmogorov, Vladimir Arnol'd, and Jürgen Moser) theorem (see Sect. 2.5). The many integrable systems now known reside in the circle in the lower left-hand corner [586, 878]; thus there are many relationships between solitons and Hamiltonian chaos, including the following.

First, the discrete system introduced by Morikazu Toda, called the Toda lattice (TL), is a soliton system in its own right (see Sect. 3.5), and a long-wavelength approximation to TL dynamics leads to the KdV equation, as Norman Zabusky and Martin Kruskal pointed out in their 1965 paper in which the term "soliton" was coined [1074]. If periodic boundary conditions are imposed on a TL and the number of elements is small, we have an example of a low-dimensional soliton system. In other words, the low-dimensional system is integrable, which is a condition for applying the KAM theorem. Now suppose that the potential function defining the nonlinear springs is slowly changed (to order ε) from the exponential form given in (3.22) to some other function. The integrability of the system will be lost, but the KAM theorem asserts that for small ε most of the solution trajectories will lie on toroidal surfaces in the phase space (see Fig. 2.4), insuring the structural stability of the soliton-like solutions. Thus the structural stability of TL solitons under Hamiltonian (energy-conserving) perturbations provides an example of the KAM theorem.

Second, as we have seen in Sects. 2.5 and 3.4, kink solutions (solitons) of the SG equation are closely related to solutions of the nonlinear pendulum problem

$$\frac{d^2u}{dt^2} = \sin u ,$$ (5.4)

which is integrable in terms of Jacobi elliptic functions [875], and this correspondence becomes exact if we replace the time variable t in (5.4) with the traveling-wave variable $x - vt$, where v is a traveling-wave speed. Considering a discrete version of (5.4), as in (2.5), we arrive at the Taylor–Chirikov map, which again demonstrates KAM structural stability but exhibits low-dimensional chaos when the parameter K, defined in (2.6), becomes of the order of one or more.

Third, in yet another example of the relations among solitons and chaos, Chris Eilbeck, Peter Lomdahl and Alan Newell have shown that a periodically driven SG system exhibits low-dimensional chaos [270]. Although the SG equation is a PDE of high (infinite) dynamic dimension, one can view the dynamics of a few emergent solitons (kinks, antikinks, and breathers) as the fundamental entities and consider the effects of a periodic driving force on their trajectories, calling KAM into play. From the interdisciplinary perspectives of nonlinear science, this study is a one-dimensional version of the low-dimensional chaos in the planetary motions of our Solar System (see Sect. 2.7).

Finally, if the standard map, given in (2.6), is modified by adding a term in the second derivative of another independent variable, it becomes a discrete version of the sine-Gordon (SG) equation, which was first studied by Yakov Frenkel and Tatiana Kontorova in 1939 as a model for the dynamics of dislocations in crystals [331]. In the limit of weak coupling (SG limit), this system is integrable, whereas the standard map is chaotic; thus these two limits are related by the KAM theorem.

5.1.3 Chaos

The upper oval, labeled 3 in Fig. 5.1, includes chaotic systems of low dynamic dimension. The reason for specifying *low-dimensional* chaos is that scientists have long been aware of practical problems in integrating the equations of motion for a system with a large number of dynamic dimensions (the 6.02×10^{23} molecules of a mole of gas, for example), although some have claimed that this integration would be possible "in principle" if computing tools of sufficient power were available. Thus it was a surprise when Edward Lorenz showed that there can be insurmountable difficulties in integrating nonlinear systems, even with a small number of phase-space dimensions (see Sect. 2.3 and Appendix A).

As we have seen in Chap. 2, these low-dimensional chaotic systems can be either energy-conserving or dissipative, and scientists who study them do not limit their attentions to one type or the other. Dissipative systems that display low-dimensional chaos include the Lorenz equations (see Fig. 2.1), the Rössler and Hénon systems, and the logistic map (Fig. 2.3), among many others. Chaotic systems that conserve energy include many versions of the N-body problem of planetary motion, the Hénon–Heiles system (Fig. 2.4), the standard (Taylor–Chirikov) map, and their analytical relatives.

Chaotic systems of large dimension give rise to the Brownian motion of small particles in a fluid, which is a physical example of the continuous but nondifferentiable curves that were discussed in detail by Benoit Mandelbrot (Fig. 2.2) [599]. Also the phenomenon of hydrodynamic turbulence, which was noted by Leonardo da Vinci as shown in Fig. 2.5, is an example of chaotic behavior in a large-dimensional system [707], to which we will return in Sect. 6.8.7.

5.1.4 Reaction Diffusion and Chaos

The right-hand edge labeled 4 of the diagram in Fig. 5.1 represents several re-lationships between dissipative low-dimensional chaos and the localized waves of activity in RD systems, but note that the KAM theorem, being restricted to Hamiltonian systems, does not apply to these models. Among such rela-tionships are the following.

First, the Verhulst equation is a model for the exponential growth of research in nonlinear science during the 1970s (see Sect. 1.2 and Fig. 1.1) and an early example of a completely solved nonlinear equation that arose in the context of a practical problem: population dynamics. As shown in Table 2.1 and Fig. 2.3, however, a discrete version of this equation displays a variety of periodic and chaotic behaviors, which were unexpected before the nonlinear-science revolution of the mid-1970s. Adding the spatially dispersive term

$$D\left(\frac{\partial^2}{\partial x^2} + \frac{\partial^2}{\partial y^2}\right)$$

to (1.2), on the other hand, yields the RD system that was proposed by Fisher in 1937 and studied by Kolmogorov and his colleagues as a model for the advance of biological genes [308, 505]. This system, in turn, is closely related to the ZF equation studied in 1938 by Yakov Zeldovich and David Frank-Kamenetskii as a model for flame-front propagation [1083] and more recently to model nerve-impulse propagation [874] (see Sects. 4.1 and 4.2).

A second example is the Swift–Hohenberg system given in (5.3) and dis-cussed above in the context of relations between soliton and RD systems. With $d = 0$, the SH equation reduces to the complex Ginzburg–Landau equation which has been studied as a nonlinear system that exhibits either filament formation or two-dimensional spatio-temporal chaos, depending on the values that are chosen for the complex parameters a, b and c [596].[3]

In RD systems of two or three spatial dimensions, finally, spatial chaos is often observed to emerge from interactions among stable solitary waves – the fell fibrillation of heart muscle being an important example, as Art Winfree has shown [1050]. Importantly, both dissipative chaos and RD solitary waves take advantage of dynamic instability (positive feedback) to establish their qualitative behaviors. In case of dissipative chaos, the attractor is said to be strange, and in the RD case a traveling-wave front emerges to balance the release and dissipation of energy.

5.2 Metatheories of Nonlinear Science

Beginning in the middle of the twentieth century, several approaches have been proposed for studies of biological and social systems that aim to describe some of the phenomena displayed in Fig. 5.1. All use the tools of physical science and are related to the nonlinear-science revolution of the 1970s, either as precursors or as contemporary activities, and there are substantial overlaps among them, suggesting the fundamental importance of the underlying ideas. Importantly, these are not special analyses for specific problems but general programs for broad areas of study; thus they should be familiar to general readers concerned with understanding physical, biological, psychological, or social dynamics. In addition to nonlinear science (NS), the seven formulations described in this section – *cybernetics* (C), *mathematical biology* (MB), *general systems theory* (GST), *nonequilibrium statistical mechanics* (NSM), *catastrophe theory* (CT), *synergetics* (S), and *complex adaptive systems* (CAS) – are the most widely known.

5.2.1 Cybernetics (C)

I first heard of cybernetics in 1950, while attending a sudent performance of Karel Čapek's play: *R.U.R. (Rossum's Universal Robots)*, which was memorable because MIT's mathematical genius, Norbert Wiener, gave an introduction that reviewed ideas from his recently published book *Cybernetics: Control and Communication in the Animal and the Machine* [1036].[4] This "new science" offered a striking example of the high technical morale that

---

[3]Opportunities to explore the dynamics of such solutions have been made available by Michael Cross at http://www.cmp.caltech.edu/mcc/Patterns/index.html

[4]In his introduction, Wiener spoke of the dawning "information age" and demonstrated a light-seeking robot which had recently been designed and built in the electrical engineering department – a small cart with an electric eye and servomotor driven wheels, all connected so it would move in the direction of maximum light intensity. When the house lights went down, he closed his remarks with a

dominated US science in the middle of the twentieth century. Both the theory and practice of communications were vastly improved over prewar levels and advances in electronics suggested the possibility of building computing machines that would free our minds from tedium, as the engines of the nineteenth century had lifted physical burdens from the backs of our grandparents.

On the crest of this technical euphoria, the Josiah Macy Jr. Foundation supported a series of high-level conferences, designed to bring disparate academics together for discussions on how the fruits of World War II research might help to solve post-war problems. Launched in March of 1946 at a meeting entitled "Feedback mechanisms and circular causal systems in biological and social systems", the aim was to study interdisciplinary areas in which closed causal loops play key roles in the dynamics, leading to systems with internal goals. (Following Plato, the term "cybernetics" derives from the Greek *kubernetes*, meaning "helmsman", and was used by André-Marie Ampère in 1834 for a science of civil government.)

In addition to Wiener, participants at the Macy conferences included psychiatrist William Ross Ashby, anthropologist Gregory Bateson, physicist and philosopher Heinz von Foerster, learning theorist Lawrence Frank, psychoanalyst Lawrence Kubie, psychiatrist Warren McCulloch, anthropologist Margaret Mead, physicist John von Neumann, neurophysiologist Rafael Lorente de Nó, mathematician Walter Pitts, and biologist Arturo Rosenblueth – a distinguished group indeed. At one of these meetings, Mead and von Neumann encouraged Wiener to write up his ideas in a book, which became his classic *Cybernetics* describing how the principles of negative feedback (NFB) are applied in the evolutionary design of biological organisms (see Sect. 6.6.3 and [189, 1036]).

Wiener was among those who introduced the concept of closed causal loops into biology, but he overemphasized the importance of NFB in relation to positive feedback (PFB). While NFB was correctly deemed necessary to explain our abilities to grasp an object and to stand erect, PFB was frowned upon as leading to spastic physiological behaviors like the unwanted *hunting* in control systems and *singing* in telephone repeater amplifiers (see Sect. 6.6.3). Overly concerned with these engineering problems, Wiener seemed unaware that the exponential growth stemming from PFB is essential for the two central phenomena of nonlinear science: chaos and emergence.

While Wiener became concerned with the problematic aspects of weather prediction in 1957 [1037] – suggesting an intuitive appreciation of chaos – he continued to ignore emergence [1038], which was then known to be an impor-

---

stirring Platonic paraphrase ("Either engineers must become poets or poets must become engineers!") and dramatically switched on a flashlight as the robot was kicked through the curtains. Unfortunately, this kick damaged one of the servomotors, so all the little thing could do was go round and round in circles, ignoring Wiener's guiding light – to the delight of his student audience.

tant biological phenomenon through the 1952 studies of nonlinear diffusion by Alan Turing [973] and by Alan Hodgkin and Andrew Huxley [440] (see Chap. 4). To its credit, however, the field of cybernetics uses closed causal loops to explain *homeostasis* by which an organism appropriately maintains its bulk, balance and body temperature while undertaking other goal-oriented activities [37], and the term has proliferated into variants like cyberspace, cyberpunk, cyborg, cyberculture, cyber rock, cyber goth, and recently even cyberterrorist – thus becoming widely known to the general public. Also, there are numerous international journals and research organizations devoted to applications of Wiener's basic ideas in various areas of the biological, cognitive and social sciences, one of which has had a strong influence on my personal scientific education.

Shortly before he died in 1964, Norbert Wiener encouraged Eduardo Caianiello (an Italian professor of theoretical physics with wide scientific interests) to found a *Laboratorio di Cibernetica* near Naples, Italy which would support a broad program of activities centered around collaborative studies of the human brain by interdisciplinary teams comprising mathematicians, physicists, chemists, engineers, physiologists and zoologists. Together with some of my electrical engineering students at the University of Wisconsin, I was fortunate to be invited to the *Laboratorio* to conduct soliton research in superconductive devices (see Sect. 6.6.7) during the 1969–70 academic year and for every summer during the 1970s, and it was there that I began experimental studies of nerve-impulse propagation on the giant axon of the squid (Sects. 4.1 and 4.2 and [893, 894]). Now an *Istituto* located in the historic Roman town of Pozzuoli, this imaginative establishment has made Wiener's dream of interdisciplinary research in biology and mathematics a reality for three generations of young scientists.

5.2.2 Mathematical Biology (MB)

A couple of years after attending Wiener's introduction to Čapek's play, I was becoming bored with an engineering job and began to think about how one might study the dynamics of living organisms. In this state of mind, it was exciting to discover Nicholas Rashevsky's two-volume opus *Mathematical Biophysics* (then recently published in a second edition) which shows how the tools of physical science can be used to discuss a host of interesting biological areas, including: cell growth and division, active transport across membranes, nonlinear diffusion, dynamics of cancer growth, self-regulation of organisms, excitation and conduction of nerve impulses, network analyses of the central nervous system, the Gestalt problem and pattern perception, locomotion, and more [790]. Reading this uniquely ambitious book encouraged me to begin graduate studies, and upon entering academia a decade later, the third edition of Rashevsky's work was my first purchase, strongly influencing my subsequent choice of professional activities.

Born in the last year of the nineteenth century in Chernihiv (Chernigov), about sixty miles north of Kyiv (Kiev), Rashevsky earned a doctorate in physics at the age of nineteen, soon publishing several of his early papers on quantum theory and on relativity in *Annalen der Physik*, the world's most distinguished physics journal in those inter-war days. Having fought with the Whites, it was expedient to leave Ukraine just as that country was entering a tragic period; thus he emigrated to the US, took a job as a research physicist at the Westinghouse Corporation in Pittsburgh, and began studying the thermodynamics of liquid droplets, paying particular attention to the concept of a maximum size. During a casual discussion, by his account, a biologist had curtly claimed that Rashevsky's studies could contribute nothing to the seemingly related problem of cell division (mitosis). Outraged by this presumption, he decided to devote his professional life to "the building-up of a systematic mathematical biology, similar in its structure and aims to mathematical physics".

At the invitation of several distinguished academics, including Arthur Compton, Ralph Lillie and Karl Lashley, Rashevsky joined the faculty of the University of Chicago in the early 1930s, where he established a mathematical biology program, in which his student-cum-colleague, Robert Rosen, reports that "he thanklessly blazed trails along which unknowing hordes plod today" [817] – a statement that bears inquiry. Trails were indeed blazed as the contents of his books and many papers show, and his program – which received start-up support for his program from the Rockefeller Foundation – was the first of its kind in the world. The adjective "thanklessly" and the phrase "unknowing hordes", on the other hand, stem from Rosen's frustrated observation that Rashevsky's contributions were and continue to be largely ignored by many biologist and scientists who work in related areas.[5]

In a major shift during the early 1950s, Rashevsky pointed out many of the shortcomings of previous research in mathematical biology and reoriented his program toward more global aspects of Life in an effort that he called *relational biology*, focusing attention on the organizational features of living creatures, their cognitive systems, and the structures of human and animal societies (see Sects. 7.2.7 and 8.3.2 and [789]). As part of this program, Rosen introduced a subtle branch of mathematics called *category theory* into mathematical biology, which emphasizes the relations among elements of sets – their organization – rather than the elements themselves and which, as we shall see in Chap. 8, has led to a novel and important definition of a *complex*

---

[5]Biographical details on Rashevsky's life are from Tara Abraham's recent historical study [8], Evelyn Keller's survey of developmental biology, accounts by Rosen [816,817,819], and an obituary in the *Bulletin of Mathematical Biology* [61]. My own dealings with Rashevsky during the mid-1960s, when he published some of my papers in his journal, revealed a kind and honest man with a sharp scientific intelligence, whom one would value as a colleague and friend; thus recent attempts to darken his memory are beyond my comprehension.

system [814,815,817,818]. At about the same time that Rashevsky was busy developing these important new activities, strangely, there was an effort at the University of Chicago to close down his program. Among the intricate reasons behind this sad situation are the following three [8].

First, the 1953 discovery of the DNA structure (by James Watson, Francis Crick, and Rosalind Franklin[6]) and the subsequent acceptance of the reductive perspectives of molecular biology seemed to most biologists to obviate the need for mathematical biology in general and for relational biology in particular – all of this would be explained by genetic codes.

Second, physical and life scientists see the relationship between mathematics and biology through different eyes. For the former, the interaction can awaken ideas for new theorems and theoretical formulations, as suggested by Stan Ulam's often-quoted dictum [408]:

> Ask not what mathematics can do for biology; ask rather what biology can do for mathematics!

Life scientists, quite naturally, tend to resent such encroachment by hordes of physicists who jabber in strange tongues and know few of the facts of Life. Ideally, biologists would have physical scientists play modest roles in their research – organizing and analyzing data, sorting out statistical implications, plotting sophisticated graphs, and so on – all activities that are appropriate for graduate student assistants but not for fully-fledged scientists. Having observed struggles over the proper role of mathematical physics in biological research playing out in five different institutions during as many decades,[7] I am persuaded that the "biomathematics problem" transcends details of institutional organization. It can, however, be overcome if researchers on both sides of the intellectual divide respect each other's integrity and make sincere efforts to understand the motivations of their opposite numbers.

Finally, Rashevsky was an outsider to the communities of both US physics and US biology; thus few were covering his back in the ugly little struggles that occasionally disgrace the halls of academe. Because he refused to fire two of his younger colleagues during the McCarthy era, for example, there was an absurd and vicious accusation that Rashevsky – who had fled communism in Ukraine – was a Red.

---

[6] Watson and Crick received the 1962 Nobel Prize in medicine or physiology for this discovery.

[7] These were the Cybernetics Laboratory of the Italian National Research Council (between the biology and physical science groups), the University of Wisconsin (between the UW Neuroscience Program and the Mathematics Research Center), the Los Alamos National Laboratory (where peaceful coexistence was enforced through funding priorities set by the Center for Nonlinear Studies), the University of Arizona (between the Program in Applied Mathematics, enlightened members of the UA Medical Center and the UA Neuroscience Program), and the Technical University of Denmark (under an interdisciplinary program in nonlinear science).

Matters became so dire in the mid-1950s (the mathematical biology program had been reduced from about thirty members to two) that Chicago neuroanatomist Gerhardt von Bonin encouraged his former colleague Warren McCulloch (then at MIT) to join him and several other prominent scientists (physiologist George Bishop, psychologist Egon Brunswik, philosopher Rudolph Carnap, mathematician Karl Menger, physician Russell Meyers, and Nobel Prize winning biochemist Albert Szent-Györgyi) in writing an open letter to the president of the University of Chicago which then appeared in *Science* and read [634]:

> We are disturbed by the drastic reductions that have been imposed on the Committee on Mathematical Biology, headed by N. Rashevsky at the University of Chicago. We wish to point out that the work of this department, the only one of its kind in the world, is of great interest and importance in our diverse fields of research, that is, biology, clinical medicine, mathematics, psychology, philosophy, and sociology. We feel that it would be a loss if that work were seriously reduced.

Under this highly visible academic onslaught, Chicago backed down and Rashevsky's program was restored to strength, but the infighting took its toll of his health, and bitterness resurfaced in 1964 when he – preparing to retire – was not consulted on the choice of his successor. With respect to this seemingly thoughtless decision, evolutionary biologist Richard Lewontin, then Associate Dean of Biological Sciences, has recently commented [556]:

> The approach of the Rashevsky school was to make simplified physical models that were supposed to capture the essence of a biological phenomenon and then describe models in mathematical terms. What Rashevsky and his school failed to take into account was the conviction of biologists that real organisms were complex systems whose actual behavior would be lost in idealizations. The work of the school was regarded as irrelevant to biology and was effectively terminated in the late 1960s, leaving no lasting trace.

Of course morphogenic dynamics are more complicated than most mathematical biophysicists would have wished (see Sect. 7.2), but the biologists' conviction that "actual behavior would be lost in idealizations" may not be entirely correct. Because the physico-mathematical properties of living systems have developed under a variety of Darwinian pressures which necessarily include the facts of physics and chemistry, an understanding of their dynamics helps to sort out some of the evolutionary causalities – as was discussed in 1917 by D'Arcy Thompson in his classic *On Growth and Form* [961] and recently by Boye Ahlborn in his *Zoological Physics* [10]. On the other hand, complex mathematical aspects of biological development have been considered in the *hypercycle* theory developed by biochemists Manfred Eigen[8] and Peter Schuster to explain the origin of life (see Sect. 7.2.6 and [269]) and physicist Walter Elsasser introduced the concept of an *immense number* (which is finite but beyond numerical control) to count possible arrangements of living organisms – a concept that has been largely ignored by mainstream biology (Sect. 7.3.4 and [287]).

That Rashevsky himself recognized the intricacies of biological development and the problematics of piecemeal mathematical studies is evident from a reading of his 1954 paper on relational biology, in which he emphasized the importance of understanding the organizational unity of living creatures (see the epigraph to Chap. 8 and [789, 790]). This perspective guided much of Rashevsky's subsequent research and led Rosen to a precise definition of a complex system in a formulation that provides room for the phenomena of Life (Sect. 8.3.4 and [817, 818]).

On the personal level, unfortunately, Rashevsky's sense of honor had been violated once too often by the University of Chicago's administration. Insisting that it was wrong to bar him and his colleagues from participating in decisions about the future of a program that he had dreamed up, founded and tended for three decades, Rashevsky chose to resign six months before his scheduled retirement, notwithstanding a written appeal from the university president to relent. In 1965 he became a professor of mathematical biology at the University of Michigan, from which he retired in 1970. As there are now dozens of centers for studies of mathematical biology throughout the world and as a host of related journals were inspired by Rashevsky's venerable *Bulletin of Mathematical Biophysics*,[9] his pioneering program in mathematical biology evidently left in its wake a vigorous international research activity that has interacted strongly with nonlinear science since the mid-1970s.

Long concern with the academic wrongs that stem from group thinking has motivated me to write at some length about Rashevsky's contributions to mathematical biology. In an attempt to achieve balance, it seems appropriate to close with some obituarial comments by three of Rashevsky's professional colleagues: Anthony Bartholomay, George Karreman and Hyman Landahl [61]:

> [Rashevsky's] work, and the work of his associates, at the University of Chicago and elsewhere reached into so many directions, including his fundamental ground breaking mathematical theories of sociology and history, that he was of necessity, a much discussed scientist, considered by other mathematical biologists to be the unquestioned all

---

[8]Eigen was awarded the 1967 Nobel Prize in chenistry for his work on fast reactions.

[9]The list includes *Bulletin of Mathematical Biology, Journal of Theoretical Biology, Journal of Mathematical Biology, Acta Biotheoretica, Rivista di Biologia, Mathematical Biosciences, Theoretical Population Biology*, and *Biofizika*, among others.

time leader in the field of mathematical biology. Like most pioneers in the history of science he has many followers as well as his share of detractors. Even under the heaviest of pressure he never deviated from his strict adherence to principle and truth. He was a man of the greatest integrity and the highest principles. He had the moral courage to admit publically that he had been wrong once he was convinced of it, truly attesting to his scientific greatness. He applied these same principles to the defence of his acquaintances whose cause he was the first to take up when he sensed the existence of an injustice, practicing his unique life style even at the risk of his own personal comfort and well being to the very end.

5.2.3 General Systems Theory (GST)

During the 1937–1938 academic year, a cultured and colorful theoretical biologist from the University of Vienna named Ludwig von Bertalanffy was invited by Rashevsky on a Rockefeller Fellowship to the University of Chicago, where he gave the first North American lecture on general systems theory (GST) [999, 1000]. With a long-time interest in the growth of biological form [998] and concern with the theoretical problems of reductionism (see Sect. 8.2), von Bertalanffy viewed GST as having broad potential for understanding the holistic nature of self-organizing phenomena in the social, biological, and physical sciences; thus he was primarily engaged with the emergent phenomena that would soon be overlooked by Wiener's cybernetics.

More generally, von Bertalanffy constructed a metatheory concerned with describing in many contexts the behaviors of *open systems*, in which a source of energy is a necessary feature of the dynamics (see Chap. 4 and Sect. 8.2.6) – a departure from typical physical theories of the day, which were energy conserving (Hamiltonian). Furthermore, he was interested in the hierarchical structures of living organisms (see Sect. 7.3.6 and [1002, 1003]) and in the ontological status of the higher levels of biological organization. Ahead of his time, von Bertalanffy was a visionary who smoothed the way for the nonlinear-science revolution of the 1970s.

The International Society for the Systems Sciences, which von Bertalanffy helped to found in 1954 – together with economist Kenneth Boulding, neuroscientist Ralph Waldo Gerard, psychologist James Grier Miller, and mathematical biologist (and former Rashevsky student) Anatol Rapoport – continues to actively:[10]

- encourage the development of theoretical systems which are applicable to more than one of the traditional departments of knowledge;

---

[10] See http://www.isss.org/

- encourage the development of adequate theoretical models in areas which lack them;
- eliminate the duplication of theoretical efforts in different fields;
- promote the unity of science through improving the communication among specialists.

On the other hand, GST was strongly opposed by certain reductive biologists. In a chapter entitled "Microscopic Cybernetics" in his book *Chance and Necessity*, for example, Jacques Monod points, perhaps correctly, to the complicated dynamics that can be displayed by combinations of interacting biomolecules, but then goes on to claim that such a network "far surpasses anything that the study of the overall behavior of whole organisms could ever hint at". Considering that dynamics tend to become more complicated, not less, as one ascends the hierarchical ladder of life, this claim is difficult to understand, never mind credit,[11] yet he went on to close the chapter with the following paragraph:

> On such a basis, but not on that of a vague "general theory of systems", [referencing von Bertalanffy] it becomes possible for us to grasp in what very real sense the organism does effectively transcend physical laws – even while observing them – thus achieving at once the pursuit and fulfillment of its own purpose.

Later comments were less tempered, including this statement published just as the nonlinear science revolution was getting under way [671]: "What I consider completely sterile is the attitude, for instance, of Bertalanffy who is going around and jumping around for years saying that all the analytical science and molecular biology doesn't really get to interesting results; let's talk in terms of general systems theory [...] there cannot be anything such as general systems theory, it's impossible. Or, if it existed, it would be meaningless." See Rosen's memoir for some further insights into the bitterness of Monod's attitude [819].

To his credit, von Bertalanffy's early emphasis on the dynamics of open systems set the scientific stage for a fundamental investigation of nonequilibrium processes in chemical and biological systems.

5.2.4 Nonequilibrium Statistical Mechanics (NSM)

Whereas Wiener had missed the possibilities for positive feedback and emergence in his cybernetics, the Russian–Belgian chemist Ilya Prigogine included them both under his program in nonequilibrium statistical mechanics (NSM), which is described in a 1962 book with that title [773].[12] The central thrust

---

[11]Perhaps Monod viewed all knowledge through the lens of his own specialty, like the Boston lady who planned her California trip "to go through the Newtons" (western suburbs of Boston).

[12]Prigogine was awarded the 1977 Nobel Prize in chemistry for his work on nonequilibrium thermodynamics and the theory of dissipative structures.

of this work was to understand the emergence of *dissipative structures* (see Chap. 4) from a fundamental perspective, showing how a unidirectional "arrow of time" (Sect. 8.2.4) in such systems can develop on an underlying basis of Newtonian mechanics in which time is bidirectional. These dissipative systems are characterized by an inflow of energy which supports steady-state emergent structures that are far from thermodynamic equilibrium – of which several examples are described in this book.

Having established a theoretical basis for emergent structures in a chemical context during the 1960s – including Turing patterns (see Sect. 7.2.4 and [973]) and Belousov–Zhabotinsky waves (Fig. 4.4 and Sect. 4.3) driven by chemical potentials, and Rayleigh–Bénard cells in fluid layers driven by a temperature difference (Sect. 6.8.3) – Prigogine and Gregoire Nicolis went on to make fundamental contributions to the development of nonlinear science during the 1970s [715]. Thus the general approach of nonequilibrium thermodynamics was applied to several self-organizing systems including atmospheric vortices (tornadoes and hurricanes) driven by temperature differences (Fig. 6.16 and Sect. 6.8.6), nerve impulses driven by a transmembrane electric potential (Chap. 4), and autopoietic living organisms (Sect. 8.3.1) and social systems (slime molds, ants, bees, humans, etc.) [716] driven by nutritional energy, confirming ideas that had been originally proposed by von Bertalanffy during his Chicago visit in the mid-1930s [1002, 1003].

5.2.5 Catastrophe Theory (CT)

As a late 1960s brainchild of the French mathematician René Thom, catastrophe theory (CT) approaches the problem of morphology (development of form) from a general study of the various ways that dynamical systems can become unstable under the continuous change of one or more of their parameters, phenomena that he called catastrophes [959]. Thus CT is in the spirit of Rashevsky's mathematical biology, although this connection is seldom noted. Perhaps because of Thom's unimpeachable mathematical credentials,[13] the CT approach to morphology was widely discussed in the 1970s, not only as part of the gathering momentum in nonlinear science research but as a distinct intellectual movement.

Briefly, Thom's mathematical contribution was to conceive of and catalog the different ways that a phase space (see Appendix A) underlying the dynamics of an open system can become globally distorted when instabilities appear with changes in system parameters. A broader scientific contribution was to suggest how these various types of instabilities might arise in the biological, cognitive and social sciences [928]. During the 1970s, CT was promoted in Britain and the US through a series of talks and papers by Christopher Zeeman [1081, 1082], whereupon the exotic names for various catastrophes caught the imaginations of science writers and the general public.

---

[13]Thom was awarded the 1958 Fields Medal for his work on topology.

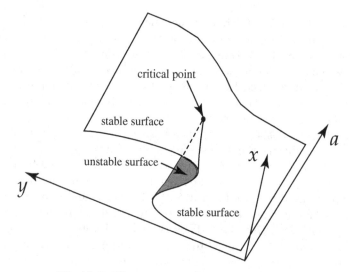

Fig. 5.2. The geometry of a fold catastrophe

The simplest of these constructions – the fold – is sketched in Fig. 5.2, corresponding to solutions of the cubic equation

$$y = x^3 + a .$$

In this figure, the surface represents stationary states of a dynamical system, the *critical point* is evidently where the *system parameter* $a = 0$, and the instability appears for $a < 0$. As y slowly increases for negative values of a, a point is reached where x is forced to jump from the lower part of the fold to the upper part. Interestingly, this construction represents the bifurcation diagram for a FitzHugh–Nagumo nerve impulse (see Fig. 4.3), and assuming a single parameter of pressure to change, it can also describe the revolution in nonlinear science (shown in Fig. 1.1) and the dynamics of a lynch mob.

Other catastrophes – related to higher-order polynomials and more system parameters – are the cusp, swallowtail, butterfly, hyperbolic umbilic, elliptic umbilic, and the parabolic umbilic, with increasing degrees of algebraic and geometric intricacy. Since the 1970s, an activity among mathematically oriented biological, cognitive and social scientists has been to decide which of Thom's types corrresponds to a particular natural instability – a new species or organism, a mental collapse, the outbreak of a war, and so on.

5.2.6 Synergetics (S)

As a theoretical physicist with a deep understanding of how quantum theory is applied to studies of condensed matter [390], Hermann Haken entered nonlinear science through his fundamental analyses of the laser (see

Sect. 6.7.1), but he soon saw that the nonlinear dynamics of such optical systems are closely connected to those of chemical reactions and hydrodynamics (Sects. 6.4.4 and 6.8 and [388, 389]). Formulated in 1969 as a program that Haken called synergetics,[14] this insight has been applied to a broad class of self-organizing systems arising in biology, psychology, economics and sociology, in addition to physics and chemistry. Importantly, Haken has developed a fairly straightforward procedure for studying instabilities that is related to catastrophe theory and which again provides theoretical support for some of von Bertalanffy's seminal ideas [263, 391–394].

A key feature of synergetic analysis is to divide the quasilinear interacting modes describing a nonlinear system near a stationary solution of the underlying phase space (see Appendix A) into m modes with positive growth constants (called the *masters*) and the remaining s modes with decay (negative growth) constants (the *slaves*); thus a general solution at the onset of an instability takes the form

$$u(\boldsymbol{x}) = \sum_{i=1}^{m} \Big[f_i(\boldsymbol{x}) \exp\left(+\lambda_i t\right) \Big] + \sum_{j=1}^{s} \Big[g_j(\boldsymbol{x}) \exp\left(-\lambda_j t\right) \Big] , \qquad (5.5)$$

where $\boldsymbol{x}$ is a multidimensional space variable, the growing (decaying) modes are indexed by i (j), and all the λs are positive numbers. Under Haken's *slaving principle*,[15] the growing modes govern the dynamics [because the first summation in (5.5) soon dominates the second one], thereby shifting the analysis to a problem of the form

$$\dot{\boldsymbol{v}} = N(\boldsymbol{v}) ,$$

where $N(\boldsymbol{v})$ is a nonlinear function and $\boldsymbol{v}$ is an m-dimensional vector from which the remaining s components of a general solution can be computed. The m components of $\boldsymbol{v}$ are called the *order parameters* of the problem, and this approach introduces a great simplification when $m \ll s$, as is often the case. Synergetic analysis has been applied to a variety of problems by Haken, his colleagues, and others, as is indicated by the 21 titles in the Springer Series in Synergetics.

5.2.7 Complex Adaptive Systems (CAS)

During the the 1980s, the US physics community awoke to the developments in nonlinear science, and some were encouraged to found the Santa Fe Institute (SFI), which is devoted to:[16]

---

[14]The term "synergetics" is coined from Greek words meaning "working together".

[15]One of Haken's former doctoral students told me that you don't really understand the slaving principle until you have done a thesis under him.

[16]See http://www.santafe.edu/

> [...] creating a new kind of scientific research community, one emphasizing multidisciplinary collaboration in pursuit of understanding the common themes that arise in natural, artificial, and social systems. This unique scientific enterprise attempts to uncover the mechanisms that underlie the deep simplicity present in our complex world.

One might question the term "unique" as this statement reflects some of the ideas that were put forward by von Bertalanffy in his University of Chicago lectures a half century earlier. Thus, the SFI has followed in the interdisciplinary intellectual tradition pioneered by Rashevsky, von Bertalanffy, Wiener, Prigogine, Thom, and Haken, establishing programs in global systems, economic markets, human language, social dynamics, human behavior, network dynamics, cognitive neuroscience, and biological computation, among others. Perhaps to distinguish itself from related efforts, SFI has chosen the term "complex adaptive systems" (CAS) to denote the area of interest, which Murray Gell-Mann, one of the SFI founders, has defined as a system that [348]:[17]

> [...] acquires information about its environment and its own interaction with that environment, identifying regularities in that information, condensing those regularities into a kind of "schema", or model, and acting in the real world on the basis of that schema.

Although Andy Warhol has claimed that there is no such thing as bad publicity[18] and the international visibility of the SFI has helped to make the general public aware of these several interdisciplinary developments, the SFI promotion of CAS has been something of a double-edged sword. Science writer John Horgan, for example, has belittled the "four Cs" of chaos theory, cybernetics, catastrophe theory and complex adaptive systems as a long-running scientific fad, under which "complexity leads to perplexity" [451, 452]. I invite the reader to make an independent judgment on this matter after reading the following two chapters, but for the present we can consider how closely these four Cs are related to mathematical biology, synergetics and to each other.

5.3 Interrelations Among the Metatheories

As precursors to and components of the nonlinear-science revolution that is the subject of this book, the metatheories mentioned in this chapter confirm – to me, at least – that much important and interesting analytical science remains to be done above the descriptive levels of physics and chemistry. The

---

[17] Gell-Mann was awarded the 1969 Nobel Prize in physics for his classification of elementary particles and their interactions.

[18] "Just check if they spelled my name right and measure the inches," he famously sighed.

	C	MB	S	NS	CAS	GST	CT	NSM
C	**12,800**							
MB	55.6	**1,830**						
S	30.8	9.59	**533**					
NS	35.4	29.8	9.85	**460**				
CAS	31.6	0.611	0.531	0.541	**400**			
GST	51.9	0.553	9.44	0.369	0.85	**221**		
CT	19.2	0.83	0.874	0.608	0.964	1.2	**207**	
NSM	0.372	0.393	0.502	0.502	0.083	0.073	0.212	**113**

Fig. 5.3. Pages on Google and interconnecting pages for metatheories of nonlinear science: cybernetics (C), mathematical biology (MB), synergetics (S), nonlinear science (NS), complex adaptive systems (CAS), general systems theory (GST), catastrophe theory (CT), and nonequilibrium statistical mechanics (NSM). (Numbers in thousands of pages were recorded in the summer of 2006)

extent to which those using the World Wide Web are aware of the above noted research programs and the degrees of overlap among them are shown in Fig. 5.3, where the numbers count the pages (in units of a thousand) that appeared on Google under the corresponding terms in the summer of 2006. Also indicated on this figure are the numbers of pages that mention pairs of terms, giving some measure of the overlaps between activities.

As the term "cybernetics" is not in fashion among US scientists, it may surprise some to see that it appears by factors of 7 to 113 times larger than those of the other six activities, but two explanations come to mind: first the term is more widely used in Asia and Europe,[19] and second, it is also employed commercially and in a nontechnical sense, as is the term "synergetics".

The largest fraction of double references is between cybernetics and general systems theory (22%), which is not surprising because many scientists were engaged in both of these activities during the 1940s and 1950s. Next largest are between cybernetics and catastrophe theory (9%), between cybernetics and complex adaptive systems (8%), between cybernetics and nonlinear science (8%), between cybernetics and synergetics (6%), between synerget-

---

[19] "Cybernetics" may have fallen out of use in the US because of its wide popularity in Russia during the McCarthy era, causing Defense Department officials to frown on project proposals employing the term. Further umbrage may have stemmed from Norbert Wiener's testy attitude toward the US defense establishment [189].

ics and general systems theory (4%), between cybernetics and mathematical biology (3%), and between synergetics and nonlinear science (2%), with the other twenty interactions being less than 1%.

In general, these modest overlaps suggest that scientists prefer single terms to describe their areas of research. Because we tend to interact with restricted cohorts, such an attitude is not unexpected, but the data in Fig. 5.3 may also reflect our subconscious wish to obscure the fact that we are developing ideas that were first proposed by Ludwig von Bertalanffy in the 1930s.

6 Physical Applications of Nonlinear Theory

> The most incomprehensible thing about
> the world is that it is comprehensible.
>
> Albert Einstein

Although several important nonlinear physical phenomena were studied before the 1970s, some (synchronized clocks, the N-body problem of planetary motion, and hydrodynamics, for example) going back to the early days of Newtonian mechanics, many more have recently been discussed as new phenomena were discovered and the mathematical connections among them realized. While not a complete survey, this chapter aims to put some of the recent results into perspective, so the reader will appreciate how broadly the nonlinear concepts introduced in the previous four chapters are now used throughout the physical sciences. Beginning with the very small (theories of elementary particles) and proceeding to the very large (cosmology), the emphasis is on verbal and visual descriptions, avoiding detailed mathematical discussions for the convenience of the more general reader.

6.1 Theories of Matter

For those who of us who accept the notion that dynamically stable "lumps" of energy can emerge from nonlinear Hamiltonian (energy-conserving) fields, it is interesting to consider this phenomenon as a theoretical basis for the elementary particles of matter. A goal of such studies is to find a nonlinear field that generates a spectrum of emergent lumps with mass values that match those found in nature. While this goal has not yet been achieved, some interesting attempts have been made, which are related to the soliton studies that were described in Chap. 3.

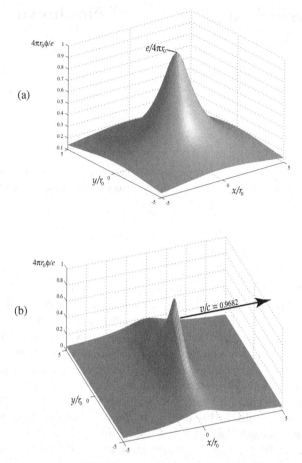

Fig. 6.1. Gustav Mie's model electron, showing the electric potential (ϕ) plotted as a function of x and y on a plane through the origin and perpendicular to the z-axis. (**a**) A stationary electron. (**b**) An electron that is moving in the x-direction with speed $v = 0.9682c$. (Note the Lorentz contraction in the x-direction by a factor of four)

6.1.1 Mie's Nonlinear Electromagnetism

Just before the First World War, a German physicist named Gustav Mie proposed a theory of matter in which the recently discovered electron emerges naturally from a nonlinear version of James Maxwell's electromagnetic equations [655]. In addition to providing an early demonstration that Maxwell's equations can be put into Hamiltonian form, Mie defined a *world function* (Φ) as a functional (function of a function) depending on electric field intensity ($\boldsymbol{E}$), magnetic flux density ($\boldsymbol{B}$) (which are measured in the laboratory) and the four components of the electromagnetic potential ($\boldsymbol{A}, \phi$), where $\boldsymbol{A}$ is a vector potential for the magnetic field and ϕ is a scalar potemtial for the electric field. Requiring Φ to depend only on the variables $\eta = \sqrt{E^2 - B^2}$ and $\chi = \sqrt{\phi^2 - A^2}$ introduces Lorentz invariance in accord with special relativity theory (SRT). (See Sect. 3.4 for a brief introduction to SRT in the context of the SG equation.)

With $\Phi = \eta^2/2$ this formulation reduces to the standard Maxwell's equations [875]; thus plane waves propagating in the x-direction satisfy the linear wave equation

$$\frac{\partial^2 u}{\partial x^2} - \frac{1}{c^2}\frac{\partial^2 u}{\partial t^2} = 0 \,, \tag{6.1}$$

where x is the direction of propagation, u represents mutually perpendicular electric or magnetic fields in the yz-plane and c is the speed of light.

Assuming $\Phi = \eta^2/2 + a\chi^6/6$ (where a is a small nonlinear parameter) led Mie to a static, spherically-symmetric electric potential of the form

$$\phi(r) \approx \frac{e/4\pi}{\sqrt{r^2 + r_0^2}} \,, \tag{6.2}$$

where $e = 4\pi(3r_0^2/a)^{1/4}$ is the electronic charge, $r = \sqrt{x^2 + y^2 + z^2}$ is a distance from the origin in standard three-dimensional space, and r_0 is both the electron radius and an adjustable parameter in the theory.

Equating the total field energy outside of r_0 to the rest energy $(m_0 c^2)$ of an electron (where m_0 is its rest mass) yields

$$r_0 \approx 2.8 \times 10^{-15} \text{ meters} \,,$$

which is the textbook value of an electronic radius. Thus Mie's nonlinear electromagnetic equations lead to the model electron shown in Fig. 6.1a, which has an electric charge of e, a radius of r_0, a rest mass equal to m_0, and an electric potential $\phi(r) \to e/4\pi r$ as $r/r_0 \to \infty$. Letting the model rotate adds spin, which then generates a magnetic moment from the rotating charge.

Just as the Lorentz transformation defined in (3.13) allows a stationary sine-Gordon (SG) kink to be boosted to the moving kink of (3.12) – leading to the Lorentz contraction shown in Fig. 3.4 – Mie's localized solution can move with any speed up to the limiting velocity of light under a Lorentz contraction. An example of this effect is shown in Fig. 6.1b, where the model electron is moving in the x-direction with a speed of $v_x/c = \sqrt{1 - 1/16} = 0.9682$, producing a Lorentz contraction in the x-direction by a factor of $1/\sqrt{1 - (v_x/c)^2} = 4$.

Although Mie showed that a model elementary particle can emerge from a nonlinear version of Maxwell's equations, his theory has several difficulties, including the following [504]. First, the choice of $\Phi(\eta, \chi)$ is not physically motivated, as it could as well depend on other powers of η and χ and products of these powers. Ideally, this nonlinear structure should be determined by empirical considerations or through the analysis. Second, the presence of nonlinear terms in the formulation means that the superposition property (independent strands of causality) of linear systems is lost near the center of the particle, making more complicated solutions difficult to compute. Third, although Mie sought to establish a connection between his model of matter and the force of gravity [504], his method of putting gravitational attraction into the theory changes electromagnetic properties of the vacuum, thereby altering the velocity of light, which disagrees with SRT. Finally, Mie's assumption of a scalar gravitational potential allows inertial and gravitational masses to differ.

Now largely forgotten, Mie's theory was esteemed in its day by Max Born, who showed the Göttingen mathematicians how Mie's treatment of energy functionals for electromagnetism was related to Lagrangian mechanics [111,504] and brought his results to the attention of David Hilbert, who used Mie's concepts in his formulation of general relativity theory (GRT) [192,434]. Albert Einstein, on the other hand, was initially unimpressed with Mie's work, put off by his inadequate treatment of gravity and so overlooking his seminal suggestions concerning variation principles and the fundamental nature of matter. Interestingly, Hermann Weyl used Mie's ideas in the early 1920s for his discussions of "fields and matter" [1031].

During the 1920s, physicists took divergent paths in these areas of research. On one hand, Einstein and his colleagues sought a formulation that would unify GRT with the Maxwell equations (MEs), just as Maxwell had unified electric and magnetic phenomena during the nineteenth century [363]. On the other hand, many physicists were involved with formulating quantum theory (QT) as a means for computing probabilities of finding atomic particles with given speeds and locations. Thus by the end of the 1920s several unifications were under active study: (i) GRT with MEs, (ii) GRT with QT, (iii) MEs with QT, and (iv) a unification of GRT, MEs and QT – all of which have deep implications for the fundamental nature of matter. Recently Hubert Goenner has written an excellent review of these unification efforts, including an extensive bibliography and many quotations from the correspondence among the main players [363]. From this reference one learns of an early suggestion by Rudolf Förster to include MEs under GRT by making Einstein's 4×4 gravitational tensor nonsymmetrical, thereby adding six additional components which can represent the electromagnetic fields $\boldsymbol{B}$ and $\boldsymbol{E}$. Although Einstein was unenthusiastic about Förster's approach, he initially liked Theodor Kaluza's idea of assuming reality to comprise four spatial dimensions plus one of time [476] – an idea that was introduced in 1884 through Edwin Abbott's delightful classic *Flatland: A Romance of Many Dimensions* [1] and has recently been revived. On physical grounds, Born began to wonder how all the energy of a particle could be contained in a small space [112], and Wolfgang Pauli worried that it was unphysical to speak of electromagnetic fields within particles because the concept of a test particle (which is used to define a field) is no longer valid. By the 1930s, importantly, Einstein had come to agree with Mie that particles should be represented by singularity-free solutions of field equations.

Mie's approach to the nature of matter was revived by Carl Anderson's discovery of electron–positron creation in 1932. In this observation, a sufficiently energetic electromagnetic wave (cosmic ray) is transformed into an electron and its positively charged sibling, a positron, establishing an empirical link between these two manifestations of energy. Born and Leopold Infeld returned to Mie's formulation, eliminating his χ-dependence and choosing $\Phi = E_0^2 \sqrt{1 + (\eta/E_0)^2} - E_0^2$, which reduces to $\eta^2/2$ for $\eta \ll E_0$. They again found a spherically symmetric model electron (as in Fig. 6.1), and a nonlinear PDE for plane waves that reduces to (6.1) for field amplitudes much less than E_0. Nonlinear solitary wave solutions of this PDE travel at exactly the speed of light, which is in accord with SRT because if these waves were to travel faster than light, they could be used to synchronize clocks in rapidly-moving frames, in violation of SRT [888, 1033]. While it would be illuminating (no pun intended!) to study the mutual scattering of interacting light beams, the estimated value of E_0 in the BI theory is about 10^{20} V/m – eight or nine orders of magnitude above the field intensities that can currently be obtained by focusing light beams from presently available high-power lasers. Thus this interesting experiment must wait for even more powerful lasers.

Erwin Schrödinger took up Born's nonlinear electromagnetic theory as early as 1935 [851], and he continued these studies through the 1940s when – as founding director of the Dublin Institute of Advanced Studies – he attempted to move physics research toward nonlinear science [674]. During this period, Schrödinger and Einstein had an extended transoceanic correspondence, in which their persistent attempts to connect nonlinear particle theories with GRT were unsuccessful. Having spent a decade of intense effort formulating GRT, Einstein was rightly concerned about the difficulties of finding the correct nonlinear particle theory among myriad possibilities; yet near the end of his life he agreed with Mie's motivating ideas in the following words [278].

> What appears certain to me, however, is that in the foundation of any consistent field theory, the particle concept must not appear in addition to the field concept. The whole theory must be based solely on partial differential equations and their singularity-free solutions. [...] If a field theory results in a representation of corpuscles free of singularities, then the behavior of these corpuscles in time is determined solely by the differential equations of the field.

Einstein's view is fully in accord with the concept of emergence as it developed during the nonlinear science explosion of the 1970s and is currently used [875].

6.1.2 De Broglie's Guiding Waves and the Double Solution

Research on nonlinear theories for elementary particles in the 1950s was largely that of Robert Finkelstein, who studied nonlinear spinor fields (vector-like entities that embody spin) as electron models [307] and of Louis de Broglie, who had first proposed that material particles can have wave properties back in 1925. After teaching physics from conventional perspectives for three decades, de Broglie – inspired by David Bohm who had recently questioned the standard interpretation of QT [99, 100] and by the above-mentioned views of Einstein – returned to an idea he had first suggested in the mid-1920s: the "double wave" theory. In a nonlinear version of this theory developed in the 1950s, de Broglie proposed that a particle should be described by two waves: a fictitious ψ-wave that obeys the linear PDE of standard QT and an objectively real u-wave that obeys a nonlinear PDE that is qualitatively similar to those of Chap. 3 [221, 222].[1]

Under de Broglie's formulation,

$$u = U e^{i\theta'} \quad \text{and} \quad \psi = \Psi e^{i\theta} \,,$$

whereupon the condition that $\theta = \theta'$ (except in a small region surrounding the particle) allows the fictitious ψ wave to guide the singularity of the nonlinear u wave which comprises the real particle. The statistical nature of standard QT enters this theory because the guiding process is assumed to be rendered stochastic by a subatomic noise field.

De Broglie's causal interpretation of QT is related to modern soliton studies in three ways. First, ascribing ontological reality to a lump of energy that is localized by a nonlinear PDE is in accord with John Scott Russell's century-old recognition that hydrodynamic waves can have particle-like solutions [833] and with other manifestations of the same idea that became evident in the nonlinear explosion of the early 1970s [888]. Interestingly, de Broglie observes that his localized u-wave "may be compared to the theory of 'solitary waves' in Hydrodynamics, which exhibits certain similarities to it" [221]. Second, the idea of a guiding wave is related to the inverse scattering transform (IST) method of soliton theory (see Sect. 3.2), under which the dynamics of a localized solution of a nonlinear PDE are determined by the solution of an associated linear PDE system [4,346,545,702,875]. Finally, a one-dimensional example of the nonlinear equation sought by de Broglie is provided by the nonlinear Schrödinger (NLS) equation (described in Sect. 3.3), where the soliton is given in (3.10). This soliton is governed by the solution of an associated IST and it evidently has both particle-like and wave-like properties. Known to nonlinear scientists since the early 1970s, such a combination of wave-like and particle-like properties in one dynamic entity (which is clearly

---

[1]It was a reading of de Broglie's book on *Nonlinear Wave Mechanics* in 1960 that brought Born's nonlinear electromagnetics to my attention and helped to awaken my interest in nonlinear wave theory.

exhibited by the NLS soliton) has long been deemed a "spooky" feature of QT, demonstrating the unfamiliarity with basic nonlinear concepts among many members of the physics community.

In 1993, Peter Holland published a careful review and analysis of the de Broglie–Bohm theory [444], discussing the connection with soliton theory and finding it a viable alternative to conventional QT. In accord with Kuhn's description of the resistance by established scientists to novel ideas, the response by the Copenhagen establishment to de Broglie's ideas, Holland writes, "was generally unfavourable, unrestrained and at times vitriolic." Werner Heisenberg, for example, was greatly disturbed by the asymmetrical treatment of position and momentum in the new theory, viewing the possibility of discerning particle orbits as "superfluous ideological superstructure" and comparing proposals to test the validity of conventional interpretations of QT as "akin to the strange hope that [...] sometimes $2 \times 2 = 5$." Such odd remarks may have discouraged young physicists from following lines of thought that challenged the concepts of conventional QT.

6.1.3 Skyrmions

In the 1960s, the Derrick–Hobart (DH) theorem was introduced [228, 438], which shows that many of the nonlinear models proposed for elementary particles in three space dimensions will be classically unstable to contraction [294, 417, 812]. In other words, localized states can release energy by becoming smaller, and in the context of (6.2), this means that $r_0 \to 0$. Of course, this contraction might be stabilized by quantum conditions (say on the mass, charge, or spin), just as are the electronic orbits in atoms.

Awareness of the classical (i.e., non-quantum) DH instability increased interest in nonlinear PDE systems with topological constraints like the SG equation, which was studied in the early 1960s by Tony Skyrme [758] and also by Ugo Enz [294], among others [875] as a one-dimensional model for elementary particles. From inspection of the mechanical model shown in Fig. 3.3, it is evident that kink solitons of SG have a finite rest energy that is permanently conserved by the fact that the pendula wind once around the central support. Skyrme originally supposed that the topological invariance of SG could be generalized to three space dimensions, and this turns out to be so for model particles, now called *skyrmions*, which are immune to the classical DH instability. The shapes of some of these topologically stable particle models are shown in Fig. 6.2 for increasing values of the *topological charge*, which is a generalization of the winding number shown in Fig. 3.3 to the three dimensions of ordinary space [605, 939]. Recently, it has been proposed that skyrmions may appear in the ground states of magnetic metals [825].

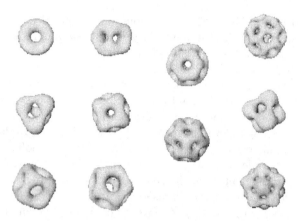

Fig. 6.2. Three-dimensional skyrmions with increasing values of topological charge. (Courtesy of Paul Sutcliffe)

6.1.4 Point vs. Extended Particles

Beginning in 1975, papers on nonlinear particle theories began to appear in *Physical Review D* (PRD), the flagship journal of elementary-particle theory in the United States. A quantification of this growing interest is provided by Fig. 6.3, which plots the annual number of papers that used the term "soliton" in the title, key words or abstract. Features of this figure can be understood by considering two complementary approaches presently taken to particle studies by the physics community.

- Point particles. In accord with Pauli's concerns, fundamental particles are modeled as mathematical points. This approach is used in the Standard Model [195], which assumes *leptons* and *quarks* as elementary particles, unifies strong, weak and electromagnetic forces, and agrees with a substantial amount of empirical data. Named for a Greek word meaning "thin", leptons include electrons, positrons, and muons, which interact with electromagnetic and weak forces. Quarks, on the other hand, are components of hadrons (from a Greek word meaning "robust"), which include protons and neutrons and interact with strong forces. Disadvantages of this approach are twofold: the masses and charges of the particles must be put into the theory as experimentally determined parameters, and unwanted infinities appear in certain analytic expressions (although these infinities can sometimes removed using renormalization theory [978]).
- Extended particles. Following Mie, fundamental fields are assumed to be nonlinear, out of which extended (non-point) particles emerge as localized (soliton-like) structures. As Einstein emphasized, this approach has the philosophical advantage that particle properties arise in a natural way from the underlying nonlinear field, offering the hope of predicting experimental mass spectra by guessing the correct nonlinear field. The problems

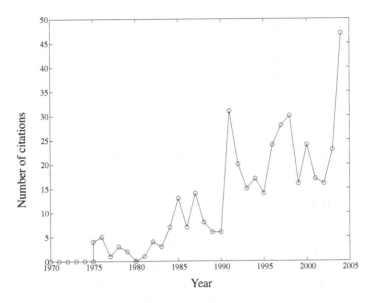

Fig. 6.3. Annual number of articles in *Physical Review D* that have used the term "soliton" in the title, abstract or key words. (Data from Science Citation Index Expanded)

with this approach are twofold: there are so many possible nonlinear field theories that one doesn't know which to use, and loss of the superposition property near the center of the particle renders detailed analyses substantially more difficult.

Figure 6.3 shows that before 1975 PRD contributors were largely unconcerned with the extended particle approach, whereas after that date such publications have been increasing in a (very roughly) linear manner at a rate of about one paper per year each year. This timing suggests that the onset of the activity was part of the explosive growth of research in nonlinear science shown in Fig. 1.1, during which several well-cited early publications on nonlinear particle models by leaders in the field made soliton studies safe for fledgling physicists [178, 332, 333, 581, 600]. Of deeper importance, perhaps, was a quantum analysis of the SG equation by Roger Dashen, Brosl Hasslacher and Andre Neveu in the mid-1970s, showing that the quantum value of a breather mass is well approximated by its classical value – thus increasing interest in classical nonlinear field theories as first approximations to their quantum counterparts [211].

As described in books by Vladimir Makhankov [594] and by Alexandre Filippov [306], several additional efforts were made during the 1980s to develop nonlinear field theories for elementary particles in three spatial dimensions, including the following:

1. The four-dimensional, self-dual Yang–Mills theory [454], a system for which plane wave reductions correspond to standard soliton equations (KdV, NLS, etc.).

2. Instantons, which are topological, solitary-wave solutions of the Yang–Mills system [453].

3. Twistor theory, which attempts to reach Einstein's decades-long goal of unifying gravity and electromagnetism through the construction of a nonlinear graviton [618, 754].

4. String theory (ST), which hopes to unify all the interactions of physics – weak, electromagnetic, strong and gravitational forces. Constructed at the Planck scale ($\sim 10^{-35}$m or about twenty orders of magnitude smaller than an electron) where gravitational forces become significant, ST presently comprises a family of formulations in which the basic particles are quantized oscillations on one-dimensional structures (strings) in spaces of higher dimensions [67]. While string dynamics may be nonlinear and therefore chaotic [66], possibly providing the stochastic underpinning needed by the de Broglie–Bohm theory to model the stochastic nature of quantum theory, Peter Woit and Lee Smolin have recently criticized the entire ST project for being insufficiently accepting of other approaches to a fundamental theory of matter [915, 1053].

Although it remains difficult to divine which of the many available nonlinear theories should be chosen for a particular model, the fact that classical nonlinear field theories give useful first approximations to particle mass spectra of quantized SG particles [211] and the development of ever more powerful computers (see Fig. 1.3) suggest that a judicious combination of the point and extended-particle perspectives will continue to provide useful information on the fundamental nature of matter. It is indeed an exciting time to be an elementary particle physicist.

But what of the quantum dynamics on the atomic scale?

6.2 Quantum Theory

As sketched in Appendix B, quantum theory (QT) is based on the Schrödinger equation (SE) for finding a probability amplitude Ψ as a function of space and time corresponding to a lossless (Hamiltonian) classical system, where the probability density (probability per unit volume of space) is then $|\Psi|^2$. Although the SE is linear, several features of QT are related to nonlinear science, some of which are discussed in the present section. Note first that QT requires a Hamiltonian formulation, so reaction-diffusion (RD) systems are not considered. Thus there is no QT for the flame of a candle, a nerve impulse, a living organism, or the human brain [954]. In general, this restriction does not cause problems because RD systems contain so many atoms that QT would not be needed even if it could be formulated.

6.2.1 Quantum Probabilities

Confusion in QT has been generated by polemics between those who view a quantum probability (QP) as describing many identical systems and those who believe that QP calculations should apply to individual systems. In other words, a QP is interpreted by physicists in two different ways.

First, QP is considered a measure of the statistical properties of a large number of identical experiments, which is called an *ensemble probability* for the following reason. Suppose we use QT to compute the probability of the outcome of an experiment. To give this statement meaning, we imagine performing identical measurements on a very large number of identical systems, called an *ensemble* – a concept that was introduced into statistical physics by Josiah Willard Gibbs [354]. Then our calculated QP is considered to tell us the fraction of measurements on the ensemble that will yield a particular value. The ensemble interpretation of QT, which was supported by Einstein, is in accord with all experimental measurements that have been made to date, and it asks for a more detailed (nonlinear) description of the system.

A second way of viewing QP is as an *individual probability*, indicating the degree of uncertainty about a measurement on a particular system, which can sometimes be calculated without reference to an ensemble. For example, consider the chances of getting an ace on a single throw of one die. From the symmetry of the cube, we know that the chances of each of the six faces coming up are equal, and as the total probability of all faces must add to unity, the probability of a particular face appearing (say an ace) is one sixth (the total probability of one divided by the six equally probable outcomes). This is a well defined concept, which is established without any reference to a large number of experiments. In his "Copenhagen interpretation" of quantum mechanics, Niels Bohr encouraged many physicists to suppose that QT computes individual probabilities. As he further claimed Schrödinger's wave function to be the most that can be known of the system, Bohr's perspective has the computational advantage of restricting analyses to the linear SE.

From a failure to agree upon or occasionally even to acknowledge differences between these two meanings of the term "probability" stems what Karl Popper has called "the great quantum muddle" [770]. For an example of the contrast between these two points of view, consider a particle of mass m moving in the x-direction with momentum $p \pm \Delta p$ and a corresponding initial uncertainty $(\pm \hbar / \Delta p)$ in its position. As the uncertainty in velocity is $\pm \Delta p / m$, the uncertainty in the position of the particle will grow linearly with time, eventually becoming much greater than $\pm \hbar / \Delta p$. If an observation is then made of the particle, its position can be more accurately determined, down to the original minimum error of $\pm \hbar / \Delta p$. From the individual probability perspective, the wave function is said to collapse in the instant that a new position measurement is made; thus the wave function of the particle seems to be influenced by observations made of it – another example of "quantum spookiness". From the ensemble probability perspective, on the other hand, no such problems arise; the new position measurement merely provides new information that updates the statistical properties of the ensemble.

6.2.2 Schrödinger's Cat

In 1935, the muddle was starkly exemplified by Schrödinger with a widely-discussed thought experiment [852], which he described as follows (in the translation of Walter Moore) [674]:

> One can even construct quite burlesque cases. A cat is shut up in a steel chamber, together with the following diabolical apparatus (which one must keep out of the clutches of the cat): in a Geiger tube there is a tiny mass of radioactive substance, so little that in the course of an hour perhaps one atom of it disintegrates, but also with equal probability not even one; if it does happen, the counter responds and through a relay activates a hammer that shatters a little flask of prussic acid. If one has left the entire thing to itself for an hour, then one will say to himself that the cat is still living if in that time no atom has disintegrated. The first atomic disintegration would have poisoned it. The [wave function] of the entire system would express this situation by having the living and dead cat mixed or smeared out (pardon the expression) in equal parts.

Failing to catch the irony of Schrödinger's example, some claim that just before the box is opened the poor cat would indeed be represented by a quantum wave function comprising in equal measures those of a living cat and a dead cat, and that this composite wave function "collapses" into either that of a live or dead cat when the box is opened and a "conscious observer" looks in. This "conscious observation" is thus thought to influence the state (living or dead) of the cat. If we suppose that QT calculates an ensemble probability, on the other hand, our physical interpretation of Schrödinger's cat experiment is quite different. We would then imagine setting up, say, a hundred such experiments, for which QT tells us that about fifty of the cats would die and fifty survive. This result would be in accord with observation were we actually to perform such a nasty experiment, and it suggests that QT provides an incomplete description of physical reality because it doesn't tell us what is going on inside a particular box.

Moore writes that "only a few commentators, the most notable being Eugene Wigner and John Neumann, have defended the uncompromising idealist position that the cat is neither alive nor dead until a human observer has looked into the box and recorded the fact in a human consciousness." Yet such a position has recently been taken by Henry Stapp, who describes the dynamics of a human brain as follows [924]:

> The parallel processors in the brain will churn out their various determinations, and the system will evolve, just like Schrödinger's notorious cat, into a superposition of macroscopically distinguishable

systems. Whereas that famous cat developed into a superposition of an "alive cat" and a "dead cat", so [a] man-in-the-box will develop, under the action of purely mechanical laws into a superposition of, for example, a "standing man" and a "sitting man".

In his examination of the cat experiment, on the other hand, Roger Penrose wonders whether the animal can actually be both alive and dead, asking [755]:

> Do we really believe that this is the case? Schrödinger himself made it clear that he did not. He argued, in effect, that the [rules] of quantum mechanics should not apply to something so large or so complicated as a cat. Something must have gone wrong with the Schrödinger equation along the way. Of course Schrödinger had a right to argue this way about his own equation, but it is not a prerogative given to the rest of us! A great many (and probably most) physicists would maintain that, on the contrary, there is now so much experimental evidence in favour of [standard quantum mechanics] – and none at all against it – that we have no right whatever to abandon that type of evolution, even at the scale of a cat. If this is accepted, then we seem to be led to a very subjective view of physical reality.

Although he thinks about this matter from the individual probability perspective, Penrose also believes that there must be something wrong with QT when it is applied to macroscopic bodies. "Common sense alone tells us that this is not the way the world actually behaves!" Ignoring the phenomenon of *decoherence* in which the integrity of an intricate quantum wave function is destroyed by random thermal vibrations [473, 954], he suggests that the SE should be replaced by some nonlinear version that is in better accord with the facts of nature [756].

After seven decades, sad to say, the quantum muddle is still with us. As I write these words, *Science* – a journal of good repute – has just published an article in which it is claimed that "Schrödinger kittens" have been produced in an experiment to create coherent optical pulses [356, 737]. Although the experimental work seems reliable, the claims of "growing the cat" through a "breeding process" are sophomoric, and there is no reference to Schrödinger's original paper nor mention of the ensemble interpretation of QT.

6.2.3 The EPR Paradox

While Schrödinger was working on his "cat paper", Einstein, Boris Podolski and Nathan Rosen wrote a related paper entitled: "Can quantum-mechanical description of reality be considered complete?" [279]. This "EPR paper" treats two particles that are moving apart with equal and opposite speeds, pointing out that according to the recognized computational rules of QT the differences between the positions of the two particles and the sum of their momenta can be measured with unlimited precision. Thus a measurement of the position of one particle gives the position of the other without disturbing it, and similarly a measurement of the momentum of one gives the momentum of the other without disturbing it. Abiding by the rules of standard QT, in other words, both the momentum and the position of a particle can be precisely measured without disturbing its momentum or position; thus both of these particle features must exist as aspects of reality, independent of observation. This result is at variance with Bohr's interpretation of QT, and it implies that QT is an incomplete description of reality. EPR close with the following short paragraph:

> While we have thus shown that the wave function does not provide a complete description of the physical reality, we left open the question of whether or not such a description exists. We believe, however, that such a theory is possible.

A model of clarity, the EPR paper landed in the capital of Denmark "like a bolt from the blue" [674, 740]. After some furious discussions with his colleagues and six weeks of intense work, Bohr responded as follows [103, 104]:

> It is true that in the measurements under consideration, any direct mechanical interaction of the system and the measuring agencies is excluded, but a closer examination reveals that the procedure of measurement has an essential influence on the conditions on which the very definition of the physical quantities rests. Since these conditions must be considered as an inherent element of any phenomenon to which the term "physical reality" can be unambiguously applied, the conclusion of the above mentioned authors would not appear to be justified.

The gist of this Proustian response seems to be that Einstein did not describe how both the position and the speed of one of the particles can be measured with the same apparatus. Without this ability, Bohr insisted, simultaneous physical reality of both quantities had not been proven. Discussions between these two acclaimed scientists continued in the same vein in correspondence and at various meetings over the next twenty years, until Einstein's death in 1955.

Perhaps the best reading on their mature views can be obtained from a volume on Einstein's thought, edited by Paul Schlipp and comprising three parts [845]. In the first section, Einstein provides a scientific autobiography, many of his colleagues comment on his views in the second section, and, finally, Einstein responds to the comments. In his autobiography, Einstein observes that the "double nature of radiation (and of material corpuscles) is a major property of reality, which has been interpreted by quantum-mechanics in an ingenious and amazingly successful fashion. This interpretation, which is looked upon as essentially final by almost all contemporary physicists, appears to me as only a temporary way out." After summarizing the main points of the EPR paper, he contrasts two attitudes as follows:

A. The individual system (before the measurement) has a definite value of [all the variables] of the system, and more specifically, that value which is determined by measurement of this variable. Proceeding from this conception, he will state: the Ψ-function is no exhaustive description of the real situation of the system but an incomplete description; it expresses only what we know on the basis of former measurements concerning the system.

B. The individual system (before the measurement) has no definite value of [the variables]. The value of the measurement only arises in cooperation with the unique probability which is given to it in view of the Ψ-function only through the act of measurement itself. Proceeding from this conception, he will (or, at least, he may) state: the Ψ-function is an exhaustive description of the real situation of the system.

Evidently A is Einstein and B is Bohr. Einstein goes on to state that the "true laws" underlying QT "cannot be linear nor can they be derived from such". But finding nonlinear formulations is very difficult, in Einstein's opinion, because there are so many possibilities that one doesn't know where to begin the search. For his part, Bohr reviews the EPR arguments and Einstein's unsuccessful attempts to pick holes in standard QT, summarizing his views as follows:

In quantum mechanics, we are not dealing with an arbitrary renunciation of a more detailed analysis of atomic phenomena, but with a recognition that such an analysis is in principle excluded. The peculiar individuality of the quantum effects presents us, as regards the comprehension of well-defined evidence, with a novel situation unforeseen in classical physics and irreconcilable with conventional ideas suited for our orientation and adjustment to ordinary experience. It is in this respect that quantum theory has called for a renewed revision of the foundation for the unambiguous use of elementary concepts, as a further step in the development which, since the advent of relativity theory, has been so characteristic of modern science.

In his "Reply to Criticisms", Einstein returns to Schrödinger's example of the cat, which he mercifully replaces with a clock that can record the times of events by making marks on a moving tape. In this context, he comments:

One arrives at very implausible theoretical conceptions, if one attempts to maintain the thesis that the statistical quantum theory is in principle capable of producing a complete description of an individual physical system. On the other hand, those difficulties of theoretical interpretation disappear, if one views the quantum-mechanical description as the description of ensembles of systems.

Although many physicists believe that Bohr won this long debate, most simply ignore the issues raised by Einstein, even though they are clearly relevant for current practitioners of nonlinear science.

6.2.4 Nonlocality and Quantum Entanglement

In considering the roles that quantum mechanical phenomena might play in biology and psychology, few terms have created more confusion in the scientific literature and unwarranted excitement among readers of the public press than the related concepts of *nonlocality* and *quantum entanglement*.[2] To avoid becoming mired in this confusion, let us consider what these terms might signify.

In condensed matter physics, the term "local", as in a "local interaction" implies talking between nearest neighbors, whereas a "nonlocal interaction" reaches beyond parochial limits. Thus, the valence bonding in a chemical molecule is local because it involves a sharing of electrons between two neighboring atoms. Electrostatic interactions among the charged regions of a large molecule (a protein, for example) are said to be nonlocal, as the forces involved extend over many interatomic distances. In a similar way, the neurons in our brains are said to interact nonlocally because their outgoing signals are carried over relatively long distances (past many nerve cells) on axons. On a yet larger scale, most would consider the gravitational influences of the sun and the moon on terrestrial tidal flows to be nonlocal.

In studies of quantum phenomena, the term nonlocality implies interaction at such a distance that causality travels faster than the speed of light and SRT is violated. Some claim that this sort of nonlocality is implied by Einstein's belief in the existence of a reality underlying quantum theory. Considering the substantial empirical evidence for SRT, this implication presents a serious challenge to Einstein's view of quantum theory.

Before we turn to quantum entanglement, let us consider ordinary, classical entanglement, which is commonly experienced. Suppose my dog has carried one of my shoes out into the back yard. If I look under the bed and find a single right shoe, then I immediately know that the dog is playing with a left shoe. Nothing strange about that, and it is easy to multiply examples. Imagine that you have two identical boxes and two balls that differ only in color (say red and black). You can mix the balls behind your back, close your eyes, and put one ball in each box. Without knowing which box has the black ball and which has the red ball, you leave one box at home and carry the other with you on a trip to a distant land. Over breakfast one morning, you open the box that you have taken along to find a red ball, and instantly you know that the box back home contains a black ball. Although the states of the two boxes have become "entangled" by the described conditions, QT

---

[2]Striking examples of how far "quantum mysticism" can be taken are presented in a recent film entitled: "What the Bleep!? – Down the Rabbit Hole".

is not involved. Identical twins are genetically entangled, as they share the same DNA code. Soldiers who survive a campaign together are psychologically entangled by their shared experiences; thus they will tend to answer certain questions in related ways. So are married couples and siblings. In the social sciences, this sort of human entanglement is an important aspect of cultural dynamics, but it has nothing to do with QT.

In the microscopic realms where QT reigns, one also finds many examples of entanglement. Consider a photon of very high energy (called a gamma ray), which may transform itself into an electron and a positron – elementary particles that are identical except for the sign of the electric charge that they carry. Now, each of these particles also carries a spin which is found to be either "up" or "down" when measured in a certain direction. As the spin of the original gamma ray is zero, it is not surprising that the spins of an electron and positron that emerge from the disintegration of the same gamma ray are entangled. If the spin of the electron were found to be up in some direction, in other words, then the spin of the positron would necessarily be down in that direction. If this were not so, conservation of total spin would be violated by the decay of a photon into a positron and an electron, making particle theorists very unhappy. This is an example of quantum entanglement, but, again, there is nothing unusual about it. It would be strange indeed if such a correlation between the spin measurements were *not* observed. So what is all the mystery and excitement about?

6.2.5 Bell's Inequality

Interest in quantum entanglement and nonlocality began in 1957 when Bohm suggested that it should be possible to actually perform an EPR-type experiment on particles that carry spin, like the positron and electron emerging from decay of a gamma-ray photon [101]. (Because spin correlation experiments differ in several ways from the original thought experiment of EPR, they are better termed EPRB experiments.) In an important series of papers published by John Bell starting in the mid-1960s, the spin correlations to be expected in EPRB experiments were calculated from the perspective of standard QT [72]. To understand Bell's results, suppose the spin of particle A (say the electron, in our above example) is measured in a direction a and the spin of particle B (the positron) is measured in direction b. Let this be done many times and compute the average of the products of the measured electron spins (in direction a) and the positron spins (in direction b). This average is called the average spin correlation $C(ab)$. QT says that the average spin correlation will be equal to the negative of the cosine of the angle between direction a and direction b (denoted $\angle ab$). When directions a and b are the same (both are in the direction of the x-axis, for example), the angle between them is zero, and the average spin correlation is -1. In other words, whenever the electron has spin up in the x-direction, the positron is found to have spin down in the x-direction, as is expected.

In order to connect with the motivating ideas of the EPR paper, Bell made two assumptions. First, he supposed that there is an underlying theory that governs the probabilities predicted by QT. To be specific, he assumed that this underlying theory involves some "hidden" variables, which comprise anything at all relevant to the underlying theory: trajectories of individual particles, initial conditions of those trajectories, internal variables, etc. Second, he assumed that the observation of the spin of particle A in direction a depends only upon direction a and the hidden variables. Thus it was specifically assumed that measurements of particle A do not depend on direction b. Similarly, Bell assumed that the measurement of particle B in direction b depends only on direction b and on the hidden variables, and not on direction a. This second assumption is called the assumption of locality, without which the results of a measurement on particle A could directly and instantaneously influence the measurement on particle B. As this would be an example of information being transmitted at a speed greater than that of light, SRT could be violated without the locality assumption.

To repeat, Bell's two assumptions were: (i) hidden variables, in accord with Einstein's belief in an underlying theory for quantum phenomena, and (ii) locality, without which SRT would be violated. From these assumptions and logical arguments stemming from probability theory, Bell derived a statement of inequality (called Bell's inequality), which can be violated by the above mentioned quantum result that

$$C(ab) = -\cos(\angle ab) \ .$$

To see this suppose a, a', b, and b' are four different angles at which average spin correlations are measured for a particular experiment. Then Bell's logical requirement is that

$$\left| C(ab) - C(ab') \right| + \left| C(a'b) + C(a'b') \right| \leq 2 \ .$$

But if the angles of a, b, a', and b' (with respect to some reference direction) are respectively 0, 45, 90, and 135 degrees, this quantity is $2\sqrt{2}$, in violation of Bell's inequality.

From Bell's logic, therefore, it seemed that at least one of the following three statements must be true: (i) There is no hidden variable formulation underlying QT. (ii) Locality (and thus SRT) can be violated. (iii) QT gives a prediction that is at variance with empirical evidence. I write "it seemed" because in the mid-1960s no EPRB experiments had been performed, but measurements in the early 1980s by Alain Aspect, Jean Dalibard and Gérard Roger (not on electron–positron pairs but on pairs of photons with correlated angles of polarization, which leads to the same quantum analysis) confirmed the above prediction of QT. Although there are some who quibble about assumptions that were made in order to interpret the experimental measurements, the data appear clean and convincing to me, yielding an averaged correlation that looks very much like a cosine function [38]. So where are we?

As empirical evidence supports the quantum prediction that Bell's inequality is violated, possibility (iii) is not so. One of the other two options seems to be required, but which one? Many physicists select (i), concluding that Bohr was right and Einstein was wrong in their celebrated debates; in other words, local hidden variable theories are dead. Others opt for (ii), inferring that violation of the limits imposed by special relativity opens the doors of physics to psychic phenomena.[3] Are these the only possibilities, or did Bell inadvertently slip an additional assumption into his derivation of an inequality that is violated both by QT and experimental observation? Some think so.

6.2.6 Joint Measurability

In an interesting and important book entitled *The Philosophy Behind Physics*, Thomas Brody has reviewed several critiques of Bell's analysis, presenting and discussing five different derivations of Bell's inequality [128]. Of these five derivations, only Bell's original analysis assumes hidden variables and only two make the assumption of locality. All of the derivations, on the other hand, assume that it is possible to measure the relevant feature (spin, photon polarization or whatever) on a single particle in more than one direction without mutual interference between measurements. Brody calls this the *joint measurability assumption* (JMA).

To understand what is involved in the JMA, suppose we measure the spin of particle A in direction a and then measure it again in direction a'. Does the fact that the spin was first measured in direction a influence the subsequent measurement in direction a'? If not, the two spin components are said to be jointly measurable, and the JMA holds. As the JMA is assumed in all five derivations of Bell's inequality cited by Brody, it is a candidate for failure in appraising the experimentally confirmed quantum prediction of the inequality.

As with entanglement, violations of joint measurability are not limited to the quantum realms. Imagine, for example, you were to give a group of students a comprehension test on passages from English literature and shortly afterwards an English vocabulary test. Could you assume that the group's collective performance on the second test would not be influenced by the fact that they had just taken the first test? Probably not, because they may have learned – or refreshed their knowledge of – the meanings of some words while taking the first test. Suppose an engineer measures the energy storage capacity of a sample of rechargeable batteries, first at a discharge rate of 20 amperes and then at a discharge rate of 10 amperes. Should she assume that the measurements at 10 amperes are not influenced by the previous measurements at 20 amperes? Probably not, because the testing at 20 amperes may

---

[3]This conclusion seems unwarranted to me, because the ratios of the distances and times involved in most experiments with extrasensory perception are small compared with the speed of light.

have aged the batteries, thereby altering their performance. In both of these examples – and others that one can readily think of – the JMA is ill-advised.

Thus one might infer from the Aspect experiments that Bell's statements (i) and (ii) are both false (there is a hidden variable theory and SRT is correct), but joint measurability does not hold in the quantum realms. To some, this is a more palatable resolution of the Bell's paradox than supposing violation of special relativity or adhering to Bohr's strict interpretation of QT. But whatever position one takes on this matter, the bottom line of the Aspect experiments is that the statistical predictions of QT have once again been confirmed, and these predictions are in accord with the ensemble interpretation of a QP.

6.2.7 Many Worlds?

In a physics doctoral thesis witten in 1956, Hugh Everett aimed to put the entire Universe under the aegis of standard QT, including cats and conscious observers. An abridged version of this work appeared in a 1957 issue of the *Reviews of Modern Physics* [297], along with some explanatory pages by John Wheeler, his mentor at Princeton [1032]. Under this ambitious formulation, an observer does not have a unique status; rather observer plus system are represented by "a *superposition*, each element of which contains a definite observer state and a corresponding system state. Thus with each succeeding observation (or interaction), the observer state 'branches' into a number of different states. Each branch represents a different outcome of the measurement and the *corresponding* eigenstate for the object-system state. All branches exist simultaneously in the superposition after any given sequence of observations." In response to a prepublication comment by Bryce DeWitt, who pointed out that in our experience only one branch actually exists, Everett added a footnote in which he asserts:

> The whole issue of the transition from "possible" to "actual" is taken care of in the theory in a very simple way – there is no such transition, nor is such a transition necessary for the theory to be in accord with our experience. From the viewpoint of the theory *all* elements of a superposition (all "branches") are "actual", none any more "real" than the rest. It is unnecessary to suppose that all but one are somehow destroyed, since all the separate elements of a superposition individually obey the wave equation with complete indifference to the presence or absence ("actuality" or not) of any other elements. This total lack of effect of one branch on another also implies that no observer will ever be aware of any "splitting" process.
>
> Arguments that the world picture presented by this theory is contradicted by experience, because we are unaware of any branching process, are like the criticism of a Copernican theory that the mobility of the earth as a real physical fact is incompatible with the

common sense interpretation of nature because we feel no such motion. In both cases the argument fails when it is shown that the theory itself predicts that our experience will be what it in fact is.

This paper was ignored by the physics community until the early 1970s, when DeWitt recalled it in an article in *Physics Today* (the general interest journal of the American Physical Society) in which he coined the term "multiple worlds" to describe the logical implication of Everett's formulation [232, 233]. Perhaps because of this catchy name, Everett's QT was widely discussed among physicists and also in the sci-fi community, including a certain John Titor, who was born in 2036 and returned to Earth in November of 2000 to confirm that the multiple-worlds picture is correct.[4]

Interestingly, just as the nonlinear-science revolution was becoming established among members of the applied mathematics and engineering communities, some theoretical physicists would jettison conservation of energy in order to preserve the linear character of the Schrödinger equation throughout the entire Universe.

6.2.8 Nonlinear Quantum Mechanics?

Although Einstein wrote to Born in 1952 that the de Broglie–Bohm formulation seemed "too cheap" [114], he often suggested that any fundamental formulation of quantum mechanics must be nonlinear [278, 845]; thus a key question becomes: Is there empirical evidence for nonlinearity in quantum theory?

In a recent attempt to answer this question, Steven Weinberg has explored the possibility of corrections to quantum mechanics at the level of atomic structure by attempting to detect a nonlinear pattern in the atomic energy levels of certain metallic ions. Detailed analysis of experimental measurements indicate that a nonlinear effect – if it exists at all – must be less than one part in 10^{20} of a typical electronic energy level in a hydrogen atom [105, 1025]. At subatomic levels of description, the effects of such nonlinear terms may be amplified by high field intensities, allowing nonlinearity to play a significant role in the dynamics, but this small amount of energy appears to be negligible in the realms of atomic and chemical phenomena. Thus the standard linear version of QT should be appropriate at and above these levels of description.

How, then, does one use the linear QT to study classically nonlinear phenomena in the border regions between the microscopic and macroscopic realms?

---

[4]See http://www.johntitor.com/Pages/Story.html

6.3 Quantum Energy Localization and Chaos

A typical nonlinear phenomenon that survives quantization is the localization of energy, which can take two forms. First are the local modes of interatomic vibration that are observed to occur on small molecules. Second, under certain circumstances such localized energy can be viewed as a *quantum soliton*, which has theoretical properties that lead to new results in quantum field theory. In addition, classically chaotic systems retain related properties under quantization.

6.3.1 Local Modes in Molecules

In molecular oscillations, the masses remain constant because relativistic velocities are not approached, but the spring constants are typically sublinear because bond stretching tends to deplete interatomic electron densities, thereby weakening the bonds. Assuming that the potential energy of an interatomic spring is $kx^2/2 - \alpha x^4/4$ where α is an anharmonic (nonlinear) parameter, the Schrödinger equation (SE) remains a linear PDE, but its energy levels are given by [875]

$$E_n = \left(\hbar\omega - \frac{\gamma}{2}\right)\left(n + \frac{1}{2}\right) - \frac{\gamma n^2}{2} , \tag{6.3}$$

where $\gamma \equiv 3\alpha\hbar^2/4m^2\omega^2$. Evidently, (6.3) reduces to Planck's law (see Appendix B) as $\gamma \to 0$, but in general this result implies the empirical relation

$$E_n - E_{n-1} = A - Bn , \tag{6.4}$$

which was first observed for interatomic vibrations by Raymond Birge and Hertha Sponer in the mid-1920s [90]. Equation (6.4) shows how level spacings decrease at higher energies, suggesting that the bond will break for $n \sim A/B$. This Birge–Sponer relation is closely followed for stretching oscillations of carbon–hydrogen (CH) bonds in benzene, where measurements of the ratio A/B and heat of dissociation have provided convincing evidence since the late 1920s that higher-amplitude oscillations (i.e., larger values of the quantum number n) are concentrated on a single bond, as shown in Fig. 6.4 [282]. Thus nonlinear local modes of oscillation were being observed and studied in small molecules by physical chemists at the same time that the implications of Schrödinger's 1926 papers were being digested by theoretical physicists.

Experimental studies of local modes in molecules were continued during the 1930s by Reinhard Mecke and his colleagues [645–648], but his work was neglected until the 1970s [429, 585]. Although this curious oversight offers additional evidence of the poor communications among early practitioners of nonlinear science, another facet of the explanation stems from the linear nature of QT. Because the SE is linear and a benzene molecule is symmetric

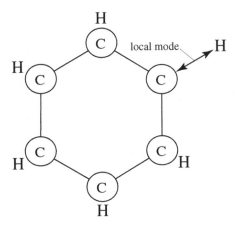

Fig. 6.4. The planar structure of a benzene molecule, showing a local mode of the CH stretching oscillation

under rotations by multiples of 60°, eigenstates for CH stretching oscillations must share the same symmetry and be spread out over the six bonds shown in Fig. 6.4, not localized on one. Following this reasoning during the 1970s, referees of manuscripts by physical chemists sometimes claimed that the experimental results showing local modes were at variance with QT and therefore wrong [428].

True as far as it goes, this theoretical argument is incomplete. At larger values of n, the physical chemist does not observe a single eigenfunction on benzene but a wave packet of them that is spread over an energy range ΔE, where [875]

$$\frac{\Delta E}{\varepsilon} \sim \frac{n(\varepsilon/\gamma)^{n-1}}{(n-1)!} \tag{6.5}$$

and ε is the energy of electromagnetic interaction between adjacent CH bonds. As $\varepsilon \doteq 4$ cm^{-1} and $\varepsilon/\gamma \sim 1/30$ for benzene, it is seen from (6.5) that $\Delta E \to 0$ very rapidly for increasing values of n. Correspondingly, a local-mode wave packet will remain organized for a time of order $\hbar/\Delta E$, which readily becomes large compared with typical measurement times. In other words, individual quantum states comprising local mode wave packets with $n > 3$ are *quasi-degenerate*, which explains the experimental observations of local modes by physical chemists without violating the physicist's linear QT. Thus does a quantum description of the dynamics smoothly approximate the classical dynamics as the quantum level becomes large.

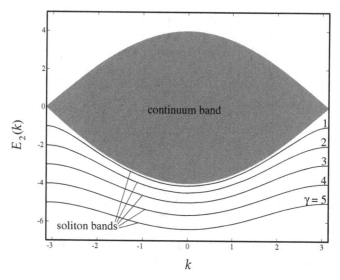

Fig. 6.5. Energy eigenvalues of the DNLS equation at the second ($n = 2$) quantum level with $\varepsilon = 1$ as $f \to \infty$. There is one soliton band for each value of γ, which is plotted for several different values of γ

6.3.2 Quantum Solitons

Consider next a chain (or one-dimensional lattice) of f classical nonlinear oscillators governed by the equations

$$\left(i\frac{d}{dt} - \omega\right) A_j + \varepsilon(A_{j+1} + A_{j-1}) + \gamma|A_j|^2 A_j = 0 , \tag{6.6}$$

where j is a site index and $A_j = A_{j+f}$ implying periodic boundary conditions. In this model, ω is the low-amplitude oscillation frequency at one of the lattice points, ε is the interaction energy between adjacent oscillators, γ measures the nonlinearity of each oscillator, and A_j is a complex mode amplitude. (The real part of A_j is proportional to the momentum of the jth oscillator and the imaginary part is proportional to its extension.) With $f = 6$, (6.6) provides a model for the benzene oscillations of Fig. 6.4 [885]. In the continuum limit ($f \gg 1$ and $\gamma/\varepsilon \ll 1$), this discrete nonlinear Schrödinger (DNLS) equation has solitary wave solutions that correspond to solitons of the continuum nonlinear Schrödinger (NLS) equation which was discussed in Sect. 3.3 [890].

Upon quantization, (6.6) has the energy eigenvaues shown in Fig. 6.5 (for a total of two quanta in the large f limit), where each eigenfunction changes by a factor of e^{ik} under translation by one unit of j (a lattice spacing). (In the jargon of solid-state physics, k is the crystal momentum of an eigenstate [390].) This figure shows both a *continuum band* (the shaded area) and a *soliton band* given by

$$E_2(k) = \sqrt{\gamma^2 + 16\varepsilon^2 \cos^2(k/2)} = E_2(0) + \frac{k^2}{2m^*} + O(k^4) ,$$

where m^* is the *effective mass* near the band center. The soliton band is characterized by two features. First, it is displaced below the continuum band by a *binding energy* $E_b = \sqrt{\gamma^2 + 16\varepsilon^2} - 4\varepsilon$. Second, inspection of the corresponding eigenfunctions shows that the two quanta are more likely to be on the same site for the soliton band than in the continuum band [875]. Quantum versions of solitons can be constructed as wave packets of states near the bottom of the soliton band.

For $\gamma \ll \varepsilon$, arbitrary n and sufficiently large f, the quantum binding energy of a soliton is

$$E_b = \frac{\gamma^2}{48\varepsilon} n(n^2 - 1) , \qquad (6.7)$$

which corresponds to the binding energy of a classical NLS soliton under the identification $n = \sum |Aj|^2 \gg 1$ [594]. In the classical DNLS system with $\gamma \gg \varepsilon$, numerical studies show that the soliton becomes pinned to the lattice. Under QT, this classical, nonlinear phenomenon is reflected by the fact that the effective mass,

$$m^* = \frac{(n-1)!\gamma^{n-1}}{2n\varepsilon^n} ,$$

becomes very large as $n \to \infty$.

Evidently, $\Psi(t)$ is a rather complicated mathematical object, and the numerical difficulty in computing it increases combinatorially with the number of quanta, but interactions between breathers have recently been studied on the quantum DNLS system [246]. A complete quantum reconstruction of a classical soliton is more complicated, as it comprises components with different values of n, each of which is a wave packet over k [875]. Soliton systems that combine a local excitation with an associated distortion of a crystal lattice are yet more difficult to solve, because interactions with an infinite number of lattice states are involved in constructing an exact solution. Thus such systems require either a product approximation based on different time scales or the truncation of an infinite dimensional matrix [390,875]. These systems include polarons, superconducting metals, and solitons in biopolymers, all of which are mentioned below.

6.3.3 Quantum Inverse Scattering

An important feature of classical soliton theory is the IST method of solution, which was mentioned in Sect. 3.2. In 1981, Ludvig Faddeev [298] and Harry Thacker [958] independently showed how this classical knowledge can be used to solve problems in quantum field theory. Their results are of interest to high-energy physicists because particle interactions in the Standard Model are based on quantum fields.

To see how this goes, start with (6.6). In the classical continuum limit, $A_j(t) \rightarrow u(x,t)$, and with appropriate normalizations u satisfies the NLS equation given in (3.9). Under quantization (as noted above), A_j becomes a *lowering operator* (b_j) that reduces the quantum level from n to $n-1$. Thus either by taking the continuum limit of the discrete operator equation for b_j or by quantizing the PDE for $u(x,t)$, one arrives at the following equation for the *quantum field operator* $\phi(x,t)$:

$$i\frac{\partial \phi}{\partial t} + \frac{\partial^2 \phi}{\partial x^2} + 2|\phi|^2 \phi = 0 \,. \tag{6.8}$$

Under the classical version of the IST method, a key function is the reflection coefficient which is constructed from $u(x,0)$ and depends in a simple way on time. From the reflection coefficient, $u(x,t)$ can be reconstructed using the Gel'fand–Levitan integral equation, thereby obtaining the solution at time t from the initial conditions at $t=0$ [875]. In a quantum version of the IST method, the reflection coefficient becomes a raising operator for wave function solutions of (6.8) [1008]. Among many useful results, this leads to a binding energy $E_b = n(n^2 - 1)/12$ which corresponds to the soliton binding energy in (6.7) [497]. In this manner, the set of classical PDEs that possess IST formulations (the classical soliton equations) leads directly to a corresponding set of solvable quantum field theories.

Thus we see that (6.6) serves as a model for the local modes in benzene (first observed in the 1920s), and it leads naturally to the quantum NLS equation – an example of a quantum field theory. Furthermore, as we have seen, solitons of the classical NLS equation have both particles and wave properties, a combination long deemed spooky.

6.3.4 Quantum Chaos?

According to standard QT, a wave function $\Psi(x,t)$ obeys a linear PDE (the SE), so chaotic behavior is not allowed for Ψ. Just as we can use the linear SE to study the quantum properties of soliton systems, however, we can also study the quantum behaviors of systems that are chaotic in the classical limit. As exact solutions of a quantum problem are difficult to obtain near the classical limit, one often turns to Bohr's semiclassical methods for such studies, but Einstein pointed out in 1917 that it is not clear how to use this approach for Henri Poincaré's irregular trajectories because they are not periodic [275]. (Evidently Einstein was one of the few physicists aware of Poincaré's results at this time.)

Although many open questions remain, some progress has been made in understanding quantum aspects of systems that are chaotic in the classical limit, including the following results [163, 260, 262]. (It should be noted that these results are not confined to quantum systems; they hold for any wave system for which the propagation of rays is chaotic in the short wavelength limit.)

- The probability of finding a nearest-neighbor energy eigenvalue spacing S for an integrable system is $\exp(-S)$ (Poisson statistics), whereas for a nonintegrable (chaotic) system this probability is approximately $(\pi S/2)\exp[-(\pi/4)S^2]$ (Wigner statistics) [612]. A qualitative explanation of this difference is that oscillation components are independent for integrable systems, allowing eigenfunctions to cross as a perturbing parameter is varied, while crossings are avoided for nearly chaotic systems because the component eigenfunctions can interact.

- The quantum mechanical eigenfunctions of classically chaotic systems become random functions with irregular nodal patterns in the semiclassical limit; however localizations (or scars) are observed in a small number of these eigenfunctions near classically unstable orbits [419].

- Quantum wave packets spread out and decay more rapidly for classically chaotic systems than for classically integrable systems. Thus some of the stability of classical solitons carries over into the quantum realm.

A perturbed hydrogen atom is a useful test bed for these ideas, because without perturbation this two-body system is classically integrable (as Newton showed) and thus chaos free [262]. Two examples are these: First, ionization of a highly-excited hydrogen atom that is irradiated by a microwave field is found to depend sharply upon the field amplitude. In the context of KAM theory (see Sect. 2.5), this phenomenon is explained by noting that tori persist at low field amplitudes, preventing the coalescence of small chaotic regions. At a critical amplitude these tori break up (see Fig. 2.4), allowing trajectories to explore the entire phase space including paths to ionization [503]. Second, the equations describing a hydrogen atom in a uniform magnetic field are no longer integrable, and the corresponding energy eigenvalue spectrum becomes scrambled, in accord with theoretical derivations of spacing statistics [335].

6.4 Chemical and Biochemical Phenomena

Proceeding up the hierarchical ladder from atomic physics to chemistry and biochemistry, we find several examples of the nonlinear phenomena, including the following.

6.4.1 Molecular Dynamics

Among the many applications of quantum theory is a 1927 proposal by Born and J. Robert Oppenheimer for calculating the force fields binding atoms into molecules, which is based upon the fact that atomic nuclei are much heavier than electrons [115]. From the SE and this work, according to Paul Dirac [242]:

> The underlying physical laws for the whole of chemistry are completely known, and the difficulty is only that the exact application of these laws leads to equations much too complicated to be soluble.

To find these "underlying physical laws", the Born–Oppenheimer (BO) approximation assumes the total wave function to be factored into electronic and nuclear components and proceeds with these steps [909]: first, choose fixed distances separating the atoms of a molecule; second, use SE to compute molecular eigenstates of the valence electrons; third, fill these states with the available valence electrons; and finally, compute the total energy of the system (valence electrons plus charged inner shells and nuclei). Repeating these four steps for many interatomic distances allows the construction of a multidimensional potential energy function which, in turn, determines the interatomic forces.

Although this program was numerically daunting in the late 1920s, analytical refinements and recent developments in computing power (see Fig. 1.3) make it possible to calculate energy functions for small molecules that depend on lengths, bending and twisting of covalent bonds, electrostatic interactions, and interatomic repulsions [627]. Presently, there are several molecular dynamics codes available as numerical packages for physical chemists to compute static structures and vibrational modes of molecules [546, 967].

Determining the structure of a molecule requires finding coordinates of the BO energy function where the total energy is a minimum (interatomic forces are zero), which is a classical (i.e., not a quantum) calculation. To obtain the relative intensities of vibrational spectra, on the other hand, QT must be used, where a multidimensional SE is written for the system of vibrating atoms. To compute the frequency and qualitative nature of a vibration, however, it is often sufficiently accurate to assume that the atomic masses obey Newton's Second Law, where the interatomic forces are obtained by differentiating the BO energy with respect to interatomic distances and orientations. This computation then involves integration of a set of ordinary differential equations.

6.4.2 Energy Localization in Biomolecules

During the 1970s, most nonlinear scientists (including me) believed that localized packets of energy had no role to play in biology. Whereas the warm-wet nature of organic tissue was known to allow localization of nonlinear diffusive activity (like the nerve impulse [440] and the heartbeat [1049]), energy-conserving processes seemed unlikely; thus the paper on protein solitons presented by Davydov at a Swedish soliton meeting in the summer of 1978 came as a surprise to me. Davydov's soliton is based on a well-defined Hamiltonian (energy functional) and modeled in the continuum approximation by the NLS equation – and so also related to (6.6) – but the physical nature of its nonlinearity is different. In energy localization on small molecules, the parameter γ represents an *intrinsic* nonlinearity arising because the force constant of a bond decreases under stretching as the electron density of the bond is depleted. Davydov's formulation, on the other hand, assumes that γ is an *extrinsic* nonlinearity, stemming from a local distortion of the protein structure [217, 875].

Extrinsic nonlinearity goes back to the idea of the *polaron*, a nonlinear entity proposed by the great Russian physicist Lev Landau in 1933 [535] and widely studied in the Soviet Union in the 1950s [748]. Under this model, an electron manages to travel through a crystal by associating itself with a moving distortion of the lattice – rather like a marble working its way down into a plate of spaghetti. Although Davydov later expanded his theoretical perspectives to include biological charge transport [218], his original picture was of a localized CO (carbon–oxygen) stretching oscillation (called an amide-I mode by physical chemists) in the peptide units of alpha-helical protein. Localized amide-I energy distorts the nearby helical structure, which, in turn, traps the amide-I energy – a closed causal loop preventing energy dispersion and offering a means for the storage and transport of biological energy. As his analysis is based on a Hamiltonian the parameters of which can be independently determined, Davydov avoided the parameter fudging that often characterizes numerical studies of biological problems.

Closely related to Davydov's basic formulation is an equation developed by Vladimir Zakharov at about the same time (and also in the Soviet Union) to describe turbulence of plasma waves [1076]. In both cases, an exact quantum analysis is more difficult than for (6.6) because it is necessary to represent the quantum character of the associated lattice. Davydov dealt with this problem by using a product wave function (similar to the Born–Oppenheimer approximation), in which the higher frequency amide-I oscillation is loosely coupled to the lower frequency lattice distortion [115, 218, 872].

On his account, Davydov was first motivated by reading the proceedings of a conference that was organized by David Green in 1973 to consider the "crisis in bioenergetics" [372]. Resolution of this crisis involved deciding whether biological energy transduction can be entirely explained by Peter Mitchell's chemiosmotic hypothesis – under which biological energy derived from hydrolysis of adenosine triphosphate (ATP) is stored in the electrostatic fields of charges separated by a membrane [664] – or whether there is some intermediate energy storage mechanism within a protein. And if so, what is it?

Of particular interest in Green's proceedings is the summary of a series of papers by Colin McClare in which he (McClare) analyzes the physics of energy transduction at the molecular level and concludes that resonant storage and transfer of energy are needed to attain the efficiencies observed in biological muscles [628–630]. For the historian of science, a useful feature of

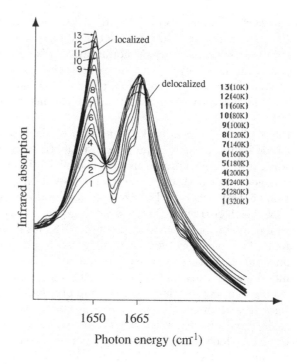

Fig. 6.6. Infrared absorption spectra of crystalline acetanilide (ACN) in the region of the amide-I (CO stretching) mode as a function of temperature. A normal (delocalized) amide-I band is at 1665 cm^{-1} and a self-localized (soliton) peak is at 1650 cm^{-1}

this publication is the transcription of discussions, from which it is clear that McClare's ideas were greeted with skepticism if not disdain by critics who claimed to have learned from physicists that resonant states must decay in a small fraction of a picosecond. Discouragement by such negative responses to his seminal ideas may have contributed to McClare's suicide early in 1977, just as our collective understanding of energy localization from the perspectives of nonlinear science was undergoing a set change [971]. Fortunately, McClare's papers have been archived in King's College, Cambridge at the direction of Maurice Wilkins, who believed that his insights would eventually be recognized.[5]

Unknown to McClare, Giorgio Careri and his students at the University of Rome were busily measuring the infrared spectrum of a model protein called acetanilide (ACN) throughout the 1970s in order to better understand the properties of natural protein. ACN forms optically clear crystals, and was used in the nineteenth century as an ingredient in patent medicines to cure headaches and fevers. (As it sometimes led to severe side effects including

---

[5]See http://www.kcl.ac.uk/iss/archives/collect/1mc30a.html

death, ACN was gradually replaced by aspirin.) In the course of their work, Careri et al. discovered the temperature-dependent band at 1650 cm^{-1} which is shown in Fig. 6.6. Notwithstanding great effort, they were unable to to find a physical explanation for this band until the early 1980s when Careri became aware of Davydov's work [152]. Although physical chemists immediately and strongly challenged Careri's assignment of the 1650 cm^{-1} band to nonlinear localization of vibrational energy, no credible counter-evidence was presented, and Davydov's theory correctly predicted both the Birge–Sponer coefficients prior to their measurement and the temperature dependence of the 1650 band intensity [872]. Recently, Julian Edler and Peter Hamm have used femtosecond pump–probe measurements to demonstrate conclusively that Careri's assignment of the 1650 band to self-localization was indeed correct [264]. These pump–probe measurements (which give null readings unless a band is nonlinear [399]) are also in accord with recent numerical calculations by Leonor Cruzeiro and Shozo Takeno [202], and they show room-temperature lifetimes of about 18 ps for NH stretching bands in ACN [265]. Thus McClare's seminal suggestion of resonant storage and transfer of biological energy has at last received firm empirical support, more than three decades after his original publications.

6.4.3 Chemical Aggregates

Beyond the level of small molecules, chemicals aggregate in several ways. From a commercial perspective, the most important may be *polymers*, which comprise long chains of repeating units (monomers) [767]. These chains can be linear (like spaghetti), branched, or crosslinked into three dimensional networks as in epoxy. Nonlinear considerations enter into autocatalytic growth processes – as in (1.2) – for these chains or networks which can depend strongly on temperature and acidity (pH). A phenomenon recently observed by John Pojman is a wavefront of polymerization, propagating through a zone of monomers like a "liquid flame" [768].

A most important problem in biochemistry is the prediction of a protein structure from knowledge of the sequence of its constituent amino acids – the folding problem [644]. Although in principle a simple task of energy minimization, analyses of protein folding are rendered difficult by the combinatorial explosion of possible structures with the length of the chain, rendering this direct approach numerically impractical. Thus a variety of heuristic procedures are presently being used with modest success [644].

Cluster coagulation is more general than polymerization, involving coalescence of rain drops, particles of smoke and dust, and bacterial aggregation, in addition to the chemical bonding of large molecules [777]. Nonlinear dynamics enter as growth laws for all of these examples, formulated by computing growth rates as functions of current and past clusters. The Verhulst law given in (1.2) is a simple example, but more complex growth phenomena are currently being studied.

Named for Irving Langmuir and Katherine Blodgett, Langmuir–Blodgett (LB) films are interesting and potentially useful structures, which are formed first as molecular monolayers on liquid surfaces [759]. Sophisticated techniques have been developed for transferring such films to crystal surfaces in connection with research in nanotechnology. Scheibe aggregates are a special type of LB film discovered independently by Günther Scheibe in Germany and Edwin Jelly in England during the mid-1930s [979]. Composed of organic dye molecules, these films are presently finding applications as efficient photon detectors, and various nonlinear models have been proposed to describe their operation, including a discrete two-dimensional version of the ubiquitous NLS equation [179].

Discovered in 1888 by Friedrich Reinitzer, an Austrian botanist, liquid crystals comprise rod- or disc-shaped molecules whose phases can be controlled by electric fields and light, leading to many odd and interesting nonlinear optical properties [718]. Although liquid crystals are biologically employed in the coloration of certain plants and insects, commercial applications were not evident to Reinitzer, but they now rank second only to the cathode-ray tube as a means for display of visual information.

6.4.4 Chemical Kinetics

Beyond the static aspects of aggregates, it is interesting and potentially important to understand the dynamics of chemical reactions. These studies go back at least to the theoretical work of Lotka in the early 1920s [578], and they have played an important role in the nonlinear science expansion of the 1970s [713, 715]. In addition to the BZ reaction described in Chap. 4, there is the Brusselator – a somewhat simpler and more flexible RD system sharing many of the same qualitative properties [712], and the Turing mechanism – an RD system in which the inhibiting component diffuses faster than the exciting component, leading to stationary patterns that may be related to those in biological organisms [110, 973]. Finally, energetic materials (explosives) provide dramatic examples of chemical kinetics [215]. These are divided into low-velocity mixtures like black powder, in which the nature of the blast depends upon the shape of the enclosure, and high-velocity explosives (nitroglycerine, dynamite, TNT and ammonium nitrate–fuel oil mixtures) in which the high speed of the reaction makes the dynamics of a blast independent of its confinement. Both types of explosion are, of course, highly nonlinear.

6.5 Condensed-Matter Physics

Comprising any of the ninety-odd atomic elements and their molecular combinations bound together by highly nonlinear valence, polarization (van der Waals) and electrostatic forces, condensed matter is a stage where many nonlinear phenomena appear, some of which were mentioned in Chap. 3. Following the Frenkel–Kontorova formulation of the sine-Gordon (SG) equation to describe the dynamics of crystal defects [331, 433, 896], for example, it was found that this same equation – augmented with appropriate dissipative and driving terms – provides a model for the motions of domain walls which separate regions of differing magnetization (or polarization) in ferromagnetic (or ferroelectric) materials [247, 976] and for the propagation of splay waves on biological membranes [304]. Other nonlinear phenomena in condensed matter include the following.

6.5.1 Extrinsic Nonlinearity

Among the earliest proposals for nonlinear phenomena in solids were Rudolph Peierls' 1930 suggestion of charge-density waves (where electrons and phonons interact to lower their total energy) [991] and Landau's 1933 concept of the polaron, in which a moving charge takes advantage of extrinsic nonlinearity (interaction with and distortion of the nearby lattice structure) to move more easily through a crystal [535, 1087]. Studied extensively in the Soviet Union in the early 1950s [748], the concept of extrinsic nonlinearity led to an understanding of the phenomenon of superconductivity, in which a pair of electrons couples (through lattice interactions) to change from two fermions into a boson-like entity which can aggregate into a macroscopic quantum (superconducting) state, sometimes called a Bose–Einstein condensate [390, 849, 966]. The dynamics of a macroscopic superconducting state, in turn, are governed by the celebrated Ginzburg–Landau (GL) equations [226], which are closely related to the NLS equation of soliton theory [597]. At a higher level of description, one finds the formation of magnetic flux vortices which allow magnetic fields to penetrate into Type II superconductors [747], and these vortices are in turn related to the fluxons that obey the SG equation on long Josephson-junction structures [58, 562] and to hydrodynamic and atmospheric vortices (see Fig. 6.18). Finally, if certain of its parameters are allowed to be complex, a GL equation can exhibit properties of an RD system [596]. Unfortunately, mechanisms for extrinsic nonlinearity are often omitted from molecular-dynamics models of condensed matter systems, limiting the range of phenomena these models can represent [967].

6.5.2 Phase Transitions

In Chap. 1 the concept of a phase transition was used as a metaphor for Kuhn's Gestalt-like paradigm shift in the perspectives of a scientific community [223]. This concept is widely employed in condensed-matter studies [1004], providing descriptions of condensation of gases, freezing and boiling of liquids, and melting of solids; the formation of domains of uniform magnetization (or polarization) in ferromagnets (or ferroelectrics); establishment of superconducting states in metals and superfluid states in liquid helium [511]; and the onset of coherent light output from a laser above a certain pumping level [922] – to name some of the more important applications.

Salient aspects of phase transitions can be understood by thinking about the properties of ferromagnetic materials such as a bar of iron or of nickel, which can be viewed as a collection of atomic magnets [1004]. If these small magnets are assumed to be uncoupled, application of a magnetic field of an external magnetic field of H would result in a magnetization M that is proportional to H and inversely proportional to absolute temperature T, because at high temperatures the magnetic orientations are almost random. Coupling among the magnets introduces a tendency for them to take the same orientations as their neighbors, whence

$$M \propto \frac{H}{T - T_c} \, ,$$

where T_c is a *Curie temperature*, below which the ferromagnetic metal can remain magnetized without the action of an applied field. (For iron, $T_c = 1043$ K, and for nickel it is 630 K.) Far above the critical temperature, i.e., for $T \gg T_c$, the coupling between neighboring atoms becomes negligible and $M \propto 1/T$. The system evidently undergoes a phase transition at $T = T_c$, below which M can be finite for $H = 0$. Why doesn't an ordinary piece of iron act like a magnet?

Two other contributions to the total free energy of bar magnet must be considered to understand its global behavior. First, there is an external magnetic field energy, which increases when the atomic magnets are all aligned in one direction to generate an external field. This external field is reduced if the interior of the iron bar is divided into small and randomly oriented domains of uniform magnetization that are separated by domain walls, across which the orientation of M changes. Second is the energy needed to create these internal domain walls. If there are too few internal domains, the external field energy grows, and if there are too many, the total domain wall energy grows. For an ordinary bar of iron that has cooled from the melt, a balance between these two opposing tendencies is established, rendering the bar almost completely unmagnetized (little or no external field is observed). If an unmagnetized bar at room temperature is placed in a sufficiently large longitudinal magnetic field H_{ext}, however, all of the internal domains are forced to become oriented in the same direction, and the internal domain walls disappear. (Under these conditions, the corresponding domain wall motions are governed by a version of the SG equation that is augmented to include both dissipative effects and the forces induced by H_{ext}.) As H_{ext} is reduced to zero, some domain walls reappear, but they are pinned to mesoscopic irregularities of the metal, leaving the residual external magnetic field of a standard bar magnet. Thus a simple bar magnet displays a variety of nonlinear phenomena, the political implications of which have not been lost upon despots.

If the natural magnetizations $M(T)$ of iron and of nickel are measured in units of their low-temperature values M_0, the resulting curves are universal functions of the *order parameter* [1004]

$$\varepsilon = \frac{T_c - T}{T_c} ,$$

and similar universal curves are observed for the densities of superconducting electrons [747]. More generally, a wide variety of critical phenomena (melting, vaporization, sublimation, etc.) display universal behaviors near critical points, where dependencies are upon powers of the order parameter and are independent of particular materials [976]. These *critical exponents* can be computed from renormalization group theory, which assumes that macroscopic behavior is independent of microscopic details; thus they can be calculated from a nested set of mesoscopic models [978]. (Similar ideas are used in high-energy physics to construct field theories with properties that are independent of the length scale chosen for short-wavelength cutoff.)

6.5.3 Supersonic Solitary Waves

To my knowledge, supersonic propagation of sound was first recorded on February 9, 1822 by Captain William Edward Parry on the second of his three arctic expeditions to find a Northwest Passage across Canada. During experiments on the speed of sound, Parry observed that the command to fire a cannon was heard "about one beat of the chronometer [nearly half a second] after the report of the gun." As the propagation distance was 1720 meters [398], these observations suggest that the high-amplitude wave of the cannon's report had a velocity about 10% above than that of the low-amplitude command to fire it. Was Parry's observation an isolated oddity, or does it suggest a general property of large amplitude waves in nonlinear media?

To understand Parry's observation, consider the spring–mass lattice shown in Fig. 3.6. If the springs are nonlinear with a certain exponential potential, the model becomes the Toda lattice (TL) having stable soliton solutions [875, 969], where (3.24) shows that these solitons travel at supersonic speeds. Although the special properties of the TL require the particular exponential form given in (3.22), supersonic solitary waves occur for a broad class of spring potentials $U(r)$, as was shown in 1994 by Gero Friesecke and Jonathan Wattis who proved the following theorem [340]:

FW Theorem. *Assume that $U(r)$ has a second derivative, with $U(0) = 0$ and $U(r) \geq 0$ in some neighborhood of the origin. A sufficient condition for solitary wave solutions of the spring–mass lattice to exist is that $U(r)$ be superquadratic on at least one side. In other words, $U(r)/r^2$ increases strictly with $|r|$ either for r between zero and some positive value R_p, or between zero and some negative value $-R_n$, where R_p and R_n could be finite or infinite.*

Furthermore, these solitary waves are supersonic, with arbitrarily large and small amplitudes, and they are either all positive or all negative (entirely expansive or compressive).

Stability of these supersonic solutions has been established in a recent series of papers by Friesecke and Robert Pego [336–339]. Among others, the conditions for this theorem are satisfied for the TL potential and for the potentials assumed in the FPU studies [305].

Thus an explanation for Parry's observation of supersonic sound follows from the FW theorem which holds for many physically reasonable models of nonlinear compression forces including those of the atmosphere. As the spring–mass model of Fig. 3.6 serves equally well for planar sound waves in a solid, supersonic solitary waves are to be expected in one-dimensional problems of condensed-matter physics [206].

6.5.4 Discrete Breathers

The supersonic waves discussed in the previous section are one-dimensional (1D) and they have no oscillatory character, but nonlinear excitations that are localized in 3D may arise if these two restrictions are lifted. Although the possibility of localized oscillatory states is implied by Mecke's observations of local modes in symmetric molecules during the 1930s [645–648] and by Careri's assignment of certain CO stretching oscillations in model proteins during the early 1980s [152], many solid state physicists continued to believe throughout the 1980s that it was theoretically impossible to have localized vibrational energy in a crystal with translational symmetry.

An important contribution to this discussion was made in a 1988 paper by Albert Sievers and Takeno, who showed how the discreteness of a molecular crystal contributes to the stability of localized nonlinear oscillations [906]. These authors informally argued that *intrinsic local modes* (ILMs) will be stabilized against radiation into phonons of the crystal lattice if overtones of the basic frequency lie within the stop bands of the lattice.[6] In 1994, Robert MacKay and Serge Aubry turned this argument into a mathematical proof for stationary (non-moving) *discrete breathers* (DBs) under the assumption of sufficiently strong nonlinearity [587]. Recently this mathematical result has generated interest in breathers within the physics community [150, 875].

Given confidence in the existence of stationary nonlinear oscillating states (DBs or ILMs), it is of interest to consider whether they can move. This is a

---

[6]Note that the terms "intrinsic" and "extrinsic" are used in two different ways. In the above analysis of local modes in molecules, an intrinsic nonlinearity is generated directly through changes in the valence bonds, whereas an extrinsic nonlinearity involves interactions with the crystal lattice. In the terminology of Sievers and Takeno, an intrinsic localization appears in a regular crystal, whereas extrinsic localization arises at the site of a symmetry-breaking dislocation or embedded atom.

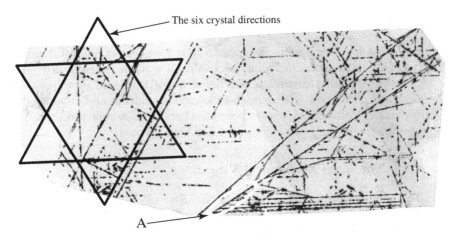

The six crystal directions

A

Fig. 6.7. A piece of muscovite mica (overall length is 23 cm). Inside the oval marked A, the tracks of several moving breathers appear to be emerging to the right from a charged-particle track. (Courtesy of Mike Russell)

more difficult question because mathematical theorems are not yet available and direct experimental observations of moving states are difficult. Nonetheless, some progress has been achieved through a combination of numerical studies and indirect empirical observations. On the numerical side, Henrik Feddersen has used Davydov's model with extrinsic nonlinearity to study solitary waves moving through a discrete 1D lattice, finding propagating solutions to a high degree of numerical accuracy over a significant range of model parameters [301]. These results are important, as they suggest that lattice discreteness is not a barrier to the formation of moving localized breathers.

Empirically, Mike Russell has proposed that many of the dark lines appearing in large pieces of muscovite mica $KAl_3Si_3O_{10}(OH)_{1.8}F_{0.2}$ provide evidence of moving breathers that were present during the cooling process [831, 832]. As shown in Fig. 6.7, mica sheets have the six-fold rotational symmetry of benzene molecules, with the six crystal axes indicated by a star imposed on the figure. Those lines that do not correspond to these six directions have been identified as charged particles generated by cosmic rays [829, 830]. Interestingly, many of the lines along the crystal axes are seen to emerge from charged particle lines, for example, those within the oval marked A in Fig. 6.7. In collaboration with Russell, Chris Eilbeck and José Marín have numerically shown that breathers localized on a two-dimensional plane can propagate for long distances along the directions of crystal symmetry, corresponding to many of the lines in Fig. 6.7 [609–611].

6.6 Engineering Applications

Problems involving nonlinear dynamics have long been of interest in many areas of applied science, including mechanical vibrations (musical instruments, water pipes and structures under the stress of wind and earthquake), and electrical oscillations (in vacuum-tube, semiconductor and superconductor systems). Several examples are presented here to show how widespread engineering applications of nonlinear science have become.

6.6.1 Nonlinear Mechanical Vibrations

The pendulum is a relatively simple nonlinear mechanical system which was used by Christian Huygens as the temporal element in his pendulum clock, invented in 1656 [529]. While indisposed and confined to bed a decade later, he noticed that two clocks in mechanical contact kept exactly the same time, because the motions of their pendula became locked together or *synchronized* [459, 934]. In addition to providing the basis for this early manifestation of nonlinear dynamics, the nonlinear oscillation of a pendulum is described by the elliptic functions which were invented in 1829 by Carl Jacobi, and these, in turn, express solutions of the KdV equation and are now a workhorse of nonlinear dynamics [145, 147, 875]. As elliptic functions are not sinusoidal, the pendulum generates overtones at multiples of its fundamental frequency – yet another characteristic nonlinear phenomenon [673].

Driven mechanical oscillations included the swaying of tree branches in a spring breeze, musical tones of a violin, screechings of loosely-mounted water pipes, and the human voice. All such phenomena are necessarily nonlinear due to effects that limit the amplitudes of their oscillations. A striking example of a wind-driven mechanical oscillation is provided by the Tacoma Narrows Bridge, which was designed to be lighter, longer, and more flexible than previous structures (the Golden Gate Bridge in San Francisco and the Bronx–Whitestone Bridge in New York), opening to traffic on July 1, 1940. From the outset, this bridge exhibited vertical oscillations with as many as seven nodal points in gentle winds, but it had remained stationary in a wind of 35 mph and survived winds of 48 mph [20, 640]. Five months after its opening in a wind of about 42 mph, the bridge began strong vertical oscillations of 36 to 38 cpm (cycles per minute) with eight or nine nodes and an amplitude of four to five feet. Interestingly, people and cars remained on the bridge, not viewing this vertical oscillation as cause for alarm, but suddenly the bridge switched to the torsional mode shown in Fig. 6.8. Rotating up to 45° with a frequency of about 14 cpm, this torsional motion definitely was cause for concern, and people scurried off the bridge as fast as they could before its collapse about 45 minutes later. Back in the 1970s, I would show a film clip of the torsional oscillation and break-up to my nonlinear waves class, describing it as one of the largest ever man-made oscillators.

Fig. 6.8. The Tacoma Narrows Bridge twisting, November 7, 1940. (Courtesy of Manuscripts, Special Collections, University Archives, University of Washington Archives, UW21413)

6.6.2 Vacuum Tube Electronics

In 1906, a decade after the discovery of the electron, Lee De Forest invented the triode vacuum tube (or audion), a device that allows the ratio of output to input power of an electric signal to substantially exceed unity. In addition to many radio and telephone applications, this amplification of an efficient cause introduced two studies of central importance in nonlinear science. The first was Adrian's discovery of the all-or-none effect (or threshold phenomenon) [9, 886], which opened new vistas in electrophysiology [874]. The second was Balthasar van der Pol's seminal study of nonlinear electrical oscillations [982]. Feeding the output voltage of a triode amplifier back to its input terminals, van der Pol arrived at an electronic system described by the equation

$$\frac{d^2y}{dt^2} - \varepsilon(1 - y^2)\frac{dy}{dt} + y = 0 \,, \tag{6.9}$$

which now bears his name.

To appreciate the dynamics of van der Pol's equation, consider several values of the parameter ε, which indicates what portion of the output signal is fed back to the input terminals:

1. For $\varepsilon = 0$, this equation describes a sinusoidal oscillation of constant amplitude which is the behavior of the linear resonant LC "tank circuit" of the amplifier output.

2. For $\varepsilon < 0$, losses predominate and any oscillation amplitude decreases with time to zero. Again, this is a linear phenomenon.

3. For $0 < \varepsilon \ll 1$, the oscillation remains approximately sinusoidal but increases slowly with time, reflecting the fact that the amplifier pumps energy slowly into the tank until the factor $(1 - y^2)$ decreases to zero. This is a weakly nonlinear phenomenon which corresponds to the Tacoma Narrows Bridge oscillation shown in Fig. 6.8.

4. For $\varepsilon \gg 1$, the dynamics is that of a blocking oscillator (or multivibrator or square-wave oscillator), switching rapidly back and forth between positive and negative solutions of $y \doteq \varepsilon(1 - y^2)y_t$ in a strongly nonlinear manner.

Thus under continuous changes of the parameter ε the qualitative nature of the oscillator dynamics becomes quite different [762].

In his 1926 and 1934 papers [982,983] – which should be read by all serious students of nonlinear science – van der Pol introduced the ideas of phase-plane analysis (see Appendix A) and of averaging methods to the engineering community [494, 558]. Along with piecewise linearization, these techniques were subsequently developed to a high degree by electrical engineers and applied mathematicians in the Soviet Union [92, 746]. Thus in the early 1960s a US engineering student with an interest in nonlinear analyses of electronic systems would study translations of Russian books that had been written a decade or more earlier by by Alexandr Andronov and his colleagues [27] and by Nikolai Bogoliubov and Nikolai Mitropolsky [98], in addition to an important book by Nicholas Minorsky [661].

6.6.3 Negative and Positive Feedback

With the invention of the audion, it seemed straightforward to design a system that would permit transcontinental or intercontinental telephone communications, but there was a hitch. About a hundred amplifying units are needed at regular intervals along the telephone line to make up for resistive line losses, and any variations in the amplifications of the individual units (caused by deterioration of the vacuums or decreases of electron-emission abilities of the hot cathodes, for example) would be raised to the 100th power at the output terminal of the system. In the mid-1920s, this was a serious problem, under intense study at the Bell Telephone Laboratories (BTL) by a number of electrical engineers, including young Harold S. Black.

While coming across the Hudson River on the morning of August 2, 1927, Black suddenly wrote the following formula on his copy of *The New York Times*:

$$G = \frac{A}{1 - \mu A} . \tag{6.10}$$

(Black's copy of the *Times* became an important patent document, establishing the date of his invention.) Now known to every electrical engineer, this seemingly simple equation embodies one of the most important inventions of the twentieth century, making transcontinental telephone possible and providing a theoretical basis for servomechanisms [127].

To understand the implications of (6.10) for transcontinental telephony, suppose that A ($\approx 10\,000$) is the (open loop) gain of three amplifier stages (audions) and μ ($= 1/10$) is the loss of a *negative feedback circuit* which connects the output of the amplifier back to the input terminals. Then to about one percent $G = 1/\mu = 10$ is the overall (closed-loop) gain of the amplifier. In other words, a gain ratio of $10\,000$ has been traded for a gain ratio of 10, with the advantage that G depends only on the passive components of the feedback circuit. This is a good trade because the large gain of the audions is cheap, whereas a stabilized overall gain (G) is absolutely essential for the transcontinental phone line to work.

Such a feedback amplifier circuit can be viewed as asking that the ratio of output to input voltage be equal to $1/\mu$. In these terms, negative feedback was used in the steam-engine governor which was invented in the eighteenth century and studied by Maxwell in the nineteenth century [528]. Following Black's invention, however, (6.10) was investigated by applied mathematicians in great detail, and the various requirements to avoid unwanted oscillations called singing (like those on the Tacoma Narrows Bridge) were thoroughly discussed in the classic book *Network Analysis and Feeedback Amplifier Design* by Henrik Bode, which first appeared as a BTL report in the early 1940s [95]. Soon the negative feedback idea was extended to servomechanisms, in which the desired value of a variable (temperature of a room, position of a boat's rudder, firing pattern of an antiaircraft gun, etc.) is fixed by requiring that the difference between the desired and actual values of some variable is forced to zero.

More generally, the concept of negative feedback introduces *closed loops of causality* into electronic systems, possibly confounding relations among causes and effects. As long as Bode's stability requirements are satisfied, a system does not oscillate, and G expresses the relationship between a simple cause and effect. But sometimes – as with van der Pol's circuit of (6.9) – oscillation is desired, and *positive feedback* is employed. From the perspectives of nonlinear science, positive feedback (PFB) is more interesting than negative feedback (NFB) for three reasons. First, NFB stabilizes a system, making the relation between input and output variables more linear. Second, nonlinearity always comes into play with PFB because the initial exponential growth must eventually be limited by some nonlinear effect, as with the saturation of nonlinear science publications shown in Fig. 1.1 or the collapse of the Tacoma Narrows Bridge. Finally, and most important, PFB is an essential element of the phenomenon of *emergence*, under which a qualitatively new entity (tornado, city, living organism, biological species, Kuhnian paradigm, etc.) comes into existence. The concept of emergence will often be discussed in subsequent chapters of this book.

6.6.4 Frequency–Power Formulas

In the early days of radio broadcasting, receivers were designed to amplify the incoming signal at its carrier frequency. These tuned-rf receivers employed two or three stages of vacuum-tube amplification, with tuning capacitors between each stage that were mounted on a common shaft for one-knob tuning. This ganged capacitor was a rather large component that was easily damaged and took up much of the physical volume of the set. Since the 1930s, however, receivers have employed the *heterodyne principle*, another example of an early nonlinear process in electronics [782].[7] Under heterodyning, the incoming (antenna) signal is "mixed" with the voltage from a variable frequency local oscillator (LO) to obtain an intermediate frequency (IF) signal at the difference between LO and antenna signals. The IF signal – usually a few hundred kilocycles – is then amplified through several vacuum-tube stages of fixed frequency; thus a large bank of tuning capacitors in no longer needed, as only the LO is tuned. Both simpler and more effective, the heterodyne process is now universally used in the design of radio receivers.

In 1956, Jack Manley and Harrison Rowe showed that lossless mixers obey restrictions on the ratios of antenna input (AI), LO and IF powers (P_{ai}, P_{lo} and P_{if}) that are more severe than conservation of energy [601]. For example, if the corresponding frequencies are respectively ω_{ai}, ω_{lo} and ω_{if} (with $\omega_{ai} = \omega_{lo} + \omega_{if}$ and $P_{ai} + P_{lo} + P_{if} = 0$), then

$$\frac{P_{lo}}{\omega_{lo}} = \frac{P_{if}}{\omega_{if}} = -\frac{P_{ai}}{\omega_{ai}}, \tag{6.11}$$

and with more frequencies present, these frequency–power formulas, also known as Manley–Rowe (MR) equations, contain additional terms [751]. As optical media are nearly lossless, MR relations also govern power flows among interacting laser beams [749]. Interestingly, (6.11) can be derived from the assumption that wave interactions in a lossless mixer involve single quanta [1028].

6.6.5 Synchronization

Although scientific interest in synchronized oscillations go back to Huygens's seventeenth century report that two of his pendulum clocks kept exactly the same time when in mechanical contact [459], modern research on synchronization began in 1934 with van der Pol's analysis of two coupled vacuum-tube oscillators, each described by (6.9) [983]. If the nonlinear parameter ε is small,

---

[7]Although credit for this invention is usually given to Edwin Armstrong in the US and to Lucien Lévy in France (both about 1917), it was in fact invented by Reginald Fessenden in 1901, who also coined the term from the Greek words "heteros" meaning "other" and "dynamis" meaning "force" (see US Patent #706,740, issued 12 August 1902).

the two oscillations are almost sinusoidal, and the oscillators must be closely tuned for frequency locking to occur. As ε increases with fixed coupling – electronic engineers have long been aware – synchronization becomes possible over a progressively wider frequency difference of the two uncoupled oscillators [821].

Motivated in the late 1950s by an interest in the phenomenon of coupling among the brain's neurons, Norbert Wiener generalized the coupled oscillator problem from van der Pol's two to many, with natural frequencies of the uncoupled oscillations described by a Gaussian probability curve [1038]. The result of the coupling, he showed, is to collapse the oscillators near the center of the Gaussian distribution into a single frequency (or delta-function distribution), leaving oscillators on the wings relatively uninfluenced. As Wiener noted, this effect is also observed in the frequency locking of generators on an electric power grid, and he suggested that it would be found in "a great many other physical cases" including planetary motions and molecular spectra. Among examples of the latter, we have noted the emergence of local modes in benzene molecules (see Fig. 6.4), for which the eigenfrequency spread of the six CH stretching modes are reduced from the nearest neighbor coupling energy to the much smaller (quasidegenerate) value given in (6.5).

A well-written survey of such physical examples has recently been published by Steven Strogatz, who describes applications to the synchronized flashing of Asian fireflies,[8] planetary motions, phase transitions, coherent quantum states (lasers, superconductors, superfluids, etc.), and biological oscillators [934]. Interestingly, Strogatz also points out that the neurological data to which Wiener referred in his *Nonlinear Problems in Random Theory* [1038] – entitled "Application to the study of brain waves, random time, and coupled oscillators" – was incorrect. Although Wiener's basic idea is sound, the dynamics of alpha rhythm in the neocortex are far more complicated than those of interacting sinusoidal oscillators.

6.6.6 Nonlinear Diffusion

With the invention of the transistor at mid-century by John Bardeen, Walter Brattain and William Shockley, not only was the vacuum tube made largely obsolete but engineering research in solid-state electronics surged.[9] In 1957, Leo Esaki invented a two-terminal solid-state device with negative differential conduction (terminal current I is a single-valued function of terminal voltage V, with $dI/dV < 0$ over some range of voltage) which allows a very simple realization of van der Pol's equation, and a corresponding superconductive device was invented by Ivar Giaever in 1960 [848].

---

[8] See http://www.pojman.com/NLCD-movies/fireflies-dsl.mov for a movie of this striking phenomenon.

[9] Bardeen, Brattain and Shockley were awarded the 1956 Nobel Prize in physics for inventing the transistor.

As an electrical-engineering graduate student in the late 1950s, these developments were of great interest, for the initial aim of my research was to study large-area negative-conductance diodes from the perspectives of linear wave theory. Nonlinear effects – it was then widely believed and taught – could be analyzed by either of two methods:

1. *Piecewise linearization*, in which solutions are obtained over the linear branches of the nonlinear device and then matched at the boundaries between adjacent linear regions.
2. *Quasilinearization*, in which a solution is first approximated by a set of linear normal modes, after which nonlinear interactions among modes are computed.

Missing from my mind was the concept of a serious search for global nonlinear solutions that cannot be well represented in either of these two approaches, but in the fall of 1960, a light switched on. Professor Jun-ichi Nishizawa from Tohoku University visited my laboratory, saw what I was doing, and commented: "We are working on similar problems in Japan, but we are using *nonlinear* wave theory." This was the first time I heard the term that became my life's work.

Under appropriate biasing, large area Esaki and Giaever diodes have "cubic-like" volt–ampere characteristics similar to the sodium ion current of a nerve fiber membrane [see (4.3)], leading naturally to schemes for electronic analogs of the nerve axon [199, 697, 857, 861]. The essential idea is to build an electronic model of the simple RD equation of (4.4) with the global solution given in (4.5), and then augment the system with a recovery variable, which returns the voltage to its original resting value.

In finding global solutions of PDEs that describe nerve-axon models, a key idea was to assume the existence of traveling waves (either periodic or localized), for which all dependent variables are functions of $\xi = x - vt$, where v is velocity parameter that is free (undetermined) at the outset of the analysis [861]. This *traveling-wave assumption* connects partial derivatives as

$$\frac{\mathrm{d}}{\mathrm{d}\xi} = \frac{\partial}{\partial x} = -\frac{1}{v}\frac{\partial}{\partial t} \, ,$$

thereby reducing the original PDE system (in x and t) to an ODE system (in ξ) that can be studied using phase-space techniques described in Appendix A [406]. In the course of such analyses, it turns out that for heteroclinic trajectories of the ODE system the speed v takes a fixed value at which energy is released by the global traveling wave at the same rate that it is consumed by dissipative processes.[10] Interestingly, Alan Hodgkin and Andrew Huxley used this method in their classic 1952 study of nerve impulse propagation on squid axons [440].

---

[10] For (6.1), the traveling-wave assumption implies $v^2 = c^2$ for waves of any shape, while for nonlinear systems that conserve energy (KdV, SG, NLS, TL, for example) v can be adjusted over a continuous range of values [861].

6.6.7 Shock Waves and Solitons

Thinking about localized traveling-wave solutions for RD systems in the early 1960s led several researchers to consider whether similar solutions can be found for nonlinear PDEs that conserve energy. Again, the growing cornucopia of solid-state electronics provided several possibilities, including the voltage dependence capacitance of a reverse-biased silicon diode (or varactor), which has been widely used since the 1950s for the frequency-locking circuits of FM radios. It is not difficult to construct an LC transmission line in which the series inductors are standard linear elements and the shunt capacitors are nonlinear varactors. Because non-dispersive lines develop shock waves, they can be used as harmonic generators, but if varactors are employed as elements in dispersive transmission lines (where the phase and group velocities are functions of frequency), one finds the KdV equation, which was discussed in Sect. 3.1. Investigations of these nonlinear electrical transmission line models of wave systems have been carried on by electrical engineers in the Soviet Union, Japan, Denmark, and the US since the early 1960s [861], and they are useful not only for scientific investigations [572] but also as teaching aids [798].

Another possibility of electronic solitons emerged in 1963 from Brian Josephson's invention of a superconducting diode for which current I is related to voltage V as [58]

$$ I \propto \sin\left(\frac{2\pi}{\Phi_0}\int V\,\mathrm{d}t\right), $$

where $\Phi_0 = 2.068 \times 10^{-15}$ volt-seconds is a quantum of magnetic flux [58, 562, 695]. In contrast to the Esaki and Giaever diodes which are nonlinear conductors, Josephson diodes are nonlinear inductors, and upon being extended in one dimension they are described by the sine-Gordon (SG) equation, for which the kink solution shown in Fig. 3.3 represents a moving flux quantum or fluxon [859].[11]

As shown in Fig. 6.9, fluxons can be used in an oscillator of technical importance. If the bias current is zero and dissipative effects are neglected, the fluxon shown in this figure will bounce back and forth in its rectangular box with a frequency equal to v/L and a voltage that can be expressed exactly as a product of Jacobi elliptic functions [875]. In general, v is determined by a balance between energy input from the bias current and dissipation, leading to an oscillator that converts direct current into oscillations at frequencies of hundreds of GHz, with this frequency controlled by the bias current [58,562]. For radio astronomy in the range of 100 to 500 GHz, the present standard comprises a superconducting heterodyne receiver with the local oscillator a long Josephson junction [747].

---

[11] Esaki, Giaever, and Josephson shared the 1973 Nobel Prize in physics for their inventions of these three tunneling devices.

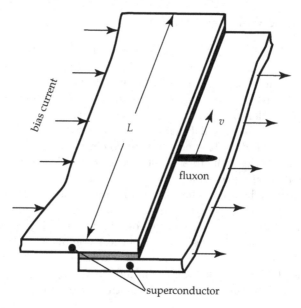

Fig. 6.9. A long Josephson oscillator in which a quantum of magnetic flux (fluxon) bounces back-and-forth between the two ends of a rectangular junction

One of the creative tensions of present nonlinear wave theory pits those who hold to the earlier view that the oscillator of Fig. 6.9 can be completely analyzed in terms of weakly nonlinear (quasilinear) interactions among the linear electromagnetic modes of the cavity against those who claim first, that the quasilinear approach is often not feasible in practice (because severe mode truncation must be employed), and second, even if it were feasible, the quasilinear approach would forgo possibilities for using perturbation theory around fully nonlinear analytic solutions [641]. In a recent paper, numerical and experimental studies of annular (ring-shaped) Josephson junctions have been conducted with sizes about equal to those of the fluxons, finding ranges of parameter values where each of these analytical approaches (quasilinear and fully nonlinear) are appropriate [698]. Moreover, Antonio Barone, who has led a Neapolitan research group in superconductor applications for several decades, recently showed that studies of the Josephson effect and the corresponding SG equation play roles in many areas including non-conventional order parameters in high temperature superconductors, macroscopic quantum tunneling, and even cosmology [55, 56].

6.6.8 Electronic Chaos

In a 1927 investigation of the nonlinear phenomenon of frequency demultiplication, written together with Jan van der Mark, van der Pol reported that an "irregular noise" was observed at certain driving frequencies [985]. Although disregarded at the time, this was the first report of low-dimensional chaos in electronics. As noted in Chap. 2, Ueda and his colleagues observed chaotic behavior in the early 1960s in periodically driven nonlinear oscillator [409] (for example, van der Pol's equation with a non-resonant periodic driving force on the right-hand side), but this observation was not widely distributed to the engineering community, perhaps because the authors were not sure that it was correct. By the end of the 1970s, as we have seen, many chaotic systems had been studied, and it was clear that chaotic circuits can easily be constructed from selections of the two-terminal elements described above: varactors or tunnel diodes along with linear circuit elements and batteries [347, 489, 952].

6.7 Optical Science

Prior to the 1960 inventions of the solid-state laser by Theodore Maiman and the gas laser by Ali Javan, optics was entirely a linear science because the electric field amplitudes available from incoherent light sources are not large enough for nonlinear effects to be observed. Predicted in the late 1950s by Arthur Schawlow and Charles Townes in the US and by Nicolai Basov and Aleksandr Prokhorov in Russia as a realization of the maser (Microwave Amplification by Stimulated Emission of Radiation) at optical (shorter) wavelengths, the laser (Light Amplification by Stimulated Emission of Radiation) could have been invented at any time after 1917, when a key paper on stimulated emission of radiation was published by Einstein [276].[12]

6.7.1 Lasers

In its simplest (two-level) form, a laser comprises a population of quantum-mechanical entities (atoms, molecules, electronic states in crystals, etc.) with an upper (excited) energy level (E^+) and a lower (ground) level (E^-). The corresponding populations (N^+) and (N^-) are coupled to an electromagnetic cavity with an excited mode of intensity I. If $N \equiv N^+ - N^-$, laser rate equations take the form [922]

$$\frac{dI}{dt} = (\alpha N - \varepsilon)I , \qquad \frac{dN}{dt} = \gamma(N_0 - N) - 2\alpha N I , \qquad (6.12)$$

where ε is the natural decay rate for the cavity energy, γ is the natural decay rate for the population inversion (N), α is a proportionality constant for stimulated emission, and N_0 is a steady-state level of inversion for $I = 0$. (Suppressed in this formulation is the pumping rate, which is proportional

---

[12]The 1964 Nobel Prize in physics was awarded to Townes, Basov and Prokhorov for developing the maser–laser principle of stimulated emission.

to N_0.) With $\alpha N > \varepsilon$, the lasing threshold is attained, and the energy of the electromagnetic field within the cavity begins to grow. For $\alpha N_0 \gg \varepsilon$ this electromagnetic field becomes very intense (in a manner that is related to the Bose–Einstein condensation of a superconducting state) and the system is said to *lase* [1068]. These equations are similar to those describing other nonlinear phenomena. For example, with $\alpha N_0 < \varepsilon$ and $I = 0$, the second of (6.12) describes a limited growth process similar to that shown in Fig. 1.1, and more generally, equations (6.12) are like those describing predator–prey dynamics of biological populations and of interacting chemical species. Thus we see again how quite different physical systems can be related through their underlying mathematical structure.

Currently, there are many types of lasers, offering a wide variety of output wavelengths and performance characteristics, including the following [922, 1019]:

1. Solid-state lasers, of which the first was Maiman's ruby (Al_2O_3) laser, where the active entities are Cr^{3+} ions and the output wavelength is red light at 694 nm. Solid-state lasers typically employ a host crystal or glass doped with active ions and are described by three- or four-level schemes, with output wavelengths ranging from 280 to 1800 nm. These lasers are usually operated in a pulsed mode in which the population inversion N is raised to a high level before the lasing action is allowed to begin, yielding an output beam of modest average power but high instantaneous power.
2. Gas lasers, of which the first was Javan's helium–neon laser, where the active entity is an excited atomic level. These lasers are usually operated in a continuous mode and offer a variety of output wavelengths from the carbon dioxide laser (10 600 nm) to the argon laser (409–686 nm).
3. Semiconductor-diode lasers in which the active entities are overlapping electron and hole states of highly doped semiconductors and the pumping (pulsed or continuous) is by direct electrical current. Familiar to us all as laser pointers, semiconductor-diode lasers are also widely used in optical communications and as optical switches. Output wavelengths range from the 330 nm of ZnS diodes to the 8 500 nm of PbSe diodes.
4. Liquid lasers (sometimes called dye lasers) in which the active entities are inorganic dyes that are pumped by intense flash lamps. The spectral range of the dyes extends from infrared to ultraviolet.
5. Free-electron lasers in which the active entity is a relativistic electron beam. These large installations are operated in tunable pulse mode with output wavelengths from infrared to X-ray.

Shortly after Maiman's invention of the ruby laser, Peter Franken and his colleagues passed its high-intensity beam through quartz, converting red light (694 nm) to blue (347 nm) and introducing *nonlinear optics* as a new realm of research in which many interesting problems have arisen [93,319,667,668,673]. Thus an aim of the laser engineer is to obtain an output beam that is uniform over the face of the cavity, which is sometimes difficult because a pure lasing mode tends to break up into independent longitudinal filaments.[13] Although a problem for the engineer, such filamentation is of theoretical interest to the nonlinear scientist. As might be expected from the discussions in Chap. 2, the laser rate equations may have chaotic solutions if the pump is modulated [582], and it was through studies of laser dynamics that Hermann Haken was led to his formulation of *synergetics* as a general description of cooperative nonlinear dynamics in physics, chemistry, biology, psychology, and sociology (to be discussed in later chapters) [391, 394].[14] Plane-wave instability (modulational instability) on optical fibers leads to solitons of the NLS equation, and NLS solitons are useful as carriers of bits of information [405, 495, 665]. Finally, advances in laser technology allow pump–probe spectroscopy in the sub-picosecond range [264, 265, 399]. Let us examine these last three phenomena.

6.7.2 Modulational Instability

Instability of a periodic wave in a nonlinear optical medium was first discussed by Lev Ostrovsky in 1966 [731] and shortly thereafter by Thomas Brooke Benjamin and James Feir in the context of deep water waves [75].[15] This phenomenon has been studied in detail by Gerald Whitham who has developed an averaged Lagrangian analysis to derive nonlinear equations for weak modulations of periodic waves – which are Jacobi elliptic functions with slowly varying parameters rather than sinusoids [1033]. Whitham's formulation is interesting as an analytic example of nonlinear phenomena at two hierarchical levels of description – the local elliptic functions and the nonlinear modulation thereof.

In the nonlinear Schrödinger (NLS) system of (3.9), a negative sign before the last $(2|u|^2 u)$ term on the left-hand side implies modulational stability; thus periodic waves of constant frequency and wavelength can propagate at a constant speed. With a positive sign before this nonlinear term, however, periodic solutions are unstable, and multi-soliton solutions will evolve from the initial conditions [729, 733, 875]. In the context of applications to pulse data transmission on optical fibers, the latter case is more interesting.

---

[13]This instability is related to the general problem of nonlinear distributed oscillators, in which van der Pol's equation is generalized to one, two, or three space dimensions ($d = 1, 2, 3$) and independent quasilinear modes are stable only for $d \geq 2$ [879].

[14]Note that Zabusky uses the term "synergetics" to mean the use of "computers as an heuristic aid to synergize mathematical progress" [1072].

[15]Inappropriately, this phenomenon is referred to as a Benjamin–Feir instability rather than a Benjamin–Feir–Ostrovsky instability because of the unfortunate tendency of western scientists to avoid reading Russian journals during the Cold War.

6.7.3 Solitons on Optical Fibers

Optical fibers offer a means for transmitting electromagnetic waves at the wavelengths of visible light that is akin to the wave guide technology developed for microwaves and radar during the Second World War. In 1973, Akira Hasegawa and Fred Tappert conceived of using the nonlinear dielectric constant in the core of these "light pipes" as the basis for soliton transmission of information [405]. Their basic idea was that each bit of information would be stably carried by a soliton of the nonlinear Schrödinger (NLS) equation [495].

Although this concept has been brilliantly realized by Linn Molenauer and his colleagues over the past three decades [665], the initial reaction of the Bell Telephone Laboratories (BTL) was unenthusiastic [1072]. In coming to this conclusion, BTL management was influenced by the incorrect belief that real optical fibers would necessarily have dispersion for which the sign before the $2|u|^2 u$ term in (3.9) is negative, allowing only "dark solitons" (blank spots within a continuous beam) to form and thus requiring too much power. As Hasegawa and Tappert were unaware in 1973 that the NLS equation was exactly integrable, two deeper problems with gaining acceptance for their proposal were: it was introduced just before the explosion of interest in nonlinear science that is illustrated in Fig. 1.1, and it was in conflict with commitments by many BTL engineers to optical communication systems based on linear waves. Again we note an example of Kuhn's claim that revolutions in science are often impeded by those with established priorities.

6.7.4 Pump–Probe Spectroscopy

From the above survey of lasers that have been developed over the past four decades, it is clear that nonlinear optics is now a sophisticated activity, offering many possibilities for novel experiments [667, 668]. Among the most important of these for nonlinear science is pump–probe spectroscopy using pulses in the femtosecond range, because this technique allows unambiguous determinations of nonlinear localization [399].

To see how this works, consider the sketch of a pump–probe experiment in Fig. 6.10a, where a key question is to decide whether or not the oscillating mode giving rise to a particular spectral line is extended over the sample (thus of small amplitude and linear) or localized over a small region by nonlinear effects. In pump–probe spectroscopy, the measurements are made both from the ground state $(n = 0)$ and from the first excited state $(n = 1)$, to which the sample is brought through the action of a pump pulse.

With the pump on as in Fig. 6.10b, the probe beam sees two upward transitions, $(0 \rightarrow 1)$ and $(1 \rightarrow 2)$, and a downward transition, $(1 \rightarrow 0)$. If the oscillating mode is spread out over the crystal (and thus of small amplitude), the measurement with and without the pump is exactly the same [875]. In other words, the net absorption from the pumped system is the sum of absorptions from the $(0 \rightarrow 1)$ and $(1 \rightarrow 2)$ transitions minus the

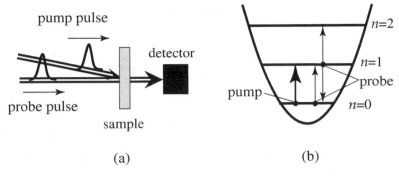

Fig. 6.10. (a) Sketch of a pump–probe setup. (b) Pump–probe measurement of an excited mode

stimulated emission from the $(1 \rightarrow 0)$ transition (minus since stimulated emission is equivalent to a negative absorption), which is just equal to the absorption from the $(0 \rightarrow 1)$ transition with no pumping. Thus subtraction of measurements with and without the pump gives zero.

If the oscillating mode is localized, however, this balance is disturbed because the spacing between the 0 and 1 levels is no longer equal to the spacing between the 1 and 2 levels (from the Birge–Sponer relation). Thus subtraction of the readings with and without the pump is no longer zero, but a value that can be used to determine the nonlinear character of the mode. As recently developed by Hamm and his colleagues, this method has been used to unambiguously demonstrate that the 1650 cm^{-1} line of crystalline acetanilide (ACN) (see Fig. 6.6) is a localized mode [264], and also to show that three of the room-temperature NH stretching lines in ACN are also localized [265].

Pump–probe spectroscopy has the following attractive features. First, the entire setup can be assembled on an optical bench, giving the the experimenter full-time access to the measurement task for as long as is needed. This is important because time-resolved measurements have previously been limited to using free-electron lasers as sources, for which beam time is limited and must be scheduled in advance. Second, the pump and probe pulses can be made very short (in the femtosecond range), allowing observation of the time course of the lattice oscillations that contribute to intrinsic localization; thus these lattice modes can often be identified [264, 265]. Finally, progress in laser technology makes it possible to perform pump–probe measurements from the far infrared to soft X-rays [283]. With these three advantages, it is expected that pump–probe spectroscopy will play a central role in the future development of experimental nonlinear science involving molecules, molecular crystals and biological molecules.

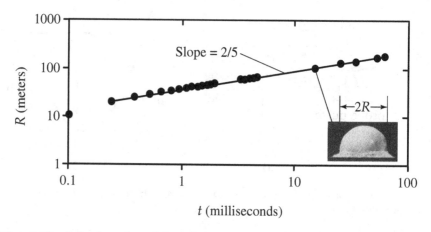

Fig. 6.11. A log–log plot of data from the Trinity explosion on 16 July 1945, showing the radius of expansion (R) as a function of time (t) during the first 62 ms. (Data from [948])

6.8 Fluid Dynamics

The oceans have frightened, fascinated and inspired our ancestors since the dawn of human experience and sound is closely tied to our speech and music; thus the dynamics of fluids have been of keen concern to Western scientists since the modern era began some four centuries ago. Studies of water waves are especially interesting because the air–water interface is both a dependent variable and a boundary condition [378]. From the perspective of Aristotle's categories of causality (see Chap. 1), in other words, the local wave height is both the effect of an efficient cause and a formal cause of that effect [29]. This sort of positive feedback loop is an essential feature in many other areas of nonlinear science, leading to the emergence of qualitatively new phenomena in a wide variety of applications.

6.8.1 Supersonic Waves

We have noted Captain Parry's observation that the "boom" of a cannon moved through the arctic atmosphere ten percent faster than sound waves of small amplitude – an observation supported by theoretical and numerical results for nonlinear spring–mass chains [336–339] and one-dimensional rods [206]. These supersonic waves (SSWs) are qualitatively similar to John Scott Russell's result for solitary water waves given in (3.1) and shown in Figs. 3.1 and 3.7a.

For a well-monitored example of a SSW, consider Fig. 6.11 which plots the radius of expansion (R) from the first atomic explosion as a function of time (t). Dimensional analysis shows that [1051]

$$\frac{R^5}{t^2} \propto \frac{E}{\rho} \, ,$$

where ρ is the density of air and E is the energy released by the blast (equivalent to about 20 kilotons of TNT). At a radius of 100 m, the hemispherical wave is expanding at 2400 mps or about seven times the speed of sound. This explosion provided a dramatic verification of the famous equation

$$E = mc^2$$

from SRT, where m is the decrease in mass of the nuclear constituents of the bomb [see (3.16)].

In the above mentioned examples – cannons, compression waves, wave tanks, tsunamis, and atomic blasts – the SSWs manage to detach themselves from their underlying linear systems. Thus SSWs differ from the bow wave of a boat and the shock waves produced by bullets and supersonic aircraft in a manner that is qualitatively similar to the difference between the near field of a radio antenna and its radiation field. Bow waves and sonic booms are tied to the objects that cause them, whereas SSWs are no longer connected to their initiating earthquakes, cannons or explosions; instead each emerges with an independent ontological status. Throughout his professional life, Scott Russell remained convinced of the importance of such solitary-wave phenomena – not only in the canals and water tanks which he had so thoroughly investigated in the 1830s – but also in the Earth's atmosphere [834].

6.8.2 Shock Waves

In contrast to supersonic waves, shock waves are experienced by the child who romps on an ocean beach. The basic phenomenon is described by the seemingly simple equation

$$\frac{\partial u}{\partial t} + (c + \beta u)\frac{\partial u}{\partial x} = 0 \, , \tag{6.13}$$

which has the solution $u(x,t) = f\big[x - (c + \beta f)t\big]$ for the initial condition $u(x,0) = f(x)$. For $\beta > 0$, higher amplitude portions of a wave travel faster than those of lower amplitude, leading to the multivalued breaking phenomenon on the front (leading edge) commonly observed for incoming waves. In more general terms, (6.13) is both nonlinear and non-dispersive, forcing all harmonics to travel at the same speed [371, 699]. Interestingly, (6.13) also describes certain traffic waves, where u represents the local density of automobiles [415, 1033]. In this application, $\beta < 0$, which causes shocks to form behind regions of increased density, as many inattentive drivers have experienced to their dismay.

In acoustics, multivalued breaking is not possible so shocks form, and their thickness is determined by detailed phenomena, which can be approximately described by the Burgers equation [699, 732, 750]

Fig. 6.12. A tidal bore (or *mascaret*) at Quillebeuf, France. The site is about 36 km from the mouth of the Seine, where tides up to 10 m occur. (Courtesy of Jean-Jacques Malandain)

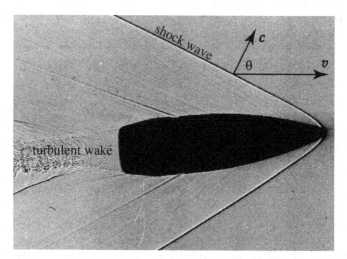

Fig. 6.13. Shadowgraph showing shock waves produced by a Winchester 0.308 caliber bullet traveling through air at about 2.5 times the speed of sound. (Courtesy of Ruprecht Nennstiel, Bundeskriminalamt Wiesbaden, Germany)

$$\frac{\partial u}{\partial t} + (c + \beta u)\frac{\partial u}{\partial x} = \delta\frac{\partial^2 u}{\partial x^2} \; . \tag{6.14}$$

Interestingly, this equation can be exactly solved for arbitrary initial conditions, just as can the KdV equation discussed in Sect. 2.2 [1033]. Hydraulic jumps are also described by (6.14), a dramatic example of which is the tidal bore (or *mascaret*) shown in Fig. 6.12. Due to dredging of the Seine in the 1960s, unfortunately, this striking nonlinear phenomenon has disappeared.

Closely related in both water waves and acoustics are the bow waves which emanate from a moving boat and the atmospheric shock waves generated by a bullet or a supersonic aircraft. In these familiar cases, which have been widely studied over the past two centuries [1033], the boat, bullet or aircraft travels at a velocity (v) that is greater than the low-amplitude wave speed (c) in the medium (water or air), and a shock cone develops which travels at speed c and angle $\theta = \cos^{-1}(c/v)$ (see Fig. 6.13) [730].

6.8.3 Rayleigh–Bénard Cells

If a pan of water is placed on the stove and the gas is turned high – as we all know – the water boils, carrying heat upward in a *convection* process involving a turbulent mixture of water and steam. If, on the other hand, the flame is very low, the water does not boil, and heat flows upward via a tranquil *conduction* process. To see what happens in between these two limits, put some peanut oil (one or two millimeters deep) in a flat-bottomed pan, shine a bright light, and slowly turn up the flame.

Using a more sophisticated apparatus, young Henri Bénard studied this phenomenon quantitatively in 1900 as part of his doctoral thesis, finding at a certain level of heating the hexagonal pattern of upward convection columns shown in Fig. 6.14a [73]. An explanation for this striking observation was provided a few years later by Lord Rayleigh, who analyzed the stability of upward heat conduction in a thin liquid layer and found that this flow becomes unstable due to buoyancy of the lower layer when the *Rayleigh number*

$$R = \frac{\alpha g \Delta T h^3}{k\nu} > R_{\mathrm{c}} \,,$$

where α is the thermal expansion coefficient, g is the acceleration of gravity, ΔT is the temperature difference between the top and bottom of the liquid, h is the depth of the liquid, κ is the thermal diffusivity, and ν is the kinematic viscosity [793]. If the Rayleigh number exceeds a critical value of $R_{\mathrm{c}} = 27\pi^4/4$, upward conductive flow of heat becomes unstable and the hexagonal pattern of *Rayleigh–Bénard cells* can form with the size of the hexagons about equal to the depth of the liquid.

For thin layers, the formation of these cells is somewhat more complicated than Lord Rayleigh had supposed because the onset of conduction instability depends also on differences in surface tension, which in turn depend on temperature at different positions of the upper surface. As differential surface-tension effects (which also lead to the "tears" that epicureans observe when they twirl their wine in the glass) had been studied by Carlo Marangoni in his 1865 doctoral thesis [606], one speaks of Bénard–Marangoni (BM) convection in thin layers when these effects dominate and of Rayleigh–Bénard (RB)

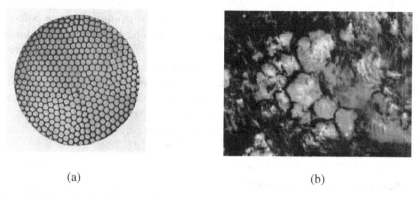

(a)   (b)

Fig. 6.14. Rayleigh–Bénard cells. (**a**) An original photograph by Henri Bénard [73]. (**b**) Clouds over the South Atlantic Ocean. (Courtesy of National Oceanic and Atmospheric Administration)

convection in thicker layers when Rayleigh's buoyancy effects dominate [169]. Since the early 1970s when nonlinear phenomena became of general interest, theoretical, numerical and experimental studies of BM and RB convections have been of great interest to the fluid dynamics community [603,604]. As is seen in Fig. 6.14b, RB convection can also be observed in atmospheric flow dynamics.

6.8.4 Plasma Waves

Sometimes called a fourth state of matter, a plasma is a fluid in which the constituent particles are charged – combining hydrodynamics with electromagnetism to form magnetohydrodynamics (MHD) [142]. Under MHD, motions of charged particles generate magnetic fields, which retroactively influence the charged particle motions; thus forming causal feedback loops that lead to many interesting and important nonlinear phenomena.

In a highly-conducting plasma, magnetic field lines (B) act like elastic strings, and these strings conduct waves (proposed in 1942 by Hannes Alfvén [16]) with low-amplitude (linear) velocity

$$v_A = \sqrt{\frac{B^2}{\mu_0 \rho}},$$

where ρ is the mass density of the plasma and μ_0 is magnetic permeability [1040]. Very nonlinear at higher amplitudes, Alfvén waves play key roles in studies of the Earth's magnetosphere, solar physics and quasars.

Plasma confinement devices for power generation by fusion are another important area of MHD research. In confined plasmas, low amplitude waves may be transverse (like ordinary electromagnetic waves) or longitudinal, with dispersion relations modified by the presence of one or more plasma frequencies:

$$\omega_{\rm p} = \sqrt{\frac{nq^2}{m\varepsilon_0}} \, ,$$

where n is the density of charged particles, q and m their charge and mass, and ε_0 is dielectric permittivity [904]. At larger amplitudes, these waves become strongly nonlinear and can be described by the KdV equation [461,1012], or a two-dimensional version of it – the Kadomtsev–Petviashvili (KP) equation [224,573,574].[16]

Recently, Leon Shohet has argued convincingly that *slinky modes* – which bedevil toroidal plasma-confinement machines through the generation of "hot spots" that degrade the vacuum – are described by the same SG equation (with damping and driving) that is used to model magnetic flux dynamics on long Josephson junctions (see Fig. 6.9) [258,641,904]. Although much empirical evidence supports this description of slinky modes [905], many plasma physicists continue to use a quasilinear approach where only the first-order interactions among linear modes are computed, thereby obscuring fully nonlinear effects. From the quasilinear perspective, for example, *mode locking* is viewed as a primary effect, whereas in a fully nonlinear analysis this phenomenon appears as a natural artifact of SG kink formation [903]. Thus the insights of modern nonlinear science have not yet been uniformly appreciated throughout the scientific community.

Finally, an intriguing plasma phenomenon is ball lightning (BL), which is largely ignored by conventional scientists because it has not yet been produced in a laboratory or studied analytically [786]. Thus its properties are derived from verbal reports, some of which are summarized as follows: BL often follows a lightning flash, is usually spherical with a diameter of tens of centimeters, has a lifetime of a few seconds, and is bright enough to be seen in daylight. Although long dismissed as a product of overheated imaginations, finding an explanation for this phenomenon presents a challenge to plasma scientists of the future. Those taking BL seriously have included Dirk ter Haar, who divided people into the following classes: (1) those who have seen BL; (2) those who know someone who has seen BL; (3) those who know someone who knows someone who has seen BL; and so on [957]. (I am in group 2.)

6.8.5 Rogue Waves

Unusual examples of nonlinear phenomena on the surface of the ocean include *rogue waves* (RWs) (also called freak or monster waves) and nearby *holes*

---

[16]Karl Lonngren has suggested that the large amplitude of the tsunami near Sri Lanka in Fig. 3.7a may be modeled by the KP equation.

(creating precipitous surface slopes), which have recently been dramatized in the film *Poseidon*, a remake of *The Poseidon Adventure* from 1972. Long dismissed by conventional ocean scientists as the unreliable tales of novelists and bibulous sailors, RWs are becoming recognized as real phenomena that can have disastrous maritime consequences, and their statistics are now being measured from a variety of data sources, including recordings from deep sea buoys, direct observations and recordings from offshore oil platforms, satellite optical imaging, and satellite radar. This is an important area of applied research, as more than 200 supertankers have been lost in the past 20 years; thus Europe has recently launched the MAXWAVE project to understand rogue waves and mitigate their effects [822].

From the statistical perspectives of linear wave theory, the sudden accumulation of localized energy in a surface wave of unusual amplitude seems unlikely, but recent observations show that RWs occur far more often than has formerly been expected [255]. Although RWs are not often closely photographed (for obvious reasons) an example is shown in Fig. 6.15a, and in Fig. 6.15b a RW is being generated in the Large Wave Channel of Hannover [843]. Guided by computer modeling, these waves can occur when slow waves are overtaken by faster waves, which then merge together in a nonlinear manner.

Present theoretical descriptions of RWs fall into three classes:

1. treating RWs as rare events of linear surface-wave statistics [315],
2. developing quasilinear (second-order nonlinear) descriptions of RWs [947],
3. formulating fully nonlinear descriptions of RWs [253, 724, 728, 1077].

All three of these approaches must consider that concentrations of energy on the ocean surface are influenced by winds, currents, and focusing caused by offshore banks; thus the region of the Agulhas current, southeast of Africa, is known to be particularly hazardous [544].

Two-dimensional surface waves on water are complicated nonlinear phenomena, but some progress has been made in understanding their dynamics by considering the NLS equation of soliton theory, which serves as a model for one-dimensional surface waves [76, 724, 728]. Using results from the IST method (see Sect. 3.2), one finds analytic expressions for unstable nonlinear solutions that rise up out of the background field and then quickly disappear, much like the ocean wave and tank wave of Figs. 6.15a and 6.15b respectively. Efficient numerical methods for computing the dynamics of such waves have recently been deveoped by Vladimir Zakharov and his colleagues [253, 1077]. To extend these results to the more realistic problem of two spatial dimensions, Alfred Osborne and his colleagues have conducted numerical studies of a corresponding NLS equation, showing that RWs and their nearby holes are characteristic phenomena for this system. A striking example of their results is presented in Fig. 6.15c, where the parameters of the model have been matched to open ocean and the axes are labeled in meters.

(a)

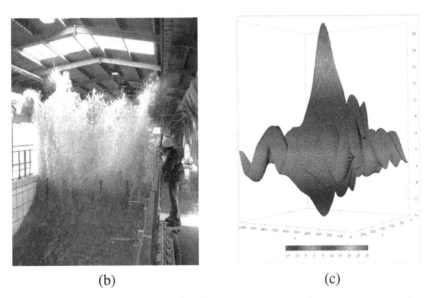

(b)   (c)

Fig. 6.15. Rogue waves (RWs). (**a**) A RW approaching the stern of a merchant ship in the Bay of Biscay. (Courtesy of NOAA Photo Library.) (**b**) A RW produced in the Large Wave Channel in Hannover. (Courtesy of the Coastal Research Centre, Hannover, Germany.) (**c**) A snapshot at the peak value of a numerical RW computed from a 2 + 1 dimensional NLS system, where the parameters are chosen to correspond to the ocean and axes are labeled in meters. (Courtesy of Alfred Osborne)

Osborne's research has shown the importance of nonlinear science in describing the dynamics of individual RWs, but is full nonlinearity needed for estimating the frequency at which RWs occur? Or does a quasilinear analysis suffice? To answer this question, Kristian Dysthe and his colleagues have numerically simulated a fairly large section of the ocean surface (about 10 000 waves) and compared the statistics of their computations with those of linear, quasilinear and fully nonlinear analyses, finding the following results [255,256,938]. First, linear statistics are not in accord with these numerical results as they underestimate the frequency of RW occurence. Second, good agreement with the quasilinear (weakly nonlinear) approach was found over most parameter ranges. Finally, a fully nonlinear description (à la Osborne) was needed in the case of long waves that are focused roughly in one direction, which required the NLS equation to reproduce the simulated statistics.

6.8.6 Coronets, Splashes and Antibubbles

Among the more famous scientific photographs of the last century is the *milk-drop coronet* which was taken in 1936 by MIT's Harold Edgerton using his newly developed stroboscopic flash tube to "stop" the lovely dynamics that transpire when a falling drop strikes a thin liquid layer. Perhaps because his image was included in the first photography exhibit at the Museum of Modern Art in New York, the coronet phenomenon is widely associated with Edgerton's name, but it was discovered and carefully studied four decades earlier by Arthur Worthington, who used sparks from a Leyden jar (an early form of capacitor, which stores electricity) to record fluid motion with flashes as short as three microseconds. By this means, Worthington obtained a series of photographs that were striking in their day [1056, 1057] and remain of scientific interest to many others, including Dror Bar-Natan who has recently recorded the images shown in Fig. 6.16.[17]

Molded by nonlinear interactions between fluid inertia and the forces of surface tension, the dynamics of a small drop falling onto a thin layer of liquid comprises several stages: first, a circularly-symmetric wave propagates outward, developing into a steepening annular ridge, much like a cylindrical soliton (see Fig. 6.16a); second, the crest of this ring become unstable to circumferential waves, with the wavelength for maximum growth determining the spacing of the jets; third, transverse localizations begin to grow (circumferential solitons riding atop a ring soliton), which can initially be described by a quasilinear analysis; fourth, the circumferential solitons elongate into jets (as in Fig. 6.16b); and finally, the jets break into droplets. This sequence is not speculation; empirically, it was observed by Worthington in the nineteenth century and by many others since the 1930s, and recently

---

[17]More of these observations can be found at http://www.math.toronto.edu/~drorbn/Gallery/Misc/MilkDrops/index.html

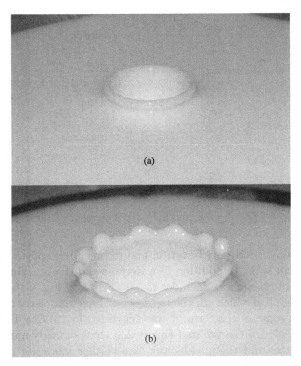

(a)

(b)

Fig. 6.16. A splashing milk drop. (**a**) Outward propagating ring wave. (**b**) Circumferential instability leads to coronet formation. (Courtesy of Dror Bar-Natan)

the entire scenario has been simulated using the Navier–Stokes equations of fluid dynamics (a PDE system that describes the dynamics of viscous fluids [12]) [1063]. And far more than scientific child's play, an understanding of drop impact dynamics is important for many technical applications, including spray cooling of hot surfaces, ink-jet printing, and solder-drop dispensing, among others.

Interestingly, the photographs by Worthington and by Bar-Natan show several variations on the theme of coronet formation, including *splashing* when drops appear at the ends of the jets [1057]. More generally, splashing comprises a broad class of phenomena that arise when a moving mass of liquid becomes too large to hold itself together by surface tension, as is commonly observed in wave breaking and at waterfalls. Howell Peregrine and Stanislav Betyaev have shown that analytical results can be obtained from simple approximations, but much remains to be done as the various components of splashing are very complex [83, 757]. In addition to the formation of fine drops, nonlinear phenomena that appear in high-energy splashes include both bubbles and *antibubbles*, in which a thin spherical film of air separates two regions of liquid (see Fig. 6.17 and [45]). At high velocity, yet another effect is *exploding water*, which bursts apart from its internal pressure.

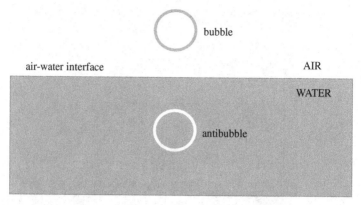

Fig. 6.17. A bubble and an antibubble near an air–water interface

Splashing is not only beautiful and technically important, but impressive on a human scale. In addition to waterfalls and the breaking of ocean waves, this class of phenomena includes rogue waves (see Fig. 6.15) and the impacts of meteors, all of which stir the feelings and imaginations of observers.[18] Using a broader sense of the term, it was suggested in 1917 by D'Arcy Thompson (in his classic *On Growth and Form*) that the seemingly miraculous creations of the potter and the glassblower "are neither more nor less than glorified 'splashes', formed slowly, under conditions of restraint which enhance or reveal their mathematical symmetry." We will return to this concept in the following chapter, where the development of biological organisms is considered.

6.8.7 Atmospheric Dynamics

The atmosphere also offers a wealth of impressive nonlinear phenomena, including vortices, first studied in liquids by Hermann Helmholtz in 1867 [421]. Most terrifying is the tornado shown in Fig. 6.18b, which usually descends from a cumulonimbus cloud and is driven by the temperature inversion of a cold front, often developing wind speeds of several hundred miles per hour. More benign are "dust devils", driven by localized heating and convection and appearing in deserts of the American Southwest on hot summer afternoons (see Fig. 6.18a). On a larger scale are hurricanes like Andrew in August of 1992, a satellite photo of which is shown in Fig. 6.18c. Called a typhoon in the West Pacific Ocean and a cyclone in the South Pacific, hurricane winds rotate clockwise below the equator and counterclockwise above with speeds of 100 to 200 mph which can last for a week or more, often with disastrous consequences – a 1737 cyclone in the Bay of Bengal claimed 300 000 lives, and an 1881 typhoon in China took a similar toll [735].

---

[18] Thus Betyaev recalls Pushkin's description of people being herded together [83]: "admiring splashes, mountains and the foam of enraged waters."

(a)

(b)

Florida

Cuba

(c)

Fig. 6.18. Natural atmospheric vortices. (**a**) A dust devil in Death Valley, California. (Courtesy of Dr. Sharon Johnson/GeoImages.) (**b**) Photo of a tornado (*right*) illuminated by a bolt of lightning (*left*). (Courtesy of National Oceanic and Atmospheric Administration.) (**c**) Satellite view of Hurricane Andrew, approaching the US Gulf Coast, 25 August 1992. (Courtesy of National Oceanic and Atmospheric Administration)

Thinking about tornadoes and hurricanes brings to mind the difficult problem of weather prediction, which was commented upon by both Poincaré and Wiener and led Edward Lorenz to rediscover George Birkhoff's irregular orbits [91,575]. Famously characterized by Lorenz as exhibiting the *butterfly effect*, the system of (2.1) with dynamics shown in Fig. 2.1 is a simple model of a weather system, which has been widely studied as an example of low-dimensional chaos since the 1970s [987].

Less problematic than weather prediction is the development of numerical models for Earth's coupled oceanic–atmospheric system, which establishes patterns of currents [13]. These nonlinear studies lead in turn to numerical models for global warming caused by anthropogenic emissions of greenhouse gases (carbon dioxide, methane, nitrous oxide, and fluorocarbons) [912]. Viewed as a parameter in a nonlinear system, a steady increase in Earth's average temperature may lead to sudden and irreversible changes in the sizes of the polar ice caps (which changes the amount of heat absorbed from the Sun), the course of the Gulf Stream, rainfall patterns, and the depth of the oceans, among others. As in any nonlinear system that nears such a major shift, several other features are expected to approach instability, leading to unusual patterns of flooding, drought, and hurricane formation. Also, less weight at the poles may induce earthquakes as Earth adjusts herself to a new set of forces. While some claim such concerns are alarmist, the nonlinear scientist does not find them so.

6.8.8 Turbulence

In 1839, a German engineer named Gotthilf Hagen and Jean Poiseuille, a French physician, independently discovered that the flow of liquid through a circular pipe at constant pressure is proportional to the fourth power of its diameter d. Thus the *Hagen–Poiseuille law* implies that the *flow resistance* Ω of a length ℓ is

$$\Omega = \frac{\ell\eta}{2\pi d^4} , \qquad (6.15)$$

where η is the dynamic viscosity. If f is a smooth flow of fluid through the pipe, energy is absorbed by the fluid at rate $f^2\Omega = f^2\ell\eta/2\pi d^4$, which increases rapidly with flow. At large enough f, smooth flow becomes unstable and turbulent flow begins, as was observed by both Hagen and Poiseuille [387,766]. Almost a half century later, Osborne Reynolds, an Irish engineer, showed that fluid flow becomes turbulent when [140,141,757,800]

$$\mathrm{Re} = \frac{4\rho f}{\pi\eta d} > 2300 , \qquad (6.16)$$

where ρ is the fluid density. From this dimensionless number (which now bears Reynolds's name), we see it is difficult to achieve turbulence for highly viscous fluids like maple syrup and motor oil.

A numerical property of the Navier–Stokes equation, fluid turbulence is widely observed – in the bore shown in Fig. 6.12, in the wake of the bullet

shown in Fig. 6.13, and in Leonardo's sketch in Fig. 2.5 – and some progress
has been made in characterizing it. Thus Andrej Kolmogorov showed in 1941
that the energy in a fully turbulent fluid should depend on wave number as
the 5/3 power [506]. Although this law is invalid in systems with two or more
spatial dimensions, many real systems are essentially one-dimensional for
which the 5/3 law holds over several decades. Shortly after the Kolmogorov
result was published, Jan Burgers proposed (6.14) as a simple model of turbulence in a bore (see Fig. 6.12). Similar in form to the KdV equation, as we
have noted, the Burgers equation can be exactly solved through a dependent
variable transformation to the linear diffusion equation [138, 750, 1033].

The many computational and conceptual difficulties associated with understanding fluid turbulence were amusingly characterized in 1932 by Horace
Lamb, the dean of fluid dynamicists, who is reported to have said [49]:

> I am an old man now, and when I die and go to heaven there are two
> matters on which I hope for enlightenment. One is quantum electro-
> dynamics and the other is the turbulent motion of fluids. About the
> former, I am really rather optimistic.

One reason behind this mystery is that fluid turbulence involves interactions among several different levels of dynamic activity, with unpredictable
emergence of new features at the time and length scales of every level – a
phenomena that philosophers call *chaotic emergence* [703, 704]. Or as one of
the pioneers in numerical weather prediction, Lewis Richardson, has put it
(in a parody of Jonathan Swift's doggerel of poets and fleas) [802]:

> Big whirls have little whirls
> That feed on their velocity,
> And little whirls have lesser whirls
> And so on to viscosity.

We will find that the concept of chaotic emergence extends beyond the realms
of fluid dynamics, offering insights into the dynamics of biological evolution,
the nature of life, and certain aspects of cosmology.

6.9 Gravitation and Cosmology

Newton's solution for the two-body problem of planetary motion led to consideration of three-body problems (the Sun, a planet and its moon, say)
and more generally to an N-body problem (the entire solar system), which
were to exercise mathematicians for some two centuries until the now famous
proof by Poincaré that the three-body problem cannot be solved by finding
integrals that reduce the dimension of the system [234]. These efforts were
not wasted, however, as many exact solutions for special initial conditions
were found [236], leading to a broad appreciation for the class of functions

that can be analytically expressed. Also there were significant developments in perturbation theory (PT), under which the solution to a nonlinear problem is progressively better approximated by a hierarchical set of tangential (locally linear) problems [496, 641, 923]. Empirical successes of PT include a prediction of the return of Halley's comet in 1758, the discovery of the planet Neptune in 1846 (from observations of unexplained variations of the orbits of known planets) [235], and ever more precise calculations of the precession of the point where Mercury's quasi-elliptical orbit comes closest to the Sun (its perihelion), which at the beginning of the twentieth century was measured to rotate by 5600 seconds of arc per century and calculated via PT to be 5557 sa/c, leaving a difference of 43 sa/c which could not be explained as the influences of other planets.

6.9.1 General Relativity Theory

According to Aristotle's geocentric cosmology, natural motion was directed toward or away from the center of the Universe (Earth), whereas Galileo and Newton supposed the laws of motion to be independent of location in an unbounded Universe. Under the special relativity theory (SRT) – proposed independently by Einstein [272] and Poincaré [765] in 1905 – and Hermann Minkowski's space-time continuum [660], it is assumed that physical laws are perceived to be the same on systems moving in straight lines at constant speeds. Just as in the Lineland example of (3.14) [1], two observers (unprimed and primed) who are experiencing uniform relative motion will agree on their respective measurements of a differential interval of spacetime ds, where

$$ds^2 = -c^2 dt^2 + dx^2 + dy^2 + dz^2 = -c^2 dt'^2 + dx'^2 + dy'^2 + dz'^2 .$$

In other words, this quadratic form is invariant under Lorentz transformations of the independent variables. To introduce curvature of spacetime, it can be written more generally in tensor notation as

$$ds^2 = g_{\mu\nu} dx^\mu dx^\nu ,$$

where a summation is taken over taken over repeated upper and lower indices and μ, $\nu = 0$, 1, 2, 3 specify the independent variables $(-t, x, y, z)$ [941]. In four dimensional Euclidean spacetime, $g_{\mu\nu} \equiv \eta_{\mu\nu} = \mathrm{diag}\,(-1, +1, +1, +1)$ – a flat metric tensor with units chosen to make the speed of light equal to one. Informally, $g_{\mu\nu}$ can be thought of as a 4×4 symmetric matrix (with ten independent components) that describes the local curvature in spacetime.

As formulated independently by Einstein and Hilbert late in 1915 [274, 396, 434], general relativity theory (GRT) follows from further assuming that the laws of physics are the same on uniformly accelerating frames, a condition that can be satisfied if the metric tensor $(g_{\mu\nu})$ is appropriately curved. Thus the Einstein–Hilbert (EH) system is the following set of nonlinear PDEs:[19]

---

[19]Questions about the priority in formulating this equation have recently been addressed by Leo Corry, Jürgen Renn and John Stachel [193].

$$R_{\mu\nu} - \frac{1}{2}Rg_{\mu\nu} = 8\pi G T_{\mu\nu} , \qquad (6.17)$$

where G is Newton's gravitational constant. Called the energy–momentum tensor, $T_{\mu\nu}$ on the right-hand side of this system is zero in a vacuum with non-zero components generated by the presence and motions of matter. (Specifically, the components of $T_{\mu\nu}$ represent the flux of the μ component of momentum in the ν direction.) The *Einstein tensor* on the left-hand side is constructed from the *Ricci tensor* $R_{\mu\nu}$ and its trace R, which depend nonlinearly on the components of $g_{\mu\nu}$ and their first two derivatives with respect to the independent variables (x, y, z, and t). Dynamics enter from the condition that small test particles will move along shortest paths (geodesics) of the curved space.[20] Thus GRT says that matter causes space to be curved and motions of matter are along geodesics of the curved space, leading to opportunities for closed causal loops. If $T_{\mu\nu} = 0$, the space can be flat ($g_{\mu\nu} = \eta_{\mu\nu}$), whereupon test particles move in straight lines with constant speed as in the empty space of Galileo and Newton. If $\eta_{\mu\nu}$ is then perturbed by a small function of the space variables $\Phi(x, y, z) \ll 1$, the EH equations approximate those for a particle moving in a potential Φ according to Newton's Second Law [158].

Just as it was difficult for astronomers of the sixteenth century to accept the geocentric cosmology of Copernicus, the concept of curved space is not easy for many of us to imagine, but a higher dimensional Euclidean alternative is also problematic. Introductions to general relativity often attempt to ease the conceptual problems by showing a locally two-dimensional spacetime closed into a sphere that is embedded in three dimensions. This works for us because $g_{\mu\nu}$ is then a symmetric 2×2 tensor with three independent components, which we can visualize as a three-dimensional figure (see [407] for some fine images). In general, the number of independent elements in a symmetric $n \times n$ array is $n(n + 1)/2$. If one considers our four-dimensional spacetime to be constraints on motions in a higher-dimensional Euclidian (flat) spacetime, $g_{\mu\nu}$ is symmetric and 4×4; thus the dimension of a flat embedding space must be ten, which is difficult (for me, at least) to conceive. Some help with this conceptualization is provided by Abbott's Flatland [1], its reduction to Lineland (exemplified in Fig. 3.3), and its subsequent generalization to Sphereland [137]. On the other hand, one can consider this example merely as a pedagogical tool, accepting or rejecting GRT as a description of physical reality based on empirical evidence.

Presently, GRT is widely accepted by the physics community for several reasons, including these [898, 1041]:

---

[20] A recent book by Sean Carroll entitled *Spacetime and Geometry: An Introduction to General Relativity* is an excellent source for details on the structure of the EH equations [158].

Fig. 6.19. Albert Einstein writing (6.18) on a blackboard. (Courtesy of Brown Brothers, Sterling, Pa)

- GRT accounts for the above-mentioned disagreement of 43 sa/c between calculated and measured values of the rotation rate of Mercury's perihelion, whereas Newton's theory does not.
- GRT correctly predicts the bending of light rays and radio waves by 1.75 sa as they pass close by by the Sun, whereas Newton's theory predicts half of that value.
- The predicted red shift of spectral lines in gravitational fields has been confirmed, as has the influence of gravity on clocks.
- Predictions of gravitational radiation are in accord with observed rates of orbital decay of neutron binary stars.
- GRT predicts the formation of black holes from the collapse of large stars, which is in accord with some astronomical observations.
- GRT rests on the assumption that gravitational and inertial masses are equal, which has been confirmed to an accuracy of about one part in 10^{13}.

If $T_{\mu\nu} = 0$, the EH equations reduce to

$$R_{\mu\nu} = 0 \,, \tag{6.18}$$

a set of ten coupled second-order PDEs with the components of $g_{\mu\nu}$ as independent variables.[21] Although far more complicated, (6.18) corresponds to the second-order nonlinear equations that were described in Chap. 3.

Under a small-amplitude (linear) approximation, solutions of (6.18) reduce to plane waves governed by the linear wave equation [69, 158]

---

[21] Taking the trace of both sides of (6.17), gives $R = -8\pi GT$ (where T is the trace of $T_{\mu\nu}$); so this equation can be rewritten as $R_{\mu\nu} = 8\pi G(T_{\mu\nu} - T/2)$.

$$\frac{\partial^2 u}{\partial z^2} - \frac{1}{c^2}\frac{\partial^2 u}{\partial t^2} = 0 \,, \tag{6.19}$$

where z is the direction of propagation and u is a component of a 3×3 symmetric, traceless *strain tensor* with only x and y components. Thus this tensor has the form

$$s_{ij} = \begin{pmatrix} a & b & 0 \\ b & -a & 0 \\ 0 & 0 & 0 \end{pmatrix} \,, \tag{6.20}$$

and the influence of gravity is not transmitted instantaneously, as is assumed in Newton's theory, but via gravitational waves.[22] Under Einstein's theory, these waves are assumed to propagate at the speed of light.

The first effort to observe gravitational waves was mounted in the 1960s by Joseph Weber, who built a detector comprising a suspended aluminum bar two meters long and a half meter in diameter, and estimated the relative change in length caused by a passing wave to be about one part in 10^{17}. To measure such a small fractional change, the bar was covered with many piezo-electric detectors (which convert mechanical distortion into electric voltage) connected in series to increase the total voltage [1016, 1017]. Although Weber reported the detection of a gravitational wave in 1969 [1018], this claim was not confirmed – nonetheless, his work inspired the construction of a number of detectors in the 1970s and 1980s, all of which failed to find gravitational waves. By the mid-1970s, the expected gravitational wave signal was revised downward to about one part of 10^{21}, requiring a new concept for the detector design [964]. To obtain higher sensitivity, present detectors have an L-shaped construction, where – based on (6.19) and (6.20) – two mirrors are expected to be pulled slightly closer together (by the a-components in the strain tensor s_{ij}) in one arm and pushed slightly farther apart (by the b-components) in the other, a small difference that is to be measured by laser interferometry.

Currently, the most important installations for detecting gravitational waves are those of LIGO (Laser Interferometer Gravitational-wave Observatory) in the United States, with two sites located at Hanford, Washington and Livingston, Louisiana, which became fully operational in November of 2005. Each of the perpendicular arms at these sites is 4000 m, leading, it is hoped, to the required sensitivity. Other sites include VIRGO in Cascina, Italy (with arm lengths of 3000 m), GEO-600 in Hannover, Germany (600 m), TAMA-300 in Mitaka, Japan (300 m), and AIGO soon to be launched in Gingin, Australia (3000 m) [149]. (Up-to-date information on these sites is readily available from the internet.) A simultaneous observation from all these installations would confirm that the recording is truly an incoming gravitational wave and not an artifact of some local seismic disturbance. Importantly, such an observation would also allow the direction of the incoming

---

[22]The term "gravitational" waves is used to avoid confusion with the "gravity" waves of hydrodynamics.

wave's source to be determined and provide a direct measurement of the speed of a gravitational wave, which up to now has merely been assumed to equal the velocity of light. As Léon Brillouin has pointed out, all we now know for sure is that gravitational waves travel no faster than light [126].

The direct observation of a gravitational wave generated by some cosmic event – a supernova explosion, radiation from orbiting neutron stars, or the collision of two black holes – is hoped for within a decade [225]. If it occurs, this event would open a new era of *gravitational astronomy* based on (6.19) and (6.20) rather than on Maxwell's equations, as are all optical, radio, X-ray and gamma-ray observations.

6.9.2 Nonlinear Cosmic Phenomena

As we have seen in Chaps. 2 and 3, the lexicon of nonlinear science includes two new concepts: chaotic (irregular) behaviors of dynamic systems and localization of dynamical variables. How does chaos fare in our Solar System? First, as was discussed in Sect. 2.7, Hyperion (a moon of Saturn) and Phobos and Deimos (moons of Mars) all tumble irregularly, and chaotic behavior is also found in trajectories of the moons of Jupiter [68, 261, 691, 1052]. Second, with the development of ever more powerful computers, it has recently become possible to integrate Newton's equations of the solar system for several million years into the past and future, leading to the discovery that the motions of the inner planets are susceptible to chaos with characteristic (Lyapunov) times of about 5 million years [261]. Thus there are difficulties in predicting the Earth's orbit beyond a few tens of millions of years, making climate reconstruction over geological time scales problematic [463, 538, 938]. Finally, the orbit of Halley's comet is chaotic (see Fig. 2.6) with a characteristic time of about 29 returns (2200 years), placing the earliest observations of this phenomenon at the edge of retrodictability [176, 261, 691].

As any description of gravitational systems (Newtonian or Einsteinian) is highly nonlinear, there are many examples of cosmic localization. On the theoeretical side, assuming plane-wave symmetry in (6.18) can reduce the number of independent variables to two, simplifying the structure of the Ricci tensor ($R_{\mu\nu}$) and allowing analytic expressions for one-dimensional singularities to be obtained [129, 602]. Also the integrable property of special soliton equations (KdV, NLS, SG, etc.) permits the analytic construction of curvature tensors that satisfy corresponding reduced versions of (6.17) and (6.18) [249, 250], which are connected with nineteenth-century studies of the relationship between nonlinear PDEs and curved surfaces [109, 117, 292, 659, 927]. For theories of localization in three-dimensional space, Vladimir Belinski and Enric Verdaguer have used the IST machinery of soliton theory (see Sect. 3.2) to generate several exact solutions of (6.18) [69], all of which they choose to call solitons. Some authors have objected to their use of this term because plane-wave solutions travel exactly at the speed of light [119], but this result is not surprising. If gravitational solitons were superluminal, the assumptions of SRT could be violated by using them to synchronize clocks in various moving coordinate systems. As we have seen in Sect. 6.1.1, this property is shared by large-amplitude solutions of the Born–Infeld system, a nonlinear version of Maxwell's equations that models the electromagnetic structure of an electron [173, 888].

On the empirical side, merely looking into the heavens on a clear night reveals many localized entities, which have teased human thought since the dawn of our species. As discussed in Sect. 2.7, the planets and their moons in our Solar System are examples of localized entities that have evolved into orbs from chaotically moving planetesimals (small chunks of matter), in an agglomerative process much like the churning of butter (see Sect. 6.4.3). Their spherical structures can be understood as a static balance between the nonlinear attractive forces of Newton's gravitational theory and the nonlinear repulsive forces of compressed matter. On a large scale, dispersed gaseous matter has recently been computed to organize itself into elliptical or spiral galaxies with masses of about 10^{11} times that of our Sun [430, 676].

In 1922, Alexander Friedmann, a Russian mathematician and cosmologist discovered an *expanding universe* solution to (6.17) [334], in 1927 Georges-Henri Lemaître proposed a model under which our Universe began with an enormous explosion [553] (later called the Big Bang), and in the 1930s Howard Robertson and Arthur Walker put the Friedmann and Lemaître theories on a sound mathematical basis [808, 1010]. Now called the Friedmann–Lemaître–Robertson–Walker (FLRW) model, this picture is widely accepted as a description of our Universe. Under the FLRW model, interestingly, the Universe may or may not be closed, depending on whether the average mass–energy density ρ is greater or less than a critical density

$$\rho_{\text{crit}} = \frac{3H_0^2}{8\pi G} ,$$

where H_0 is the observed rate of expansion (Hubble parameter). Although ρ is difficult to measure with precision [281], recent data suggest that ρ/ρ_{crit} is close to unity, an observation that generates uncertainty of our future [520].

More generally, the Einsteinian revolution has replaced Newton's action-at-a-distance forces by mass-generated variations in the geometry of space-time, as described by the EH equations of GRT – one of the more complicated nonlinear systems yet formulated [158, 964]. From this new perspective, gravity is no longer viewed as a force, but a parameter related to the local curvature of space-time geometry, and its influence has changed from an efficient to a formal cause. Although Newtonian mechanics were widely accepted for about two centuries and have led to striking advances in many realms of science, it is amusing to note that Einstein's mechanics re-establish two key Aristotelian ideas: motion is again determined by the local nature of space (its curvature) and there is a possibility that our Universe is finite.

6.9.3 Black Holes

The saga of black holes began in the eighteenth century when John Mitchell and Pierre-Simon Laplace independently used Newton's theory to compute the radius of a mass that is small enough to prevent light from escaping its gravitational field [518].[23] Just after the publications of GRT by Einstein and Hilbert in late 1915 and just before his untimely death at the Russian front in 1916, Karl Schwarzschild assumed a static and spherically symmetric solution of (6.17) and showed that the curvature tensor near a mass m becomes singular at the radius [855]

$$r_s = \frac{2Gm}{c^2} . \qquad (6.21)$$

Thus if a mass is somehow compressed to within its *Schwarzschild radius*, it will be impossible for anything to escape from the surface. (The Schwarzschild radius is about 1 cm for a body with the mass of Earth and about 3 km for the mass our Sun.) Whether collapse to such a radius is possible has been vigorously debated through much of the twentieth century by some of the best minds in physics.

As our Sun steadily radiates energy, it must eventually cool and then shrink, because its heat (the thermal agitation of its constituent atoms) is presently balanced against inward gravitational forces. In 1931, young Subrahmanyan Chandrasekhar proved that our Sun will eventually collapse to a *white dwarf* with a radius of about 5000 km, at which the quantum forces of compressed electrons will resist the inward pull of gravity [168]. Furthermore, he showed that there is a maximum mass (about 40% larger than that of our Sun), above which compressed electrons are no longer able to resist gravity, thus allowing further collapse. Although this concept of a maximum mass was publicly dismissed by Arthur Eddington (his research supervisor), Chandrasekhar's results have been confirmed both numerically and empirically, and are now a bedrock of modern cosmology. Sadly for Britain, Eddington's rejection caused Chandrasekhar to leave Cambridge University for the University of Chicago, and he ceased doing research on gravitational collapse for almost four decades.[24]

What will a star with a mass greater than 1.4 times that of our Sun eventually become? In 1934, Walter Baade and Fritz Zwicky proposed that it would collapse into a *neutron star* with a radius of about 16 km [41, 42]. This idea received support from Landau, who suggested in 1938 that a typical star might have a neutronic core [536]. Disliking singularities, Einstein thought that extreme gravitational collapse was unphysical, and in 1939 he

---

[23]The history of gravitional singularities has been surveyed in an excellent book for the general reader entitled *Black Holes and Time Warps: Einstein's Outrageous Legacy*, by Kip Thorne [964].

[24]In 1983 Chandrasekhar was awarded the Nobel Prize in Physics for this work.

claimed that it could not happen under GRT [277]. Inspired by Landau's work, however, Oppenheimer and his students began to study neutron stars in detail, concluding in a paper with George Volkoff that – like a white dwarf – there is also a maximum mass for a neutron star, lying somewhere between a half and several solar masses [726]. Beyond that maximum, a star should collapse to the Schwarzschild radius, generating an internal singularity in a process that was described in some detail by Oppenheimer and Hartland Snyder [725].

During the Second World War, these problems were set aside as physicists concentrated on the military needs of their respective nations, but with the return of peace, Dennis Sciama in Britain, Yakov Zeldovich in Russia, and John Wheeler in the US, among others, returned to the subject of stellar collapse with fresh insights into the nature of nuclear interactions, hosts of eager students, and vastly improved computing power. Founder of the Relativity and Gravitation Group at Cambridge University, Sciama was a former student of Dirac and the mentor of Stephen Hawking. Zeldovich – whom we have met in Chap. 4 as he who formulated and solved (4.4) for the nonlinear diffusion of a flame front – was convinced that a star with a sufficiently large mass will collapse to its Schwarzschild radius, whereas Wheeler, an influential Princeton physics professor, was initially skeptical of this idea. By the early 1960s, however, Wheeler agreed with the pre-war estimates of neutron-star formation collapse made by Oppenheimer and his students, and in 1967 he named the resulting entity a black hole (BH)[25] Thus began the Golden Age of black-hole research, during which Roger Penrose used topological methods to show that a singularity will appear even when the collapsing form is not spherically symmetric [753], leading the discovery that – according to the EH equations – this singularity will oscillate chaotically within the BH [70, 71, 663]. It was also learned that the EH equations could predict all the properties of a BH from its mass, rate of spin, and electric charge [964]. From this perspective, a BH is like an enormous three-dimensional soliton, bearing the same relation to (6.17) as the electron model in Fig. 6.1 does to Mie's nonlinear electromagnetic equations. In the mid-1970s, Chandrasekhar returned to BH research, writing a masterful survey of the theory in the early 1980s [170].

But does the BH really exist in nature, or is it – like a unicorn – only a figment of physicists' imaginations? Although they don't emit or reflect light, Zeldovich aggressively sought empirical evidence for BHs, initially looking for

---

[25]Although of different cultures, social backgrounds and personalities, Wheeler and Zeldovich were entangled (in the sense of Sect. 6.2.4) through working on A-bomb projects during the War and then on the H-bomb. Together with Andrei Sakharov, Zeldovich had independently invented the same approach to H-bomb design that Stan Ulam had come up with to save Edward Teller's "super" project in the US. Recall that Ulam is he of the "FPU problem" which led to many soliton discoveries during the mid-1960s (see Sect. 1.3).

stars that wobble, which would indicate that they might be binary stars in orbit around BHs. A careful search of the astronomical data brought up about a dozen candidates, but none provided convincing evidence, as the dark sister might be a white dwarf or a neutron star. Undaunted, Zeldovich next showed that a BH should emit X-rays as its large gravitational field cuts through the charged gases emitted from a nearby star. From a broad international effort to construct space-borne X-ray detectors with sharp angular resolution, the modern era of astronomy began, with X-ray and radio observations joining classical optical astronomy, and it was soon discovered that there are many intense sources of X-rays and radio waves, some of which may be related to BHs.

First proposed as a BH candidate by Charles Bolton in 1972 was Cygnum X-1, the dark companion to a star called HDE 226868, which has a mass of about 30 times that of our sun [118]. From the rotation dynamics of this pair, Cyg X-1 has a mass of about 7 times that of our Sun, which is too large to be a white dwarf or a neutron star. Of course this assignment was hotly contested, leading to a famous wager between Hawking and Thorne [964], but by the 1990s there was wide agreement that Bolton was right. Presently, there are many BH candidates for which the evidence seems to be equally strong.

6.9.4 Tests of GRT

If BHs do exist – as many physicists now assume [964] – there remains the question of whether (6.17) describe them correctly. During the Golden Age when enthusiasm for GRT was gathering momentum, Brillouin [126] and Robert Dicke [238,239] independently called for a reevaluation of the evidence supporting this theory. As these are both talented scientists who have made substantial contributions to diverse areas of physics, their concerns should be carefully considered, and many have been addressed by Clifford Will [1041] and Irwin Shapiro [898], among others.

First, Brillouin cited assertions that GRT gives incorrect values for the rates of perihelion advance of Venus and Mars. In response, it can be pointed out that GRT correctly accounts for the rates of perihelion advance for both Mercury and Earth [184,185], and also for the asteroid Icarus [897]. The rate for Venus is not accurately known because her orbit is almost circular, and the influence of GRT on the Martian rate is small because his orbit is far from the Sun, avoiding regions where space is curved. The best current values are given in Table 6.1, and these show (to me, at least) an acceptable agreement between measurements and calculations based on GRT [82,184,898,1022].

During the 1960s, Dicke and Mark Goldenberg claimed that the diameter of the Sun at its equator is about 52 km greater than at the poles [240,241], an *oblateness* that would account for about 3 sa/c of Mercury's perihelion advance [1041]. If correct, this is problematic for GRT because from Table 6.1 the needed correction is almost exactly 43 sa/c. Stimulated by the possibility

Table 6.1. Measured and GRT values for rates of perihelion advance

Planet	Observed (sa/c)	Error (sa/c)	GRT (sa/c)
Mercury	43.11	$\pm$ 0.45	43.03
Venus	8.4	$\pm$ 4.8	8.63
Earth	5.0	$\pm$ 1.2	3.84
Mars	–	$\sim \pm$ 1.5	1.35
Icarus	9.8	$\pm$ 0.8	10.3

that the GRT paradigm would need to be changed, many papers were written on the subject over two decades. The present consensus, however, is that our Sun is not substantially oblate, so the entire 43 sa/c can be ascribed to GRT [1041].

Second, Brillouin questioned Eddington's claim – based on observations from Principe (an island off the West Coast of Africa) and Sobral (a city in Northern Brazil) during a total eclipse in May of 1929 – that he had confirmed the GRT calculation of the bending of a star's rays near the Sun, thereby validating one of Einstein's predictions and making him (Einstein) famous. To the contrary, Brillouin cited "individual errors of 100% and averaged errors of 30%", and he observed that the "theory is not safe because it assumes an ideal vacuum near the sun's surface". From Principe, however, the Eddington team found an average value of 0.91 ± 0.18 times the GRT prediction from only two reliable photographic images (because of cloudy weather) and 1.13 ± 0.07 from 8 images taken at Sobral [254].

Of course, such measurements were repeated during subsequent Solar eclipses (in 1922, 1929, 1936, 1947, 1952, and 1973), but the accuracy remained about the same (of the order of 10%) – in approximate accord with GRT. After 1969, bending measurements could be made yearly using radio waves from quasars (QUASi-stellAR radio sources) that were passing behind the Sun. These radio measurements began with a precision of about 8% in 1970 (after correction for the Sun's corona), which steadily improved to a present value of 0.1%. Thus current gravitational bending measurements are in accord with GRT [1041].

Third, Brillouin questions Einstein's interpretation of the *red shift* of an atomic clock in a gravitational field, claiming: (i) that the observed shift is due to the influence of gravity on photons that travel from the clock to an observer, and (ii) suggesting the possibility of a frequency shift caused by a gravitational potential that was overlooked by Einstein [125]. Brillouin's first concern seems to be a different point of view without operational significance. In the late 1970s, Einstein's prediction of the gravitational slowing of clocks was carefully checked by a joint effort involving the Smithsonian Observatory in Cambridge, Massachusetts and the Marshall Space Center in Huntsville, Alabama. This project used a space-borne hydrogen maser that was launched vertically upward to 10 000 km, finding agreement with predictions of GRT to an accuracy of about one part in ten thousand [995].

Can other theories of gravity explain the same empirical observations? Prominent among several alternate formulations advanced over the years, is a scalar–tensor theory proposed by Carl Brans and Dicke in 1961 [121], which deals with an elephant in the living room of GRT – the problem of locally sensing the presence of the Universe. To grasp this difficulty, assume (6.17) and imagine that there is only a single test particle in an otherwise empty universe. How does this particle know whether it is at rest, moving with steady speed, or accelerated? Or how do we know whether our Earth is rotating and the stars are standing still, or – as Aristotle supposed – Earth is stationary and the stars are rotating?

In a 1953 paper, Sciama responded to this question by defining motion with respect to an average over all the matter in the Universe, which suggested a scalar gravitational potential $\Phi = -c^2/G$ [856]. This result plus a dimensional argument led Dicke to the conclusion that [237]

$$\frac{MG}{Rc^2} \approx 1 \,, \tag{6.22}$$

where R is the radius of the known Universe and M is its mass.[26] Assuming that G is a function of position, Brans and Dicke augmented the standard variables of GRT with a scalar field $\phi \equiv 1/G$ to obtain a theory incorporating a fixed parameter ω and influencing measurements as follows.

1. The gravitational constant is multiplied by a factor of $(4+2\omega)/(3+2\omega)$. As this correction can be included in the definition of G, no change in measurements of the gravitational red shift is anticipated.
2. The GRT contribution to a planet's rate of perihelion advance is multiplied by the factor $(4+3\omega)/(6+3\omega)$, which explains the reduction of Mercury's rate from 43 sa/c to 40 sa/c for $\omega \approx 7$ if one believes the Sun to be oblate (see above).
3. The angle of light deflection near the Sun is multiplied by the factor $(3+2\omega)/(4+2\omega)$, which puts a lower bound of about 500 on ω at the present precision of 0.1%. At this value of ω, predictions of the Brans–Dicke theory are numerically identical to those of GRT.

Assuming that GRT is numerically correct and BHs do exist in nature, it is of interest to integrate (6.17) under various initial conditions and see what they say. Although numerical solutions of these equations are computationally demanding, a group at NASA's Gravitational Astrophysics Laboratory has recently published full three-dimensional calculations, from which one frame

---

[26]Although neither M nor R are accurately known, taking R to be 14 billion light years and M as 10^{80} hydrogen atoms [955] gives a LHS value of 0.93. More generally, M comprises baryons (about 3 or 4%), nonbaryonic matter (about 26%) and about 70% dark energy, which is probably vacuum energy [929, 955].

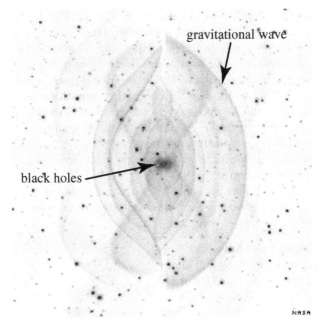

gravitational wave

black holes

NASA

Fig. 6.20. A full three-dimensional computation from the EH equations of gravitational waves radiating from two inwardly spiraling black holes [47]. (Courtesy of NASA's Gravitational Astrophysics Laboratory)

is shown in Fig. 6.20 [47]. The initial conditions for this computation were two inwardly spiraling black holes, which radiated about 4% of their mass as gravitational waves – encouraging news for those who watch and wait at LIGO's laser interferometer.

6.9.5 A Hierarchy of Universes?

Finally, we may observe from (6.21) and (6.22) the remarkable fact that the Schwarzschild radius for the mass of the known Universe is about equal to its size. What does this mean? Are we living in a big black hole?

Just as an atom (with its relatively light electrons orbiting around a heavy nucleus) seems something like a small replica of our Solar System, could the above-mentioned chaotic oscillations of singularities within a black hole replicate the dynamics of our Universe? Did the Big Bang occur when our Universe formed within (or popped-out from) a bigger universe? Could our Universe be drawing mass from a larger universe and passing it on to smaller ones? Might our Universe, in other words, be but one in a branching hierarchy of universes? To visualize the idea we can visit Abbott's Lineland [1], where a closed universe is represented by a circle and the corresponding hierarchy might be the Mandelbrot set shown in Fig. 6.21 [227, 598].[27] In the two dimensional spaces of Flatland and Sphereland [1, 137], these circles would be rotated into spheres, on which smaller spheres would be randomly tangent – topologically equivalent to the branching tubes in our arteries, nerves and lungs, or to the linked pads of the little prickly-pear cactus tree that grows outside my office window.[28] To check this hypothesis in four-dimensional spacetime, it would be necessary to integrate (6.17) with appropriate initial conditions. This would be a daunting task, especially as the computation would seem to be chaotically emergent (see Sect. 6.8.7 and [703, 704]) because the times and the locations at which new black holes pop out would not be under numerical control.

As with Lamb and his abhorrence of fluid turbulence, I hope for enlightenment in this matter upon meeting my Maker and offer another version of Richardson's parody of Swift:

> Big black holes have little holes
> That suck out mass and energy,
> And little holes have lesser holes
> And so on to infinity.

Let us turn next to the biological sciences, where other nonlinear hierarchies and examples of chaotic emergence are to be found.

---

[27] In the context of Chap. 2, this figure is generated by the discete map $z_{n+1} = z_n^2 + c$, where $z_0 = 0$ and c is a point in the complex plane. If $|z_n|$ remains finite (is bounded) as $n \to \infty$, the point c is colored white.

[28] Interestingly, cosmologists have recently engaged in such metaphysical speculations [280, 565, 566, 956].

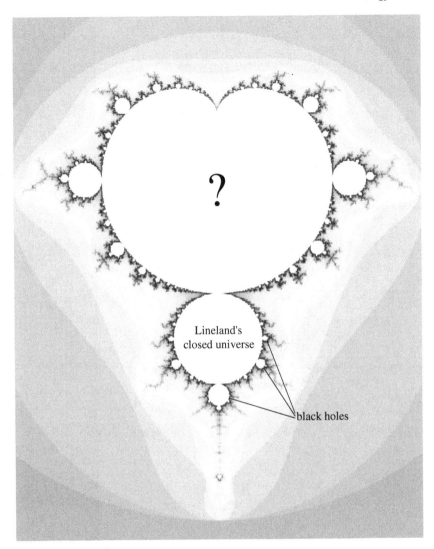

Fig. 6.21. A Mandelbrot diagram that can be interpreted as hierarchical universes in Abbott's Lineland

7 Nonlinear Biology

Biology does not deny chemistry, though
chemistry is inadequate to explain biological phenomena.

Ruth Benedict

In the previous chapter we looked at examples of nonlinear dynamics in the
physical sciences, where Albert Einstein was surprised that such phenomena can be understood. Here we consider some nonlinear formulations of
Life's dynamics, beginning again with the small (biochemistry) and working
through the growth of form and population dynamics to several aspects of
neuroscience. Along the way we will notice a distinctive feature of biological sciences – the number of possible entities in any species (proteins, DNA
molecules, organs, neural structures, and brains, and so on) is very large. Indeed, this number is often *immense* in the sense defined by physicist Walter
Elsasser,[1] meaning that available possibilities, although finite in number, cannot be physically realized; thus the sets of items being studied in the life sciences are *heterogeneous*, rather than *homogeneous* as in the physical sciences.
To keep account of these many possibilities, living systems are conveniently
organized into hierarchical levels, interacting nonlinearly among themselves
like those of Horace Lamb's hydrodynamic turbulence (see Sect. 6.8.7).

Although the dynamics of living organisms are far more nonlinear than
those of inanimate matter, biologists have not been as exposed to the
nonlinear-science revolution of the 1970s as have physical scientists, because
they tend to ignore or reject physico-mathematical formulations of Life's dynamics. For this reason, many biologists still turn to reductive explanations
based on well-defined causal chains, thereby missing the implications of *emergence* for understanding their phenomena.

---

[1]In this chapter, the professions of scientists will be noted as they are introduced,
because this information is relevant to understanding the role that physical science
has played in the formulation of mathematical biology.

7.1 Nonlinear Biochemistry

The sad story of biophysicist Boris Belousov's discovery of an autocatalytic chemical reaction being rejected by the Soviet scientific establishment (see Sect. 4.3 and [252,1048,1084]) was not the only example of chemists and biochemists with blinkered conceptions of causality in the 1970s. In Sect. 6.4.2 it was mentioned that biophysicist Colin McClare may have taken his life in response to British rejection of his proposals for resonant storage of biological energy [629, 629, 630], and physicist Giorgio Careri's experimental evidence [152, 872] for physicist Alexander Davydov's theory of nonlinear localized states in protein (solitons) [217,218] was strongly resisted by US physical chemists in the mid-1980s. Thus the scientific suspicion or rejection of nonlinear localization in biochemistry was a worldwide phenomenon.

7.1.1 Fröhlich Theory

Resonant storage of biological energy – as opposed to the localization of dynamic activity by reaction-diffusion processes – was proposed in 1968 by Herbert Fröhlich, a condensed-matter physicist who constructed a theory of collective biological oscillations that was related to his seminal studies of Bose–Einstein condensation and superconductivity [342, 343, 977]. Unfortunately, McClare seems to have been unaware of Fröhlich's work, which offered support for his ideas and would have encouraged him by showing that not everyone rejected the possibility of resonant energy storage in biology.

Observing that living organisms remain in stable ordered states while far from thermal equilibrium and that both cells and large biomolecules have unusual dielectric properties (large dielectric constants), Fröhlich outlined a general theory of collective biochemical effects to be anticipated in cell membranes and in the interior of a hydrogen-bonded protein molecule [343]. As in laser dynamics (see Sect. 6.7.1), he showed that if energy is supplied at a sufficiently high rate, a collection of oscillators will become synchronized in a single, strongly-excited mode, which may be useful for biological processes [343]. This is more a *metatheory* than a specific explanation, however, because it points to a general family of possibilities rather than a particular phenomenon.

Fröhlich visited the University of Wisconsin in the mid-1970s to give a talk at biochemist David Green's Enzyme Institute, and both his ideas and his appearance were impressive – with his long, white hair and intense eyes, he impressed me as a modern version of Beethoven, who happened to be interested in science rather than music. In the early 1950s, Fröhlich had famously used the methods of quantum field theory to describe holistic phenomena in condensed-matter physics, and two decades later he was taking the same approach in biology. In this realm, he explained in his UW lecture, energy for a nonequilibrium state is typically obtained through the hydrolysis of adenosinetriphosphate (ATP) to adenosinediphosphate (ADP) through the reaction

$$\text{ATP}^{4-} + \text{H}_2\text{O} \quad \longrightarrow \quad \text{ADP}^{3-} + \text{HPO}_4^{2-} + \text{H}^+ + \Delta E \,, \qquad (7.1)$$

where (in various energy units)

$$\Delta E \approx 10 \text{ kcal/mole} = 6.76 \times 10^{-20} \text{ joules} = 422 \text{ meV} = 3409 \text{ cm}^{-1}$$

is the amount of energy released by the reaction under typical physiological conditions [317] and $\approx$ indicates that the precise value depends on temperature, acidity, and concentrations of the reagents.

The proposals of Fröhlich and McClare received empirical support in 1978 from observations of sharp resonance peaks in microwave-induced *action spectra* measured on the growth rates of yeast.[2] As these spectral peaks were too narrow to be explained as thermal effects from the microwave heating, it was suggested that a version of Fröhlich's mechanism might be playing a role in yeast growth [381, 382]. Also inspired by Fröhlich's theory, biophysicist Sydney Webb published measurements of *laser-Raman spectra* from synchronously dividing colonies of *Escherichia coli* [1015], which were soon confirmed on colonies of *Bacillus megaterium* by a group in the Soviet Union [52].[3]

7.1.2 Protein Solitons

In the summer of 1978 at an international soliton conference in Gothenburg, Sweden, Davydov presented his theory for energy storage and transport via nonlinear Schrödinger (NLS) solitons in alpha-helical regions of protein. As biophysical theories typically have several undertermined parameters, it was a surprise for me to learn that his formulation was based on a Hamiltonian (energy functional) in which the values of every parameter had been fixed through independent experiments. Could it be, I wondered upon hearing this talk, that these solitons which I had been studying for over a decade play functional roles in living organisms? In those days, it was difficult for Soviet and US scientists to communicate, so Davydov asked me to carry several of his reprints back to the University of Wisconsin and discuss them with Green, which I was pleased to do. Throughout an afternoon of vigorous discussion, however, I couldn't get Green to understand what Davydov was proposing.

---

[2]Action spectra are obtained by irradiating different lots with different wavelengths and then plotting the rates of growth against wavelength.

[3]In measuring Raman spectra, a beam of laser light is passed through the sample, after which the outgoing beam is examined in a spectrograph. Slight decreases or increases in wavelength indicate emission or absorption of energy by the sample, at sum or difference frequencies according to the Manley–Rowe relations discussed in Sect. 6.6.4.

He kept coming back to the nerve pulse as a metaphor, which misses the point: Davydov's soliton is an energy-conserving entity (see Chap. 3) rather than a reaction-diffusion wave (Chap. 4).

During a visit to the applied mathematics group at the Los Alamos National Laboratory a few months later, David McLaughlin and I used a large (for then) ODE solver that had recently been developed by Mac Hyman ("so scientists can do science rather than numerical analysis") to integrate Davydov's dynamic equations. We studied an alpha-helical chain of 600 amino acids, which was modeled by 1200 ODEs, and our modest aim was to generate some computer images that would help biochemists see what Davydov was proposing. During this numerical exercise, an interesting threshold effect appeared that was related to protein parameters [460]. Becoming hooked on the problem area, I began to study energy localization in biomolecules both theoretically (using perturbation methods) and numerically – a research activity that would continue for more than a decade [869, 872].

Upon being appointed as director of the newly-established Center for Nonlinear Studies at Los Alamos in 1981, I was able to fund and guide numerical and experimental work in this area, and, surprisingly, found that some of the laser-Raman spectra reported by Webb were almost identical to those that I had independently computed for Davydov's solitons on alpha-helix regions of protein [868, 1015]. Thus I immediately visited Webb at the Max Planck Institute in Stuttgart, where fortunately Fröhlich was also in residence. After studying the details of these measurements and calculations, Fröhlich strongly supported the assignment of Webb's measured spectrum to internal vibrations of Davydov's alpha-helix solitons [571]. In retrospect, this interest is not surprising because, as biophysicist Jack Tuzyński soon pointed out, Davydov's soliton formulation can be viewed as a specific example of Fröhlich's metatheory [974].

For a qualitative appreciation of a Davydov soliton, consider Fig. 7.1, which shows the arrangement of peptide group (HNCO) atoms in a short section of alpha helix, a secondary structure in many natural proteins. Under Davydov's theory, localized oscillations of carbon–oxygen (CO) stretching vibration (called the Amide-I mode) induce a local distortion of the crystal lattice, and this distortion acts as a potential well to trap the CO oscillations; thus there is a closed loop of causality, from which a soliton emerges. In the lowest order of approximation, the dynamics of this soliton are described by the NLS equation (see Sects. 3.3 and 6.3.2 and [869])

$$\left(\mathrm{i}\frac{\partial}{\partial t} - \omega_0 \right) u + \varepsilon \frac{\partial^2 u}{\partial x^2} + \gamma |u|^2 u = 0 \,, \tag{7.2}$$

where ω_0 is the oscillation frequency of a CO stretching mode, ε is a dispersion parameter arising from electromagnetic (dipole–dipole) coupling between neighboring CO oscillators, and γ is a parameter of extrinsic nonlinearity arising from the fact that hydrogen-bond stretching slightly alters the CO

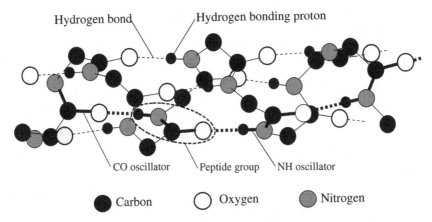

Fig. 7.1. A short section of alpha helix. The *dashed lines* indicate relatively weak hydrogen bonds. Peptide (HNCO) groups are shown for each amino acid, but residues are not shown. (To aid in seeing it, the bonds of one peptide channel are bolder)

oscillation frequency. The spectral lines that I had numerically computed involved lateral oscillations among the three channels of the form [868]:

$$\cdots H - N - C = O \cdots H - N - C = O \cdots H - N - C = O \cdots ,$$

where the bonds for one of these chains are drawn more heavily in Fig. 7.1.

Although the evidence for a connection between Webb's measurements and Davydov's soliton theory seemed convincing (to me, at least), an independent experimental confirmation was needed. Thus a project was begun in which Scott Layne (a medical doctor and CNLS postdoctoral researcher) grew synchronously dividing colonies of *E. coli* and Irving Bigio (a senior optical scientist) measured their Raman spectra. Whereas Webb had employed a conventional scanning spectrometer, Bigio used a recently developed *optical multichannel analyzer* which records the entire spectrum simultaneously. After several months of effort, these experiments failed to confirm Webb's observations.

But Webb had seen *something* – what could it have been? A possible explanation was a strong transient fluorescence that Bigio and Layne observed during cell division. Thus what appeared to Webb as spectral lines in his scanning observations might have been fluorescence appearing during the times of synchronized cell division, a distinction that could not have been made without using a multichannel analyzer to record the spectra [543]. Instead of confirming the motivating theory, therefore, these experiments revealed a new phenomenon that also fell under the aegis of Fröhlich's metatheory, but Webb disagreed. In an attempt to resolve the issue, we invited Webb to visit the Los Alamos experiment for several weeks, and at his suggestion Layne and Bigio repeated their measurements on synchronized colonies of *B. mega-*

terium (no small task) with the same results: no Raman spectra but a rise in fluorescence during synchronized cell division [542].

This experience convinced me that physical scientists must be very careful in experimenting with living organisms to exclude artifacts arising from poorly understood features of such complicated systems. Although convincing to me when it was first recognized, the correlation between Webb's measured and published spectra and my independent calculations of interchannel oscillations (based on Davydov's theory) has not been explained and is currently ascribed to happenstance [571, 868, 1015].

At about the same time, however, supporting empirical evidence for Davydov's soliton was offered by Careri's infrared absorption and Raman-scattering studies of acetanilide (ACN), a small but protein-like molecule that can be grown in conveniently large crystals of optical quality.[4] Suggestively, ACN crystals have linear peptide chains (as in the alpha helix of Fig. 7.1), making it a model protein, and Careri's experiments, which were carried on at the University of Rome throughout the 1970s, had revealed an anomalous spectral line at 1650 cm^{-1} that had not been explained (see Fig. 6.6 and [152]). These experimental results plus related numerical studies done together with applied mathematician Chris Eilbeck (who was visiting at Los Alamos during the early 1980s) suggested that the 1650 line was indeed evidence for the existence of a localized mode in ACN [271]. Several additional experiments and theoretical studies were carried out, and a series of informal papers appeared in *Los Alamos Science* (spring 1984), along with numerous refereed publications [872]. This highly-visible theoretical, numerical, and experimental effort ignited a surge of activities throughout Europe and the US during the 1980s, both supporting and challenging the assignment of Careri's 1650 line to a local mode in ACN and the extension of this idea to solitons in natural protein.

In the early 1980s, most physical chemists did not believe that a crystal with translational symmetry can have local modes of vibration; thus it was claimed that the 1650 line should be assigned to Fermi coupling (a mixing of other modes as described in Sect. 6.6.4), even though this possibility, among several others, had been considered and eliminated by Careri and his coworkers during the 1970s as they searched for a viable explanation [152, 469]. The possibility that the 1650 line could be caused by a change in the crystal structure was eliminated through a careful X-ray diffraction study at Los Alamos, and the claim that it could be assigned to a second-order phase change of the crystal was eliminated through the neutron scattering studies caried out by Mariette Barthes and her French group [60].

Prior to its measurement, interestingly, (7.2) and the data from Fig. 6.6 allowed us to predict the Birge–Sponer relation for temperature-dependent

---

[4]Chemically related to acetaminophen (Tylenol), acetanilide was used as an analgesic in the nineteenth century before being supplanted by aspirin, which is less toxic.

Table 7.1. Overtone spectrum for crystalline ACN

Number (n)	Predicted	Measured	Fitted
1	1650 ± 1	1650 ± 0.5	1 649.3
2	3254 ± 4	3250 ± 1	3 249.2
3	4812 ± 9	4803 ± 3	4 799.7
4	6234 ± 16	6304 ± 5	6 300.8

overtones of the 1 650 line to be (see Sect. 6.3.1 and [891])

$$\omega_n = \omega_0 n - \gamma n^2 = 1\,673n - 23n^2 \ .$$

In determining the parameters of this relation, $\omega_0 = 1\,665 + 2J$, where $J = 4$ is the nearest neighbor coupling energy (calculated from Maxwell's equations [152, 271]) and $\gamma = \omega_0 - 1\,650$, as required by (7.2).

Table 7.1 presents a comparison of predicted and measured values for the first four temperature-dependent lines, where the errors in the predicted values are roughly estimated by assuming an uncertainty of ± 1 cm^{-1} in the value of γ.[5] The last column of Table 7.1 shows that the measured lines fit well to

$$\omega_n = 1\,674n - 24.7n^2 \ ,$$

implying errors of about 1 cm^{-1} in ω_0 and 1.7 cm^{-1} in γ.[6]

Further analysis led to a quantitative expression for the temperature dependence of the 1 650 line intensity, $W(T)$, in terms of γ and the energies (ω_1) of M lattice vibrations associated with the self-localization [872, 887]. Thus it was predicted that

$$\frac{W(T)}{W(0)} = \left\{ \frac{\exp\left[(-\gamma/M\omega_1)\coth(\omega_1/kT)\right]I_0\left[(\gamma/M\omega_1)\text{csch}(\omega_1/kT)\right]}{\exp(-\gamma/M\omega_1)} \right\}^M , \tag{7.3}$$

where γ, ω_1 and kT (the thermal energy per degree of freedom) are all in the same energy (frequency) units. Equation (7.3) agrees well with Raman measurements of $W(T)$ for $\gamma = 24.7$ cm^{-1}, $M \geq 5$, and $\omega_1 = 71$ cm^{-1}, which is a reasonable effective value for a lattice vibration.

---

[5]Measuring this overtone spectrum was not as easy as it might seem because the $n = 4$ line is weak, requiring the growth of a special crystal of ACN that was several centimeters long.

[6]Such errors are not unexpected, because there could have been an error of ± 1 cm^{-1} in computing the nearest-neighbor coupling energy J, as it is not clear how many next-nearest-neighbor interactions to include due to uncertainties in dielectric shielding [152, 271].

In 1986, Denise Alexander, a gifted research student working together with condensed-matter physicist James Krumhansl, published an important survey of self-trapping in molecular crystals, pointing out that Davydov's soliton theory was similar to a polaron theory that had been proposed by physicist Theodore Holstein back in 1959 [15, 445, 446]. Importantly, Alexander also proposed that the NH stretching oscillation (see Fig. 7.1) be considered as a candidate for biological energy storage and transport, in addition to the CO stretching oscillation which was originally considered by Davydov. As she was beginning her academic career at Columbia University in New York in 1987, sadly, Alexander was hit by an automobile and killed; thus her idea of using the NH oscillation was largely forgotten, perhaps because the structure of this band was then poorly understood. If she had not died at this time, the force of her intellect would – I feel sure – have gained an earlier acceptance for self-trapping of biological energy via the NH stretching mode.

These many theoretical, experimental, and numerical studies were discussed in detail at a NATO-sponsored conference on "Davydov's soliton" that was held on the Northwest Coast of Denmark in the summer of 1989 [180]. Under a bright Nordic Sun in the dog days of the Soviet Union, Davydov and his colleagues were brought together with most of the western scientists working on protein solitons, and – like unacquainted families meeting at a wedding – the two groups at first regarded each other warily from opposite sides of the lecture room. Stimulated by Viking hospitality, however, friendships soon formed and led to many fruitful research interactions over the following years. Most but not all left Denmark convinced that local modes in ACN are well established and that corresponding local modes in natural protein are possible [180, 872].[7]

During the 1990s, theoretical analyses of energy localization in ACN far outnumbered empirical studies, but the experimental picture was pinned down early in this century through the femtosecond pump–probe measurements of physical chemists Peter Hamm and Julian Edler, which are described in Sects. 6.4.2 and 6.7.4. In addition to proving conclusively that the Amide-I (CO stretching) mode is localized [264], this work sorted out the structure of the NH stretching band (see Fig. 7.2), which is more complex than the CO stretching band (shown in Fig. 6.6) [265]. In the NH case, the delocalized line at $3\,300$ cm^{-1} is accompanied by three lower energy satellites that are localized through interactions with lattice phonons. (Note that the temperature sensitive line at $3\,250$ is the first overtone of the $1\,650$ line.)

From the perspectives of biology, however, the central issue is whether such localized states perform a biological function on real alpha helix, and it

---

[7]Although he was not present at the Danish meeting, optical scientist Robert Knox kept abreast of the events, proposing a pump–probe method to measure the time-of-flight of a vibrational wave packet on alpha helix using chromophores to insert and sense vibrational energy that was included in the conference proceedings [180, 500].

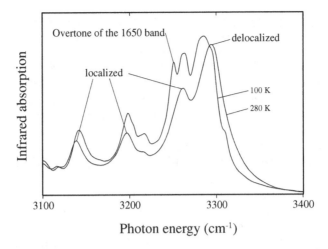

Fig. 7.2. Infrared absorption spectra of crystalline ACN in the NH stretching region at two different temperatures [265]. (Data courtesy of Peter Hamm)

is important for such applications that the three localized NH lines are strong at room temperature.[8] In addition, as is seen from (7.1), the quantum energy of an NH line is about equal to that released by hydrolysis of a molecule of ATP; thus any of the three localized satellites is a more favorable candidate for a biological soliton than the CO line at 1 650, which was originally chosen by Davydov [217, 218].

From Hamm's measurements, excitations of the three localized NH modes become spectrally dark in about 1 ps, followed by a slow return of the energy to the ground state in about 18 ps. Interestingly, biophysicist Robert Austin and his colleagues have recently published free-electron-laser observations of similarly long-lived CO modes in myoglobin, a globular protein that is largely composed of alpha helix [1060]. Although Austin's measurements were at a temperature of 20 K, they suggest that self-trapping should be expected in a natural protein as well as in ACN.

With reference to Fig. 7.1, the current status of self-trapped states in ACN and in natural protein can be summarized as follows:

- Self-localization of CO (Amide-I) oscillations. Although the nonlinear self-localization originally proposed by Davydov has been confirmed at low temperature by pump–probe measurements in ACN, this phenomenon may not be biologically relevant for natural proteins at normal temperature because of the weak nonlinearity. On the other hand, numerical stud-

---

[8]This can be qualitatively understood because the same proton (hydrogen nucleus) participates in both the NH oscillation and in the hydrogen bond that forms the alpha helix (see Fig. 7.1), leading to a much stronger coupling to lattice distortions (γ) than for the CO (1 650) line.

ies by theoretical biologist Leonor Cruzeiro and physicist Shozo Takeno suggest that elevated temperature may lead to a form of *Anderson localization* [24] where adjacent CO oscillators become largely detuned and quanta jump from one CO oscillator to another [201, 202], a picture that is in accord with empirical observations of Edler and Hamm [264]. As the quantum energy of a CO oscillator is about half of that released by the hydrolysis of an ATP molecule, however, a localized state comprising two quanta would be needed to store and transport this amount of energy. (This is the 3 250 line shown in Fig. 7.2.)

- Self-localization of NH oscillations. As originally proposed by Alexander and overlooked by Davydov and the rest of us, the NH line is a promising mechanism for biological energy storage and transport in natural protein for three reasons: it persists at room temperature (see Fig. 7.2), the lifetime may be long enough for a biological function, and each quantum of NH oscillation can transport the entire energy released by the hydrolysis of an ATP molecule; thus only a single-quantum soliton needs to be considered. (As I write in May of 2006, Hamm is attempting to implement Knox's time-of-flight measurement to observe soliton motion on a natural protein [400].)

7.1.3 Biological Applications of Protein Solitons

While delving into studies of protein solitons at Los Alamos in August of 1983, I came across an article by biophysicist Sungchul Ji on the *conformon*, a polaron-like structure for storing and transporting biological energy or charge in protein that had been independently proposed a decade earlier by three different biophysics groups: Mikhail Vol'kenstein and his colleagues at the Institute of Molecular Biology in Moscow [996], Gabor Kemeny and Indur Goklany in the Department of Biophysics at Michigan State University [487, 488], and David Green and Ji in the University of Wisconsin's Enzyme Institute [373]. As the conformon concept was close to that that of Davydov's soliton, I couldn't understand how Green had not been able to grasp Davydov's ideas during our long afternoon of discussion in the fall of 1978; thus I immediately telephoned him, only to sadly learn that he had died a month earlier. In subsequent discussions, Ji explained that Green had been so intimidated by the negative response to McClare's presentations at his (Green's) 1973 conference on "The Mechanism of Energy Transduction in Biological Systems" (see Sect. 6.4.2 and [372]) that he abandoned the conformon concept completely – a striking example of the Kuhnian barriers to be overcome in the 1970s [515]. As all of these proposals for energy storage or charge transport in protein (Vol'kenshtein, Kemeny/Goklany, Green/Ji, Davydov, and Careri) were advanced to account for observed phenomena in living organisms, let us consider some of the suggested applications.

Muscular Contraction

In the early 1970s, electron microscopy had shown that skeletal muscle contracts not by changing the lengths of individual biomolecules but by sliding thick filaments (comprising myosin) past thin 'filaments (comprising actin), and it was supposed that a force is generated through the making and breaking of cross bridges from the head groups of myosin molecules to actin filaments [457, 458]. In this picture, myosin heads: (i) swing out, (ii) attach to the actin, (iii) bend and therefore generate a force, and finally (iv) detach from the actin to begin another cycle.

In Davydov's first paper on solitons in alpha-helix protein [217], a main application was to provide an alternate explanation for the sliding filament theory, avoiding the ad hoc nature of the above picture which he dismissed in private conversations as the row-boat theory of muscular contraction. Instead he proposed that groups of solitons are launched from the myosin heads and travel along their alpha-helical tails, forming *varicosities* on thick filaments that push against neighboring thin filaments because of the solitons' kinetic energy. As he wrote in 1991 [218]:

> According to this model, the heads of myosin molecules attach and detach further from thin filaments, as in the model of formation and breakdown of bridges. The origin of this motion is not, however, due to lengthening, rotation, and contraction of the lengths of the heads [i.e., the "row-boat" theory], but due to the movement of solitons inside a thick fibre [alpha-helical myosin tails] which is accompanied by the motion of bent regions of helical molecules. The kinetic energy of the solitons thus transforms into the energy of contraction, or results in a tension if the muscle is under load.

There are two problems with this theoretical picture: First, only a small fraction of a soliton's total energy is kinetic; most is rest energy stored in quanta of the CO oscillation [869]. To explain the high efficiency of skeletal muscle, therefore, it is essential to assume that the soliton's rest energy exerts a force between the myosin and actin, as was proposed by Ji in his 1974 version of the conformon theory of muscular contraction [468]. Second, Davydov's theory is not in accord with the best current knowledge of how myosin behaves, under which the row-boat theory is still afloat [309, 846, 981].

Active Transport Across Nerve Membranes

Discovered in 1882 by Edward Reid – an English physiologist in unhappy exile at the University of Dundee – active transport is a process under which ions are pumped uphill (in the direction of higher ionic concentrations), using ATP hydrolysis as a source of energy [795, 796]. As physician Stanley Schultz has recalled in an interesting historical review, this important phenomenon was ignored or considered suspect by biologists until the middle of the twentieth

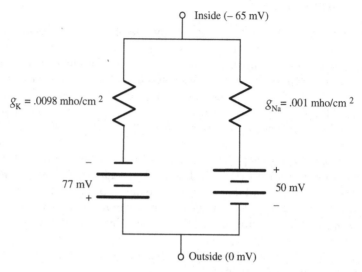

Fig. 7.3. An electrical circuit model of a unit area of squid nerve membrane in its resting state

century, because it had a whiff of vitalism [854], but it fits naturally into the family of nonlinear processes discussed in Chaps. 3 and 4.

Returning to the discussion of a squid nerve in Sect. 4.1, recall that the concentration of sodium ions is greater outside the axon than inside, whereas the concentration of potassium ions is greater inside than outside. In the electrical model of Fig. 7.3, the 50 mV (77 mV) battery on the right (left) represents the tendency of sodium (potassium) ions to diffuse inward (outward), leading to a resting potential of -65 mV inside the axon with respect to the outside [874]. Under this model, g_{Na} represents the electrical conductance (or permeability) per unit area of the membrane for sodium ions, and g_K is the corresponding conductance per unit area for potassium ions. (Note that $g_{Na} + g_K = g$, which is given in Table 4.1.)

In a squid nerve, the tendency of potassium ions to diffuse outward (because of their higher inner concentration) is approximately in balance with their tendency to be conducted inward (because the electric field across the membrane is directed inward). With sodium ions, on the other hand, the outer concentration is greater than the inner concentration; thus sodium ions want both to be conducted inward and to diffuse inward, a tense situation held modestly in check because $g_{Na} \ll g_K$. As was described in Sect. 4.1, it is the release of inward sodium current by the nonlinear properties of the membrane (g_{Na} takes a larger value) that leads to the emergence of a nerve pulse.

Over the long term, sodium ions are maintained in their nonequilibrium resting state by an enzyme known informally as the *sodium pump* and more precisely called Na/K-ATPase, which is an *intrinsic membrane protein* (embedded in the nerve membrane) that moves sodium ions from inside the nerve axon to outside. This enzyme was first isolated in the 1950s by the Danish biochemist Jens Skou and his colleagues, who collected nerves from the legs of some 25 000 crabs [907, 908].[9] As each hydrolysis of an ATP molecule pumps three sodium ions outward (requiring $3 \times 50 = 150$ meV of energy) and two potassium ions inward (requiring $2 \times 77 = 154$ meV of energy) for a total of 308 meV [567, 584, 666], we see from (7.1) that the needed energy is comfortably supplied by ATP, but how does it work? What is the mechanism of active transport? How are sodium ions actually pushed outward against their concentration gradient?

To answer such questions in detail, it is necessary to know the enzyme structure, and this is more difficult to determine for an intrinsic membrane protein than for extrinsic proteins which are more soluble in water and thus more readily crystallized. Nonetheless, approximate X-ray images of Na/K-ATPase have recently been obtained [1086], from which it is seen that ten segments of alpha helix cross the membrane, suggesting that charge-carrying conformons or (which is about the same thing) charge-carrying versions of Davydov's solitons may be transporting sodium ions against their gradient of concentration. Under this scenario, the kinetic energies imparted by ATP hydrolysis to three sodium solitons and two potassium solitons would be about 422 meV. As we have noted, this is sufficient to recharge the sodium and potassium batteries shown in Fig. 7.3, with the excess (about 118 meV) accounting for ohmic losses in g_{Na} and g_K.

In addition to Na/K-ATPase, it is now known that several other ions are actively transported across cell membranes in a similar manner [436], including: (i) the pumping of calcium by Ca-ATPase [930] (which replaces sodium as the active ion in some nerves), (ii) pumping of protons (hydrogen ions) in mitochondrial energetics by F1F0-ATPase as ADP is converted back to ATP [244], (iii) acidification of our stomachs by H/K-ATPase [781, 835]. As shown for Ca-ATPase in Fig. 7.4, all of these transmembrane proteins have about ten cylinders of alpha helix spanning the membrane, suggesting that something like Lev Landau's polaron (see Sect. 6.4.2) is transporting the charge. Recalling that a polaron is said to move like a marble working its way down through a plate of spaghetti, this model seems apt, but the details are surely more intricate. Whether the polaron/conformon/soliton picture will help biochemists understand active transport as the ATPase structures become better defined remains to be seen.

---

[9]In 1997, Skou was awarded the Nobel Prize in Chemistry for this work.

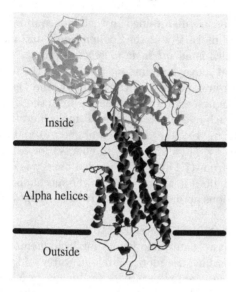

Inside

Alpha helices

Outside

Fig. 7.4. The structure of Ca-ATPase. (Courtesy of David Stokes [930])

The Turning Wheel

In 1965, a positive feedback model was proposed by biochemists Jacques Monod, Jeffries Wyman and Jean-Pierre Changeux for the *allosteric* action of enzymes, in which the binding of a ligand at one site induces a conformational change that enhances the attraction at another site, leading to a sigmoid response function of the reaction [670, 935].[10] Building on this picture a decade later, Wyman proposed the concept of a *turning wheel* as a general formulation for the steady-state operation of an enzyme [350, 435, 1059].

A simple example is shown in the kinetic diagram of Fig. 7.5a, which describes the conversion of a substrate (ligand) L to its product L'. (In this figure, an appropriate rate constant is attached to each arrow; thus the diagram represents a system of ODEs.) It is assumed that the concentration of L is kept constant, and the enzyme has only two conformational states, M_1 and M_2, each of which can bind to the substrate L.[11] Under the desired steady-state solution of these ODEs, there is a counterclockwise circulation in the diagram, and the product substrate is created at the rate

$$\frac{\mathrm{d}[L']}{\mathrm{d}t} = k_{\mathrm{L}}[M_1 L'] . \tag{7.4}$$

---

[10]Monod received a Nobel Prize in Physiology or Medicine for his work on genetic control of enzymes.

[11]If there are more than two enzyme conformations, the square in Fig. 7.5 can be generalized to a cube or hypercube. As Wyman limited his analysis to one-step transitions, the diagonal transitions $M_1 \rightleftharpoons M_2 L$ and $M_2 \rightleftharpoons M_1 L$ are not included in his scheme.

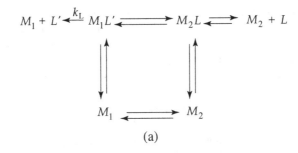

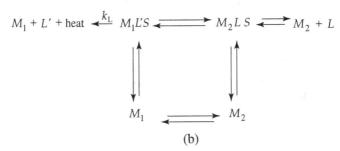

Fig. 7.5. (a) The kinetic diagram for Wyman's turning wheel [1059]. (b) A soliton-driven turning wheel [153]

In the row-boat model of muscular contraction, for example, L and L' would be ATP and ADP respectively, while M_2 is the unbent conformation of the myosin head group and M_1 is the bent conformation; thus (7.4) gives the rate at which a myosin molecule is doing work.

Motivated by the experimental observations of CO stretching oscillations in ACN, Careri and Wyman (CW) have suggested that Davydov's soliton might help to keep the wheel turning [153]. Their basic idea is shown in Fig. 7.5b, where the soliton S, is assumed to be:

> [...] an energy packet trapped [...] inside a portion of a protein matrix, with the possibility of storing energy without dissipation until it decays into heat as a result of a strong perturbation from some other portion of the same protein.

Thus the soliton forms as the enzyme binds to the substrate L, and it has a long lifetime in conformation M_2 but a short one in M_1. This short lifetime drives the unidirectional output reaction producing L', and the energy of the soliton accounts for the energy difference between L' and L, the remainder being dissipated as heat. In a somewhat different language, this scenario corresponds to the conformon picture of muscular contraction that was first formulated by Green and Ji in 1973 [373].

The key assumption of the CW picture is that the substrate binding can give rise to the formation of a long-lived vibrational soliton. To make this assumption plausible, CW consider the binding of negative phosphate groups at the ends of alpha helices, finding that the binding of one negatively charged group involves an attractive energy of about 13 kcal/mole, whereas the energy to create a CO stretching mode soliton is about 5 kcal/mole and about 10 kcal/mole for an NH stretching mode. Further, CW proposed that rapid compression of the end of the helix may induce soliton formation.

In a later paper [154], CW extended these ideas to a suggestion for a prebiotic mechanism driven by solar energy (rather than ATP) that is based on the CO stretching overtone bands of ACN shown in Table 7.1. Before the development of chlorophyll, an amide *protoenzyme* may have developed with infrared properties similar to ACN. For harvesting sunlight, the fundamental (1 650) band is ruled out of consideration because of the high absorption by the HOH bending oscillation of water, and the first overtone (3 250) is similarly ruled out because of high OH stretching absorption in water. But the second CO overtone ($n = 3$) at 4 803 falls in a range where water has a window for mid-infrared transmission. Although the above-mentioned temperature problems for a CO soliton are only modestly less severe in the prebiotic application than for warm-blooded creatures, it is not interesting to turn to the NH stretching oscillation, because the NH frequency and its overtones lie close to those of water's OH stretching oscillation.

7.1.4 DNA Solitons and the Hijackers

As conviction grew among biologically oriented members of the nonlinear science community that solitons play functional roles in proteins, it seemed natural to suppose that an extended biomolecule like DNA (deoxyribonucleic acid) would also support solitary waves that might be biologically useful – for example in the "unzipping" process of the double helix by RNA polymerase that is needed to read out the genetic code. There are actually two scientific questions here. Within the realms of condensed matter physics, the first question is: Do synthetic double helices (see Fig. 7.6) support one or more solitons? If the answer to this physics question is positive, a second question is functional: Do soliton-like objects on biological DNA play useful roles in any living organisms?

The physics question was broached in 1980 by Krumhansl, Alan Heeger[12] and three of their biochemical colleagues (Walter Englander, Neville Kallenbach, and Samuel Litwin), who suggested the following discrete version of the sine-Gordon (SG) equation (3.11) to explain experimental observations of unusually rapid deuteration of internal hydrogen bonds in DNA [291]:[13]

---

[12]Heeger was awarded the Nobel Prize in Chemistry in 2000 for his discovery and development of conductive polymers.

[13]In (7.5), u_i is an angle of rotation of a base about its sugar-phosphate backbone axis, K is a torsional spring constant between adjacent base pairs, M is a moment of inertia of a rotating base pair, T is a constant of the restoring torque, and d

$$K \left(\frac{u_{i+1} - 2u_i + u_{i-1}}{d^2} \right) - M \frac{\partial^2 u_i}{\partial t^2} = T \sin u_i \,, \qquad (7.5)$$

which is also the dynamical equation for the mechanical model of Fig. 3.3. Noting that the kinetics of this exchange data on normal (B-type) DNA indicated that regions of about ten base pairs open under thermal equilibrium, these authors proposed each open region to comprise an SG-like kink and an antikink (see Sect. 3.4 and Fig. 3.3).

With reference to the upper diagram of B-type DNA in Fig. 7.6, the idea of Englander et al. was that several adjacent sets of hydrogen bonds between the base pairs are broken by thermal agitation, allowing the base pairs to swing outward in a coupled manner and offer themselves for deuteration in an exchange experiment. Merely a glance at Fig. 7.6, however, shows that real DNA is far more complicated than can be modeled by (7.5), allowing for many different sorts of nonlinear wave propagation, and even these atomic structures do not include the associated molecules of water, which are important factors in transitions between A-DNA and B-DNA.

In a book by biophysicist Ludmilla Yakushevich entitled *Nonlinear Physics of DNA*, the various soliton theories are organized in a hierarchical manner (shown in Fig. 7.7), proceeding from the simplest type-1 models (compression waves on a single continuum or discrete longitudinal rod) through several intermediate models (pairs of rods, pairs of discrete rods, helical rods, helical discrete rods, and so on) to the most complicated type-5 model (realistic molecular dynamics simulations) [967, 1061, 1062]. From this book one sees how enthusiastically the nonlinear science community was studying DNA in the 1980s. As we have seen in Chap. 1, the nonlinear science revolution had spawned many centers of nonlinear science, staffed with biologically oriented – if not well-informed – physicists and applied mathematicians who were tenured, adequately funded and eager to exercise their tools on a problem as interesting and important as the dynamics of DNA. Although some in the biological community were unhappy about the new reality, there was little that could impede this growing research momentum. It was no longer possible to close down programs, as was almost done to Nicholas Rashevsky's effort in the 1950s (see Sect. 5.2.2); thus the nonlinear science of DNA would succeed or fail on its merits. Unfortunately, two early examples of the physico-mathematical approach to DNA were based on artifactual data.

In 1984, electrical engineers Glen Edwards and Chris Davis, working together with biochemist Jeffrey Saffer and biologist Mays Swicord, published measurements of two peaks of resonant absorption of microwaves between 5 and 10 GHz in aqueous solutions of strands of natural DNA (from *Es-*

---

is the longitudinal spacing between adjacent base pairs (see Fig. 7.6), all of which can be measured, albeit with varying degrees of accuracy [1061]. If distances are normalized to units of $\lambda_0 = d\sqrt{K/T}$ and time to units of $\sqrt{M/T}$, then solutions varyingly only slightly over a longitudinal distance d are described by (3.11), and the limiting (characteristic) speed is $d\sqrt{K/M}$ [859, 861].

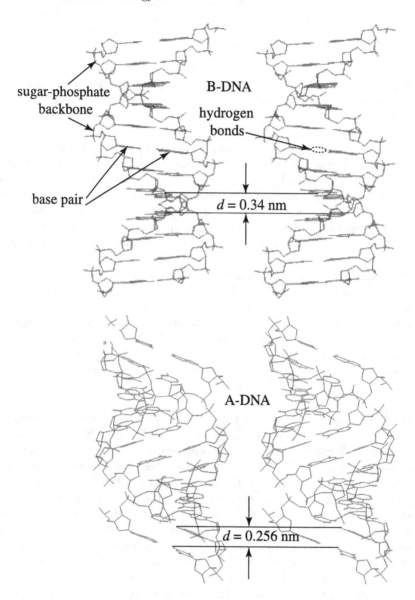

sugar-phosphate
backbone

B-DNA

hydrogen
bonds

base pair

$d = 0.34$ nm

A-DNA

$d = 0.256$ nm

Fig. 7.6. Stereographs of stick models of DNA. *Upper*: B-type (aqueous) DNA, which has 10 base pairs per turn. *Lower*: A-type (dehydrated) DNA, which has 11 base pairs per turn. The diameters of these helices are about 20 nm [190, 626]

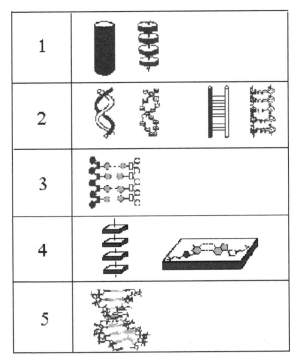

Fig. 7.7. A hierarchy of DNA models from a simple rod (type-1), through double rods (type-2), base-pair interactions (type-3), and base-pair atomic models (type-4), to a full atomic model (type-5). (Courtesy of Ludmilla Yakushevich [1061])

cherichia coli) of known lengths [266, 267]. Both circular and linear DNA strands were studied, and the observed frequencies indicated a sound speed of 1.69 km/s, which seemed in accord with an independently measured speed of 1.9 km/s for longitudinal wave propagation in strands of B-type DNA (from calf thymus) by biophysicist Stuart Lindsay and his coworkers [395].

These empirical results aroused keen interest in the nonlinear science community for two reasons. First, they were obtained by careful experimentalists and appeared in two journals of high reputation (*Physical Review Letters* and *Biophysical Journal*), leading *Nature*'s editor, physicist John Maddox, to ask if physicists are about to "hi-jack" DNA [590]. Second, it was to be expected from the steric resistance between adjacent base pairs that the longitudinal restoring force will be nonlinear in the same manner as in the Toda lattice potential (3.22) of Sect. 3.5 (compression forces are stronger than expansion forces) [692, 693]. Although some physical scientists disagreed with this nonlinear interpretation of the microwave resonance data [989], it supports the most basic description of DNA solitons: longitudinal compression waves on a uniform rod, shown as the type-1 model in Fig. 7.7 [870, 871].

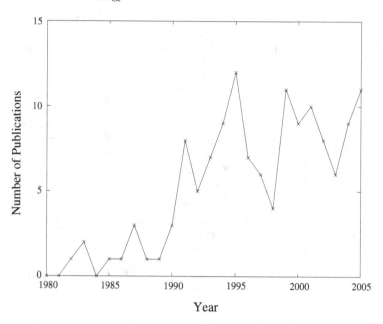

Fig. 7.8. Papers that use both the words "soliton" and "DNA" in their titles, abstracts or lists of key words

Disappointingly for soliton enthusiasts, the microwave resonance data soon appeared to be invalid, for in 1987 two independent groups reported their failures to confirm observations of the absorption peaks [316, 344, 345],[14] and in the same year, the measurements of hydrogen–deuterium exchange rates in DNA used by Englander et al. to construct (7.5) were also shown to have been incorrectly measured [384].[15] These two empirical lapses were promptly noted by molecular biologist Maxim Frank-Kamenetskii,[16] who crowed in *Nature* that: "the hijackers have failed again".

But lack of particular evidence for them is not evidence that solitons do not exist in DNA, and the hijackers' morale was little curbed by these empirical setbacks, as is evident from the publication data shown in Fig. 7.8. Publications during the 1980s include physicist Sigeo Yomosa's type-3 *base-rotator* model [1064, 1065], which was refined by physicists Shozo Takeno and

---

[14] As with Webb's Raman-scattering measurements on *E. coli*, the experimental situation was again sorted out in 1993 by Bigio at Los Alamos, who used a more accurate measurement technique (based on optical sensing of dielectric heating) to establish beyond reasonable doubt that the measurements of Edwards et al. were artifacts, possibly stemming from resonances in a microwave connector [88].

[15] In a personal communication in September of 2006, Kallenbach agreed that the data used by Englander et al. was incorrect.

[16] Maxim D. Frank-Kamenetskii is a son of David Frank-Kamenetskii, whom we met in Sect. 4.1 in connection with the ZF equation (4.4).

Shigeo Homma [942] and Chun-Ting Zhang [1085], and type-1/type-2 Toda lattice models (see Sect. 3.5) by physicists Virginia Muto and Peter Lomdahl and applied mathematician Peter Christiansen and Jesper Halding [692–694]. In addition, physicist Mario Salerno used a version of (7.5) with values of T depending on the site n in order to represent the presence of a specific DNA code [838]. Thus T_n for a C-G base pair with three hydrogen bonds is 50% larger than for an A-T base pair with only two bonds, and it was found that an initially static soliton can either remain static, oscillate or begin to move in one of the two possible directions [839, 840].

Building on previous DNA melting sudies by physicist Earl Prohofsky [776], physicists Michel Peyrard and Alan Bishop set the question of soliton propagation aside and concentrated on the thermodynamics of strand sepa-ration, using a double-rod model of type-2 in Fig. 7.7 [760]. In their exact analysis of a simple energy functional to capture the essence of DNA strand melting, the only nonlinear term is for the potential energy holding the base pairs together with a sum of Morse functions

$$PE = \sum_n D(e^{-ay_n} - 1)^2 ,$$

where n is an index counting base pairs and the adjustable constants (D and a) can be chosen to mimic the hydrogen bonds between base pairs (see Fig. 7.6). From this simple model, Peyrard and Bishop obtained an-alytic expressions for the average hydrogen-bond length and for its average square as functions of temperature, and they also derived a discrete nonlinear Schrödinger (NLS) equation (Sects. 3.3 and 6.3.2) for which stationary soli-tons represent "bubbles" of strand melting as the temperature is increased. While commending the work of Peyrard and Bishop in a 1989 *Nature* note, Maddox stated [591]:

> In this business, there is more value in a problem that can be solved than in an attempt to specify at the outset all the complications of an ideal solution.

These words would have warmed Nicholas Rashevsky's heart, as he made the same point very clearly a half century earlier in the preface to the first edition of his classic *Mathematical Biophysics* [790].

With discrete versions of the SG equation providing models of topological solitons in DNA and the discrete NLS equation of Peyrard and Bishop mod-eling thermal denaturation by a nontopological soliton, the nonlinear science revolution of the 1970s became relevant to DNA research in the 1980s, but more realistic type-5 models continued to remain of central interest.

An early response to this interest was offered by biophysicists Asok Baner-jee and Henry Sobell in 1983 [51], who proposed more complex types of non-linear DNA excitations to explain the relationship between premelting and melting. Figure 7.9 shows the atomic structures of these excitations, which

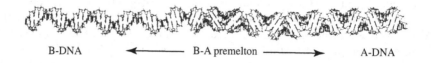

B-DNA   ←——————— B-A premelton ——————→   A-DNA

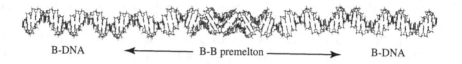

B-DNA   ←——————— B-B premelton ——————→   B-DNA

Fig. 7.9. Molecular structures of topological (B-A) and nontopological (B-B) premeltons in DNA. (Courtesy of Henry Sobell)

are of two kinds: *B-A premeltons* (these are topological and similar to the SG kink in Fig. 3.3), and *B-B premeltons* (which are non-topological like the SG breather in Fig. 3.5). A key feature in both types of premeltons is the presence of an intermediate structural form in their central regions that differs from either A- or B-DNA. Called *beta*-DNA, this form is both metastable and hyperflexible, and contains a regularly alternating sugar-puckering pattern along each sugar-phosphate backbone.[17] Beta-DNA is connected to either B- or A-DNA by boundaries possessing a gradation of structural change that comprises a modulation in alternate sugar puckering present within each polynucleotide chain.[18]

B-B premeltons are proposed to arise at specific DNA regions to nucleate site-specific DNA melting. These *nuclease hypersensitive sites* allow drugs and dyes to *intercalate* into DNA – a process that allows flat planar organic molecules to slip between base pairs when binding to DNA. Additionally, hypersensitive sites are known to exist at the ends of genes in naked DNA and DNA in chromatin (a tightly-bound complex between histones and DNA within the living cell) [162, 467], suggesting that they act as attachment sites for the RNA polymerase, first to form a weak binding complex and subse-

---

[17]This alternation is between the C2′ endo conformation present in B-DNA and the C3′ endo conformation present in A-DNA [801].

[18]The method of Linked-Atom Least Squares (LALS), used by Sobell and his colleagues to create Fig. 7.9, was devised in 1978 to refine helical structures of both natural and synthetic DNA fibers against the paucity of X-ray fiber diffraction data usually available [914]. The procedure allows one to obtain near perfect stereochemistry for a molecular structure with the best fit possible between the observed and calculated structure factors. Details of these calculations and related papers can be found at `http://members.localnet.com/~sobell/`, and final coordinates for the structural intermediates connecting B-DNA with A-DNA have been published [918].

quently to form a tight binding complex [167]. The attachment requires the enzyme be able to spontaneously melt DNA in these regions [917].

According to this more detailed soliton theory, B-A premeltons act as moving phase boundaries, connecting B-DNA with A-DNA during the B-to-A structural phase transition. Under suitable thermodynamic conditions, phase boundaries within B-B premeltons appear and begin to move apart to form longer and longer core regions, whose centers then modulate into A-DNA structure. This necessarily involves the formation of B-A premeltons and A-B premeltons, which continue to move apart, allowing long regions of A-type DNA to appear with A-A premeltons embedded within.

7.1.5 The Coils of Chromatin

As the DNA in your genome comprises about three billion base pairs, each separated by a third of a nanometer, its total length is about a meter, all of which resides within a compact form known as *chromatin* (because it colors on staining). At first guess, one might suppose the DNA to be wound up like a ball of yarn, but chromatin turns out to be organized into the hierarchical series of structures shown in Fig. 7.10, allowing several opportunities for nonlinear wave propagation.

Counting the right-handed double helix as the first stage in this hierarchical ordering, the second is the *nucleosome* comprising 146 base pairs of DNA wound around a core of histones in a left-handed toroid [508]. The histones are small, positively-charged proteins (called H2A, H2B, H3 and H4), which are spatially related by two-fold symmetry [579, 803]. In the presence of an additional protein (H1) complexed to linker DNA (additional DNA that exists between nucleosomes [637]), DNA is further compacted into *tetranucleosomes* that form a 300 angstrom *superhelix*, which can readily be seen by electron microscopy [251, 841, 960].

In contrast to this description of inactive chromatin, active chromatin is unraveled and spread out to permit transcription of the genetic code. Although not yet understood in detail, gene expression is known to involve as many as a hundred different proteins, each playing a role in regulating gene expression [920]. These proteins, called transcriptional factors, replace many of the histones found in inactive chromatin, a process called chromatin remodeling [77]. Additionally, the left-handed toroidal superhelix present in inactive chromatin disappears, probably switching into its topologically equivalent right-handed interwound superhelical form [200].

This interconversion could play a key role in signaling the onset of gene expression. Bending strain energy present in the left-handed toroidal structure can be expected to reappear as torsional unwinding energy in the right-handed interwound structure, giving rise to the premelting regions described in Sect. 7.1.4. These, in turn, could determine the sequential binding of transcriptional factors along DNA, a process eventually guiding the RNA polymerase to the promoter to begin the process of genetic transcription.

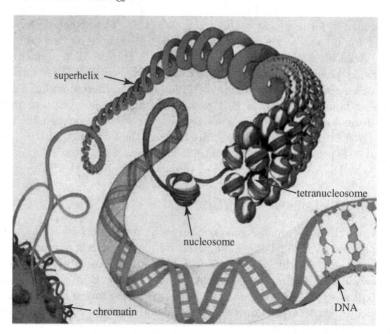

Fig. 7.10. The hierarchical structure of chromatin. (Courtesy of Ludmilla Yaku-shevich [1061])

7.2 On Growth and Form

The processes by which chemical stuff becomes organized into living creatures – called *morphogenesis* – is one of the most mysterious and important in biology. Prospective students of this subject are encouraged to begin with a beautifully written work by the zoologist and polymath D'Arcy Thompson entitled *On Growth and Form*, which first appeared in 1917, was greatly enlarged in 1942, and has been available in an abridged edition for the past four and a half decades [961]. In this book, which inspired Rashevsky to develop his physico-mathematical approach to biology, Thompson made the following introductory comments.

> How far then mathematics will suffice to describe, and physics to explain, the fabric of the body, no man can foresee. It may be that all the laws of energy, and all the properties of matter, and all the chemistry of all the colloids are as powerless to explain the body as they are impotent to comprehend the soul. For my part, I think it is not so. Of how it is that the soul informs the body, physical science teaches me nothing; and that living matter influences and is influenced by mind is a mystery without a clue. Consciousness is not explained to my comprehension by all the nerve-paths and neurones of the physiologist; nor do I ask of physics how goodness shines in one

man's face, and evil betrays itself in another. But of the construction and growth and working of the body, as of all else that is of the earth earthy, physical science is, in my humble opinion, our only teacher and guide.

Limits of physical science are taken up in the following chapter, but attention here is given to seeing how far the ideas of nonlinear science can take us toward understanding the determinants of biological shapes, beginning with the concepts of force and energy. Just as a sphere is the natural shape of a planet, a soap bubble and a star, energetic concepts bear *some* relevance to biological form, as is evident from elementary considerations: the weight of an African elephant cannot be supported by the spindly legs of a sandhill crane, not does a skater insect share the mass of a humpback whale, because the latter swims *in* water (supported by its buoyancy) while the former walks *on* it (held up by surface tension). Such forces and constraints have acted throughout the course of evolution, allowing some shapes and proportions to emerge while suppressing others, but how can this action be described?

7.2.1 The Physics of Form

Thompson begins his discussions of analytical morphology with a survey of the classical surfaces of revolution, which were studied by nineteenth-century mathematicians and now provide a basis for the theory of solitons (see Sect. 3.5 and [117, 659]). Not suprisingly, many of the figures he includes on the shapes of varicosities in tubular structures look like solitons, as they must in order to minimize the sum of nonlinear localization energy and nearest-neighbor interactions. Thus the sleek "bell shape" – with which we became familiar in Chap. 3 – appears in many biological contexts, including filamentary algae, protozoa, vascular aneurysms, improperly spun spiders' webs, and even the shapes of the glass microelectrodes that are used to measure intracellular electrical potentials [961].

A more dramatic example of biological form is the crown of *Hydra vulgaris* (a freshwater polyp) shown in Fig. 7.11, which, Thompson observes, is similar in shape to the milk-drop coronet that was discussed in Sect. 6.8.6 (Fig. 6.16).[19] Named for a fell monster of Greek mythology whose nine heads would immediately regenerate after being lopped off – thereby taxing the ingenuity of Hercules – *Hydra* is a tiny freshwater cousin of the jellyfish. With stinging tentacles and a tubular body, comprising two layers of cells, *Hydra* boasts a ring of nerve cells just below the tentacles, one of the earliest neural nets to evolve some 700 million years ago [656]. The foot at the base of its body is usually attached to vegetation, leaving the tentacles free to sting and capture passing prey.

---

[19] I feel close to this little creature, as it has long been studied at the *Laboratorio di Cibernetica* near Naples, Italy, where I often visited during the 1970s (see Sect. 5.2.1).

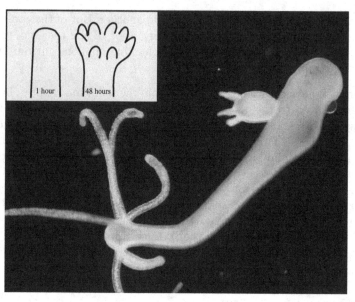

Fig. 7.11. Photograph of *Hydra vulgaris*. The body is about 0.5 mm in diameter. (Courtesy of Paola Pierobon.) The *inset* is sketched from [447]

Interestingly, *Hydra* shares the ability to regenerate lost limbs with its Attic namesake, as is indicated by the inset to Fig. 7.11. Sketched from a photograph in reference [447], this inset shows the regrowth of tentacles two days after being cut in half, which is not unlike the milk-drop coronet in Fig. 6.16. Although it would "seem venturesome", in Thompson's words, to claim a theoretical connection between these two phenomena, "there is something to be learned from such a comparison". Referring to Arthur Worthington's seminal studies of falling liquid drops [1056, 1057], Thompson asserts that many aspects of *Hydra* are "features to which we may find a singular and striking parallel to the surface tension phenomena of the splash". Among others, these parallels evidently include the cylindrical shape of the body, and the "jet-like row of tentacles" that grow from the upper edge of the living cylinder, but how far do these comparisons go? What is the connection?

Tubes are no surprise, of course, as they appear widely throughout both the physical and the biological realms, but the cylinder in Fig. 6.16 is formed by an outwardly propagating ring soliton, whereas the cylinder of *Hydra* is a protoplasmic structure that has evolved to satisfy its biological needs; thus there is little theoretical connection between them. Making a parallel between the jets of the milk-drop crown and the growing tentacles of *Hydra* seems at first even more tenuous. The liquid jets, as we have seen in Sect. 6.8.6, can be numerically computed as solutions of the Navier–Stokes equations of fluid dynamics [1063], whereas the genome of *Hydra* plays a role in the formation of its tentacles [447,548,570,656]. Has Thompson's ample imagination run away with him here? Is physical science failing as "our only teacher and guide"?

From the perspective of many experimental biologists, the answer to this question is an unqualified "yes". Physical scientists are stepping beyond appropriate professional boundaries by presuming to consider (say) the regrowth of tentacles on a severed *Hydra*, because this problem is far more complicated than they imagine. As is slowly being revealed through many careful experiments, growth processes are structured around *signaling pathways*, in which genes that are "turned on" in one cell produce *morphogens*, which diffuse to nearby cells, causing the same or other genes to turn on, producing other morphogens, and so on. Although these pathways are not yet understood for *Hydra*, it is claimed they must be before the entire process of generating a global morphology can be explained [447,548,656]. Hans Meinhardt disagrees with this perspective.

Trained as an experimentalist in high-energy physics, Meinhardt turned in the late 1960s to theoretical biology and began collaborating with Alfred Gierer, an experimental biologist with a group working on *Hydra* at the Max-Planck Institute for Developmental Biology in Tübingen [368]. In 1972 – at the dawn on the nonlinear science revolution – these two scientists published a paper which developed the mathematical ideas from Alan Turing's now famous paper of 1952 (see Sects. 4.3 and 7.2.4) in a biological context and then applied them to pattern formation in *Hydra* [355]. Surprisingly for this time, they showed that a pair of coupled nonlinear diffusion equations (for activator and inhibitor concentrations) successfully modeled the onset of tentacle regrowth from a smooth structure and was in accord with available biological evidence.[20] From a variety of empirical studies, it turns out that nonlinear diffusion as originally proposed by Turing does indeed serve as a "minimal model" for the diffusion of morphogens through signaling pathways [650,651].

This is not to claim that it is unnecessary to work out the genetic bases for the signaling pathways – for it certainly is – but the situation somewhat parallels that in electrophysiology, when Alan Hodgkin and Andrew Huxley published their classic paper (also in 1952) on nerve-pulse propagation (see Sects. 4.1 and 7.4.1 and [440]). At that time, the biomolecular basis for the sodium and potassium ion currents that activate and inhibit transmembrane voltage were not known, so this was also a "minimal model" based on a global theory of nonlinear diffusion. In contrast to the reception of the developmental biology community to Turing's proposal, however, the neuroscience community recognized the importance of the Hodgkin–Huxley work, rewarding them with Nobel Prizes and using their model in many global empirical studies. Of course it is interesting and important to learn how membrane proteins are organized to form ionic pores, and much progress has been made in this area over the past half century [436], but it would have been a poor

---

[20] "Surprisingly" because this important paper was turned down by three biological journals [368].

research strategy for electrophysiologists to wait until this area of science was completely understood before employing nonlinear-diffusion formulations to study nerve-pulse dynamics [874]. Similarly, it is a poor research strategy for developmental biologists to continue to insist that understanding the growth of form must wait until the details of their efforts are completely sorted out. We will return to this vexing question in Sect. 7.2.7.

Armed with these insights, let us return to Thompson's assertion – made in 1917, recall – that "there is something to be learned" from comparing the growth of fluid jets on a milk-drop coronet (Fig. 6.16) with the regeneration of tentacles on a severed *Hydra* – dynamic processes that differ by about five or six orders of magnitude in their time scales. In both cases, a circumferential wave instability leads to patterned growth; thus a mathematical link between these quite different dynamic processes may be found upon examining the instability equations that govern the onsets of growth: fluid jets in one case and fleshy tentacles in the other. As René Thom has pointed out in the context of catastrophe theory (Sect. 5.2.5), there is only a limited number of ways that a mathematical system can become unstable [928, 959, 1082], and the instability class could be the same in these two cases.[21] Because Meinhardt has been concerned with the global features of developmental biology, therefore, a new branch of this study has opened, which fits comfortably within nonlinear science.

Accepting that *Hydra* is alive and the milk-drop coronet (Fig. 6.16) is not, how can we decide what shapes contain Life? Although living organisms have not yet been – and may never be – constructed from lifeless constituents in a laboratory, a system need not be alive for strikingly life-like forms to emerge. This was shown by Stéphane Leduc, a pioneering biophysicist who developed a "synthetic biology" over a century ago [547] – during the same years that Worthington was studying the shapes of splashes. In the chemical garden shown in Fig. 7.12, which appeared as the frontispiece to his book *The Mechanism of Life*, a solution of sodium silicate (commonly known as water glass) is seeded with salt crystals, whereupon a sequence of reaction-diffusion phenomena involving chemical reactions, osmotic forces and buoyancy results in the observed growth [160]. Although Leduc overstated the case for the relevance of these effects in present-day biology [484], his life-like shapes may have played roles in the emergence of Life from the hot chemical soup of the early oceans – perhaps by helping to form a physical environment in which the protoenzyme of Careri and Wyman (discussed near the end of Sect. 7.1.3) could have prospered.

Employing a variety of nonlinear physical and chemical phenomena – some of which may have been known to Isaac Newton in the course of his alchemical studies [359, 705] – Leduc's life-like creations were the subject of intense discussions among physical and biological scientists during the early decades

---

[21] An interesting thesis problem in mathematical biophysics would be to study and compare these two instabilities from the perspective of catastrophe theory.

Fig. 7.12. A chemical garden based on sodium silicate (water glass) [547]

of the past century, in part because it was not then clear to the biological community whether Life is entirely based on the laws of physics and chemistry. In accord with Henri Bergson,[22] vitalists believed that a "Life force" (*élan vital*) acting beyond these laws was needed to explain the empirical evidence [78], while physicalists asserted that the laws of physical science will eventually offer a sufficient basis for biology.[23]

According to the physicalists, then, how did Life on Earth begin? At the birth of Life some four billion years ago, they would claim, all of Leduc's nonlinear chemical reactions and others that have and have not yet been discovered plus the broad spectrum of nonlinear physical phenomena associated with highly energetic splashing were available to act as midwives (see Sect. 6.8.6 and [1056, 1057]), including but not limited to bubbles, antibubbles, (Fig. 6.17 and [45]), vortices, and foam, and other features of turbulence. Out of this literally unimagined physical and chemical turmoil – it is claimed – there emerged a handful of protobiological molecules that could reproduce themselves with variations and thus embark on the road to Life [269].

Since the tree of Life took root and began to evolve, the process of natural selection – discovered independently and jointly announced by Charles Darwin and Alfred Wallace in 1858 [130, 209] – has been an efficient cause of incremental shape changes, but the Darwin–Wallace theory is qualitative. From Thompson's mathematical studies, incremental evolutionary changes can be described by nonlinear *topological transformations* of which he gives many examples. (To grasp the nature of a topological transformation, imagine that the shape of a creature is drawn on a rubber sheet, which is then distorted until the original shape matches that of a related creature. The mathematical

---

[22]Bergson received the 1927 Nobel Prize in Literature.

[23]The original interest in Leduc's work by the biological community and its later demise have been thoughtfully discussed by Keller in a recent book [484].

transformation between the original and final sheets then defines a mapping from one shape to another.)[24] Using such transformations, Thompson has shown how the forms of various biological entities – fish, mammalian skulls, shellfish, etc. – are related to one another, thereby obtaining quantitative descriptions of the actions of evolution. Interestingly, the Poincaré conjecture, which was recently proven by Grigori Perelman, asserts that such topological mappings can also be done in three dimensions [842] – a result that is evidently important for the application of Thompson's ideas to biological structures.[25]

One of these nonlinear mappings is shown in Fig. 7.13, where the fossil record provides the shapes of the pelvic bones of earth-bound *Archaeopteryx* (a late dinosaur) and of flying *Ichthyornis* (an early bird) during the Cretaceous period (about 100 million years ago). From mechanical considerations, the walking creature needs more vertical support than the flying one, as is seen from the shapes of their respective pelvic bones (1 and 5). Determined by Thompson through a mathematical process of interpolation between the topological transformation from *Archaeopteryx* to *Ichthyornis*, the intermediate stages (2, 3, and 4) are theoretical predictions of fossils yet to be discovered.

From the perspectives of nonlinear science, evolutionary changes in the shapes of organisms are jointly caused by influences of both natural selection and the ever-present laws of physics and chemistry. Thompson offers several examples of physical influences, relating the skeleton of a four-legged animal to the shape of a cantilever bridge, the human femur to the shape of a crane head, and the metacarpal bone in a vulture's wing to a truss bridge, among others. Many more examples have been provided by Boye Ahlborn in his recent book *Zoological Physics*, which evolved from notes for a course developed for and presented jointly to students in zoology and physics [10]. As this book assumes little knowledge of mathematics and explains all of the details, it should be of great interest to those biologists who would understand how the laws of physics have influenced the course of Life's evolution.

In the Aristotelian language of Chap. 1, therefore, we can say that *efficient* causes of shape changes act through evolutionary pressures and are mediated in detail by changes in genetic (DNA) codes and at a more general level by reaction-diffusion models, whereas the laws of physics (gravity and mechanics in the case of Fig. 7.13) can be described as *formal* causes of morphogenesis. Both of these types of causes are highly nonlinear, as is most of biological dynamics.

---

[24]Whereas linear topological transformations merely describe uniform stretchings in one or more directions, nonlinear topological transformations induce arbitrary global distortions while remaining locally smooth.

[25]In August of 2006, Perelman refused to accept a Fields Medal for this work.

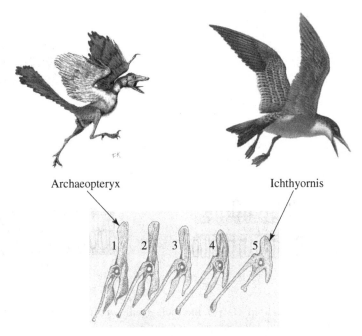

Fig. 7.13. Transformation of the pelvic bones between *Archaeopteryx* and *Ichthyornis*. (The bird images are reproduced with permission of T. Michael Keesey and the pelvic bones are from Thompson [961])

7.2.2 Biological Membranes

When teaching about cell membranes, I like to begin with a short discussion of the "black" soap film, which was known to Isaac Newton in the course of his alchemical studies [359, 706]. Such a film is called black because it doesn't reflect light, and it doesn't reflect light because it is much thinner than a wavelength of light, and it is so thin because of its molecular bilayer structure, as is the membrane of a living cell. Interestingly, this bimolecular property follows directly from energetic considerations.

Although more sophisticated procedures are perhaps appropriate for lecture demonstrations [131], a black soap film can easily be made and observed in the kitchen, with nothing more than a small length of thin wire [965]. Here's what you do. Put some warm water in a glass with a squirt of liquid dish soap, and then bend the end of the wire into a small loop, about a millimeter or so in diameter. Dip the loop into the water to form a thick soap film and watch this film under a bright light. (If you are nearsighted like me, you can hold the loop close to your eye, otherwise a small magnifying lens might be helpful.) Soon colored bands are seen (indicating that the film thickness has reduced to the order of a wavelength of light), and after a

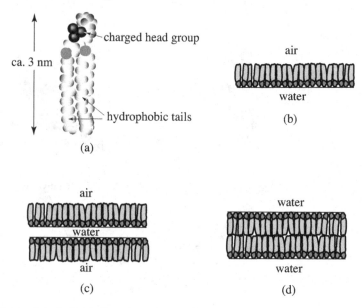

Fig. 7.14. (a) A lipid (fatty) molecule (redrawn from Goodsell [366]). (b) A monomolecular lipid layer on a water surface. (c) A "black" biomolecular soap film. (d) A biological lipid bilayer

minute or so black spots appear in the film, which swirl around and become larger, eventually filling the entire wire loop. That's all there is to it![26]

The energetics of a black soap film and of a cell membrane can be understood with reference to the structure of an oily (lipid) molecule, shown in Fig. 7.14a. This molecule comprises a relatively long *hydrocarbon tail* with a charged *head group* attached to one end. In air, the tail appears inert, but there are electrical field lines emanating radially from the head group to create an energetic electric field, just as with the model electron that we considered in Sect. 6.1.1. This external field energy is greatly reduced if the head group is close to water, because the dipolar molecules of water rotate themselves to cancel the field. Thus if a thin lipid film is spread onto the surface of water, the molecules will orient themselves as in Fig. 7.14b, with heads down and tails up, because this arrangement has the lowest possible energy. The black film that (I hope) you have just made consists of two such orientations, one down the other up, with both sets of heads facing a thin layer of water, as shown in Fig. 7.14c.

In the early 1960s, Paul Mueller, Donald Rudin, Ti Tien, and William Wescott showed the scientific world that a biological lipid bilayer can be reconstructed using the apparatus shown in Fig. 7.15 [683, 965]. In this ex-

---

[26]I have read that such black spots can form at the top of a soap bubble, but I've never observed this.

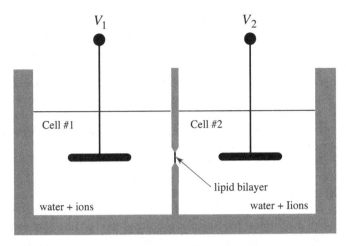

Fig. 7.15. An experiment for making physical measurements on artificial cell membranes. (The two electrodes are connected to voltages V_1 and V_2)

periment (which Newton *didn't* try!) there is a small hole between the two chambers that can be covered with a lipid bilayer using a technique that is similar to your formation of a bilayer soap film. Touching the hole with a camel's hair brush that has been dipped in lipid will first give a relatively thick film, which soon collapses into the energetically more favorable bilayer of Fig. 7.14d.[27] Again, the lipid molecules are arranged to minimize total energy, but now with their head groups facing outward, toward the water. Thus it is possible to construct artificial membranes just like those of our bodily cells, and the fact that these membranes can be so readily constructed supports the speculation of the previous section that Life emerged in the turbulent Hadean oceans.

During experiments on artificial membranes, the two chambers of the vessel in Fig. 7.15 are held at different electric potentials (V_1 and V_2) and they are filled with aqueous solutions containing different ionic concentrations, allowing biophysicists to study the mathematical relationships among transmembrane voltage ($V_1 - V_2$), concentrations of various ions, and transmembrane ionic currents – the key players in cell membrane dynamics [874]. These studies have included visual observations of bilayer formation, quantitative measurements of electrical conductivity and capacitance of a lipid bilayer, measurements of resting potential as a function of ionic concentrations, and observations of the influence of intrinsic (embedded) membrane proteins on membrane properties. Three fruits of these studies are the following [465]. First, the electrical capacitance of a lipid bilayer is about one microfarad per square centimeter, which agrees with both theory and measurements on

---

[27] A short film showing the formation of a lipid bilayer is available on the Internet at www.msu.edu/user/ottova/soap_bubble.html.

the membranes of living cells [440]. Second, the electrical conductivity (or ionic permeability) of a pure lipid bilayer is very small, corresponding to that of a good insulator such as quartz. Finally, membrane permeability to ionic current flow is very sensitive to the presence of intrinsic proteins. If certain proteins are dissolved in the lipid bilayer, in other words, membrane conductivity increases by several orders of magnitude, proving that ionic current is carried across cell membranes by proteins. These observations lead directly to the battery–resistor model of a cell membrane that was used in Fig. 7.3 to describe active transport.

In the biophysics of a cell membrane, therefore, we find confirmation of Thompson's claim that "the laws of energy, the properties of matter and the chemistry of the colloids" can explain much of Life's dynamics. Indeed, with a proper choice of embedded membrane proteins, even the switching action of a nerve membrane has been reproduced [684]; thus this essential feature of an animal organism can be constructed from its chemical components, a point we will revisit in Sect. 7.4.1. Finally, it is expected that experiments using apparatus similar to Fig. 7.15 will help biophysicists sort out the nature of active transport across cell membranes – as mediated by intrinsic proteins like the Ca-ATPase in Fig. 7.4 – perhaps leading to an understanding of the role played by solitons in this essential process.

7.2.3 Leonardo's Law

Among the many gems in the notebooks of Leonardo da Vinci is his comment that:

> All the branches of a tree at every stage of its height when put together are equal in thickness to the trunk below them.

This seminal observation was followed by the admonishment [804, 1029]: "Let no man who is not a Mathematician read the elements of my work." Accordingly, the widespread phenomenon of tree formation can be described by a general *branching law*, which gives the relation between the diameter d_0 of a *parent branch* and $d_1, d_2, \cdots, d_n$ of *daughter branches* as

$$d_0^\Delta = d_1^\Delta + d_2^\Delta + \cdots + d_n^\Delta , \qquad (7.6)$$

where Δ is a *branching exponent*. Figure 7.16 shows the geometry for two daughter branches of equal diameter.

Fluid flow through a branch is expected to be problematic if the daughters are too small in relation to the parent, an effect that we have all experienced when a wide auto highway divides into smaller ones. With Leonardo, one might expect the flow through a branch to be most efficient when the total area of the daughters is equal to that of the parent, which implies $\Delta = 2$; so two equal daughters would have diameter

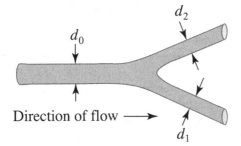

Direction of flow ———→

Fig. 7.16. A branching fiber

$$d_1 = d_2 = \frac{d_0}{\sqrt{2}} = 0.707 d_0 \ .$$

For the bifurcation of our main abdominal aorta into two daughter arteries, Thompson records measurements that indicate a daughter-to-parent diameter ratio of 0.7 ± 0.02, in accord with the equal-cross-section assumption that $\Delta = 2$ in (7.6). For much of our arterial system, however, the optimum ratio of daughter-to-parent diameter is larger than that implied by $\Delta = 2$, because – from the Hagen–Poiseuille law of fluid dynamics (see Sect. 6.8.8) – fluid flow resistance is proportional to $1/d^4$. As the arteries get smaller, in other words, flow resistance increases, and the combined area of the daughters must exceed that of the parent in order for blood to flow smoothly into them.

Inspired by Thompson in 1926, biologist Cecil Murray defined and used a *principle of minimum work* to optimize the main factors influencing oxygen transport from heart to muscles in humans. If the blood vessels are too small, he showed, then the heart must work too hard to pump the blood through them, but if the vessels are too large, the body must produce more blood than is needed [686, 687]. Balancing these two factors implies an optimum rate of blood flow[28]

$$f \propto d^3 \ ,$$

which leads to (7.6) with $\Delta = 3$. With this branching exponent, (7.6) is called *Murray's law* in the physiological literature, and it tells us that if a parent bifurcates into two equal daughters, then

$$d_1 = d_2 = \frac{d_0}{\sqrt[3]{2}} = 0.794 d_0 \ .$$

An analysis published by Rashevsky in 1960 confirms Murray's law for all branches of the human arterial system except the largest [790], as does a

---

[28] Assuming no turbulence and (6.15), the power (energy per unit time) expended per unit length of tube is $f^2 \eta / 2\pi d^4 + B d^2$, where the first term accounts for viscosity (η) losses and B is a constant that measures the energetic cost of maintaining the blood per unit volume and per unit time. Minimizing this sum implies that $f = d^3 \sqrt{(\pi B / \eta)}$.

recent study by environmentalist Page Painter and physicists Patrik Edén and Hans-Uno Bengtsson [739]. Empirical bases for the many arterial branching studies have been reviewed by biologist Thomas Sherman, who also proves that branching according to Murray's law gives the least global resistance for a fixed volume [900].[29]

Returning to the Plant Kingdom which inspired Leonardo da Vinci, Murray devoted the summer of 1926 to a study of nine types of botanical trees (ash, aspen, beech, butternut, cedar, hickory, hornbeam, oak, and maple), comprising 116 trees in all which were randomly selected from the neighborhood (Grindstone Island, New York), rejecting only those that appeared to have been recently injured. Measurements were made from trunks to twigs, with diameters ranging from 9 cm (trunks) down to 0.4 mm (stems of leaves). From these data he found close accord with (7.6) if $\Delta = 2.49$ [689]; thus for bifurcations into two equal daughters,

$$d_1 = d_2 = \frac{d_0}{\sqrt[2.49]{2}} = 0.757 d_0 \ ,$$

which approaches the equal-area ratio.

Yet closer to Leonardo's original observation is the work reported in 1964 by Kichiro Shinozaki, Kyoji Yoda, Kazuo Hozumi and Tatuo Kira, forest ecologists whose imaginations were not tied into the branching concepts associated with Murray's law [901, 902]. Seeking a means for estimating the biomass in a forest, they made extensive measurements of the supporting and leaf structures in a variety of trees, finding that:

> The amount of leaves existing above a certain horizontal level in a plant community was always proportional to the sum of the cross-sectional area of the stems and branches found at that level.

This observation led to their *unit-pipe model* of a tree, where each pipe sustains and provides mechanical support for a certain amount of leafy superstructure. On the other hand, biologists Katherine McCulloh and John Sperry and biomathematician Frederick R. Adler have recently observed that water transport systems in vines follow Murray's law ($\Delta = 3$), which they claim is expected for conduits that do not provide mechanical support [635], and more recently they have found agreement with Murray's law in branchings of whisk fern [636]. Evidently there is more to be done on the botanical problem suggested by Leonardo some five centuries ago.

---

[29]The basis for using the term Murray's law for (7.6) with $\Delta = 3$ is uncertain. In his first 1926 paper [686], often given as a reference for this designation, Murray shows that optimum flow is proportional to d^3, but he doesn't state (7.6). In a later paper [688], he mentions the branching condition $d_0^3 = d_1^3 + d_2^3 + \cdots + d_n^\Delta$ as a result that was "obtained by Miss E. Hendee and Miss M.E. Gardiner in this laboratory." In Thompson's book, this formulation is described as [961]: "an approximate result, familiar to students of hydrodynamics."

In 1962, physicians Ewald Weibel and Domingo Gomez published a detailed empirical study of branching in five human lungs (ages 8 to 74), finding bifurcations with $\Delta = 3$ [1021]. A few years later, Theodore Wilson, a mechanical engineer, independently developed an argument similar to that of Murray [686] to show that $\Delta = 3$ in the bronchial tree in accord with Weibel and Gomez [1047]. Branching in mammalian lungs has been reviewed by physicists Bruce West and William Deering [1030], who note that mathematician Benoit Mandelbrot has constructed a fractal model of a lung with $d_1 = d_2$ at each stage of the tree and $\Delta = 3$ [599, 1029]. Although air is cost free for an organism, Sherman has pointed out, one should also expect Murray's law to hold for the lungs because the cross-sectional area of the tube walls is proportional to their total cross section [900]. Finally, it has recently been shown that Murray's law holds also for the vascular trees of our kidneys [721].

All of the above examples (arteries, botanical trees, lungs, and kidneys) involve fluids (blood, sap, air, and urine), and they all seem to evolve toward morphologies in which the energetic cost of producing or filling large cylinders is balanced by the metabolic energy needed to push fluids through small tubes, leading to (7.6) with $2 \leq \Delta \leq 3$. There is another branching structure in animal organisms that works on an entirely different principle – the nerve fiber, which was introduced in Chap. 4.

Although a parent nerve fiber (incoming dendrite or outgoing axon) often branches into daughters as in Fig. 7.16, what moves through the branch is not fluid but a localized pulse of electrical activity, as shown in Fig. 4.1; thus the optimum morphology of the branch is expected to differ from Murray's law. Surprisingly, (7.6) still holds but with a smaller branching exponent. There are two ways of seeing this, one from the perspectives of linear dynamics and the other from nonlinear dynamics.

The linear approach was proposed in 1959 by Wilfred Rall, a neuroscientist who was motivated by the problem of signal transmission through dendrites. As dendrites were considered to be passive electrical structures in those days, he viewed optimum branch design as a problem of admittance matching, which is familiar to the communications engineer [783–785]. Because the *characteristic admittance* of a passive fiber is proportional to the 3/2 power of its diameter, it follows that a branch will be optimally designed for signal transmission if (7.6) holds with $\Delta = 3/2$, which makes the admittance of the parent equal to the sum of the admittances of the daughters. This matching of parent to daughters is called *Rall's law* by electrophysiologists. Under it the bifurcation of a parent fiber into two equal daughters would have

$$d_1 = d_2 = \frac{d_0}{\sqrt[1.5]{2}} = 0.630 d_0 ,$$

which is significantly smaller than the ratios found for optimum transmission of fluids through a branch.

From a nonlinear perspective, nerve pulse propagation is an active process as described in Chap. 4, so it is important to consider the *threshold condition* that must be met in order to establish a pulse on a daughter fiber. Interestingly, this condition requires a certain amount of electrical charge, which in turn is proportional to the 3/2 power of the fiber diameter, and the charge carried on the leading edge of a pulse on the parent fiber has the same dependence [863, 874]. Optimum branch design, therefore obtains when the leading-edge charge carried into the branch by a pulse on the parent fiber is just equal to the sum of the threshold charges of the daughters, implying (7.6) with $\Delta = 3/2$.

Presumably Rall's law would be selected by evolution whenever optimum transmission through a branch is desired. To quantify this optimum design condition, the concept of a *geometric ratio* (GR) was defined by Rall (for two daughters) as [783–785]

$$GR = \frac{d_1^{3/2} + d_2^{3/2}}{d_0^{3/2}} . \tag{7.7}$$

For most efficient pulse transmission through a branch, one expects $GR = 1$, because if the GR is too large, pulses won't be launched on the daughters, and if it is too small, energy will be wasted.

As a test of this speculation, GRs have been calculated for the first branchings of 109 giant axons of the squid (*Loligo vulgaris*, taken from the Bay of Naples between December of 1979 and May of 1980), with the result [893,894]:

$$GR = 1.017 \pm 0.029 .$$

Similar observations were made by neuroscientist Dean Smith, who measured

$$GR = 1.01 \pm 0.15$$

on branchings of excitor motor axons in the crayfish (*Orconectes virilis*) [913]. The fact that these two independent observations both give $GR = 1$ within experimental uncertainty, provides substantial support for Rall's law where efficient pulse transmission through a branch is desired.

In Sect. 7.4.2, we will consider cases in which branches are performing digital logic in dendritic trees; thus GRs significantly greater than unity are needed, and Rall's law does not apply. Yet another situation where Rall's law seems not to hold is in axons of the central nervous system, where speed of transmission through the branch is a major design consideration. Using a minimization analysis related to that of Murray, neuroscientists Dmitri Chklovskii and Armen Stepanyants have recently balanced propagation delay against volume, finding Murray's law ($\Delta = 3$) and offering some empirical confirmation [686].

Finally, a note about nomenclature. Considering that physiologists use the term "Murray's law" for (7.6) with $\Delta = 3$ but not for other values of Δ [900]

and the electrophysiologists use the term "Rall's law" for $\Delta = 3/2$, there is need for a term to designate (7.6) with an arbitrary branching exponent. For this more general usage, the term "Leonardo's law" is hereby proposed.[30]

7.2.4 Turing Patterns

Turing's now-famous paper on biological morphogenesis [973] was mentioned in Sects. 4.3 and 5.2 in connection with the growing awareness during the 1970s of reaction-diffusion processes among the nonlinear science community, and in Sect. 7.2.1 for stimulating Meinhardt and Gierer to develop a numerical model for the regrowth of tentacles on a severed *Hydra* [355]. Like Lorenz's now-famous chaos paper (see Fig. 1.4), which was neglected for over a decade and then discovered during the mid-1970s, Turing's publication has had a curious citation history, as we see in Fig. 7.17.

After being ignored by the science community for a decade and a half, Turing's approach to biological morphogenesis was noticed around 1970, after which the annual number of citations has continued to grow, until presently it is one of the most highly-cited paper in all of nonlinear science. But who is reading it? Not developmental biologists, it seems, for as Richard Lewontin has recently pointed out [556]:

> The irony is that while Turing's model turns out to be correct in its simple outline form, it did not play a significant part in the production of modern molecular developmental biology. Developmental geneticists now study how the cell's reading of different genes produces spatial patterns of molecules in embryos, but that detailed and messy description, involving large numbers of genes and proteins, owes nothing to Turing's model.

While Lewontin's judgement is surely correct in describing activities in his field, he seems unaware of the role that approximate theories can play in developing scientific understanding of natural phenomena. As Rashevsky pointed out in the 1930s [790], the science of statistical mechanics was developed under the assumption that atoms are little spheres, although everyone knew this was not so, and the reaction-diffusion theory for nerve pulses that was developed by Hodgkin and Huxley also in 1952 assumed nonlinear properties for squid membrane that had no theoretical basis at the time [440] – but that's the way reductive science works. Chemistry would be in sorry shape if chemists had decided to wait for the high-energy physicists to develop a complete theory of the elementary particles so atomic structure could be put on a sound basis before beginning their work, and developmental biologists quite properly are not sitting on their metaphorical hands until chemistry is a

---

[30]Note, however, that this general law differs from Leonardo's incorrect claim that the speed of a falling body is proportional to time, which has also been called Leonardo's law.

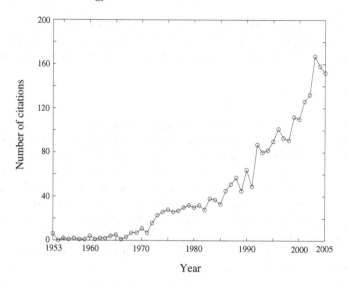

Fig. 7.17. The number of citations per year to Turing's 1952 paper on biological morphogenesis [973]. (Data from Science Citation Index Expanded)

finished science. The efforts of all scientists – from string theorists to cultural anthropologists – involve making assumptions concerning the nature of entities at lower levels of description, theorizing about how these might interact, and comparing the conclusions of their theories with empirical evidence. We will return to these matters in a later section, for our present concern is to understand who *were* writing the papers indicated in Fig. 7.17, and why.

Of the 2468 citations to Turing's paper through September of 2006, the greatest number (152) appeared in the *Journal of Theoretical Biology* (no surprise), but the second largest number (115) were in *Physical Review E*, a journal that was launched by the American Physical Society in 1993 to accomodate the avalanche of new manuscripts in nonlinear science. Overall, physics journals have published at least 19% of the references to Turing's paper – the largest fraction for any professional field. On the other hand, only one reference appears in the *Cold Spring Harbor Symposia in Quantitative Biology*, which is surprising as the stated aim of these symposia is "to consider a given biological problem from its chemical, physical and mathematical, as well as from its biological aspects" [403], but this lack of interest accords with Lewontin's appraisal of the attitude of developmental biologists toward morphological theories not grounded in genetics.

Carefully written for a biological audience, Turing's paper on morphogenesis is highly original, as is all of his work [439], introducing reaction-diffusion phenomena into biology at a time when it was scarcely appreciated in any area of science. Briefly, he considered the presence of two *morphogens* in an organism, where one morphogen is an activator and the other is an in-

Fig. 7.18. A dappled Turing pattern (drawn after a figure in [973])

hibitor [973]. These morphogens represent *form producers* or *evocators* in the theory [1009], which by diffusing into a tissue "somehow persuade it to develop along different lines from those which would have been followed in its absence". If the diffusion constant for the inhibitory morphogen is smaller than the activator diffusion constant, there can be nonlinear waves of activity, as is the case for the Hodgkin–Huxley theory of nerve pulses which was also published in 1952. If the inhibitory diffusion constant is larger, on the other hand, inhibitory morphogens can move out faster and surround activator morphogens, leading to the dappled pattern of stationary regions shown in Fig. 7.18. Stationary patterns evidently serve as models for the coloration of animals, like (say) a Holstein cow or the wings of a butterfly, and following Turing's suggestion, such a formulation was used by Gierer and Meinhardt to model the regrowth of *Hydra* tentacles [355] (see Sect. 7.2.1).

Although readily demonstrated on a computer, it was not until the 1990s that stationary Turing patterns were reproduced in the laboratories of physical chemists [165,554]. An explanation for this delay is that in chemical solutions all molecules tend to diffuse at about the same rate, making it difficult to establish conditions under which inhibitory molecules diffuse significantly faster than activators. In successful laboratory demonstrations of stationary Turing patterns, activator molecules were eventually bound to large color indicators, significantly reducing their diffusion constants [110].[31]

---

[31] Oddly, the lore has developed that Turing's theory of morphogenesis applies exclusively to the development of stationary patterns, in which inhibitory morphogens diffuse significantly faster than activator morphogens – a misapprehension that arose during the surge of interest in Turing's formulation by the physics community during the early 1990s. Coming into nonlinear science during the 1980s, a full decade after the revolution described in Chap. 1, physicists may have felt a need to define areas that they could dominate. I recall commenting in a session on Turing models at a physics conference during the 1990s that Turing's problem is similar to that introduced by Hodgkin and Huxley, also in 1952. "No, no," was the

In developmental biology, which is the real problem that Turing took up in 1952, the morphogens are enzymes (proteins) which diffuse from one cell to another, activating or inhibiting genes to produce other morphogens, allowing much more freedom in the realization of different diffusion constants than in the aqueous solutions of physical chemistry. Thus a challenge for modern molecular biologists, as they develop their "messy descriptions, involving large numbers of genes and proteins", is to uncover situations in which activator morphogens diffuse more slowly than inhibitors and then to explore – as in the Gierer–Meinhardt model for regrowth of *Hydra* tentacles [355] – the dynamic implications of this difference.

7.2.5 Buridan's Ass, Instability and Emergence

We saw in Sect. 7.2.1 that morphogenesis begins with an instability, after which a new form appears, usually with a higher (more restricted) symmetry than the lower symmetry of the environment from which it emerges. In the severed *Hydra* body of Fig. 7.11, for example, there is continuous rotational symmetry about the body axis after 1 hour, but symmetry only upon rotation by 45 degrees after 48 hours, as 8 equally spaced tentacles have appeared. Similarly, the milk-drop coronet in Fig. 6.16 repeats upon rotation by about 360/21 degrees, corresponding to the 21 jets in Fig. 6.16.

A classical scenario of something new *not* coming about is the stupid ass of fourteenth French cleric Jean Buridan, which stands *exactly* between two *identical* piles of hay. When it thinks of eating the first pile, the second pile seems more appetizing, and when it thinks of eating the second, the first looks better, so the poor beast dithers to death, unhappily preserving the original (lower) symmetry of the problem [455]. (Buridan's metaphor goes back to Aristotle's discussion of the natural location of Earth, where he writes of a man who "though exceedingly hungry and thirsty, and both equally, yet being equidistant from food and drink, is therefore bound to stay where he is" [32].)

From the perspectives of nonlinear science, the emphasis given by philosophers to Buridan's ass and to Aristotle's hungry man is like worrying about a sharp pencil that is balanced *exactly* on its tip: How can the lower rotational symmetry of the upright pencil be broken? This misplaced concern shows that the thinking of many remains guided by linear dynamics, according to which a symmetrical solution to a symmetrical system must remain so. Under realistic dynamics, instabilities can appear with growing solutions that have symmetries more limited (higher) than those of the initial conditions. The subsequent growth of these unstable solutions is eventually checked by nonlinear effects, whereupon we say that new entities with higher symmetries

---

immediate response, "the Hodgkin–Huxley problem is very different." But moving patterns may also be useful in morphogenesis.

have emerged.[32] In the case of the ass, one can suppose that it doesn't keep its head exactly still but moves momentarily closer to one of the piles, causing that pile to seem more attractive, causing the ass to move closer to that pile, and so on. Eventually, the original mirror symmetry of the problem is lost as the ass reaches one of the piles of hay, eats it, and belches, happily avoiding starvation.

Instabilities, in turn, arise from *closed causal loops*, as shown in Fig. 7.19, with A causing B and B, in turn, causing A. If one assumes that $B = xA$ and also that $A = yB$, where x and y are numerical constants, then $A = xyA$ and $B = xyB$, so that xy necessarily equals unity, which seems trivial. But this result holds only if the causalities act *instantaneously*. As we have seen in the previous chapter, however, an instantaneous causal response is not possible as this would require an infinite speed of information flow and thus would contravene the special theory of relativity, under which information can move no faster than the speed of light.

In general, therefore, causality must allow for a time delay, which can be introduced by letting x and y be functions of frequency (think of the gain of your stereo amplifier), whereupon Fig. 7.19 refers to each frequency component. The relationship between B and A is then given by the feedback formula that was discovered by young Harold Black on the morning of August 2, 1927 and introduced in (6.10) [127]:

$$\frac{B}{A} = \frac{x}{1 - xy} \, .$$

In this ratio, it is supposed that A is a *cause*, B is an *effect*, and B/A is a *response function* of frequency. For $|xy| < 1$, the right-hand side of this equation is finite and well behaved. With $|xy| > 1$, interestingly, $1 - xy$ can be zero at frequencies that correspond to those of instabilities; thus the variables in a corresponding physical system (voltages, say, or morphogens) may grow exponentially with time, oscillate, or both. As is well understood in the engineering literature and taught to undergraduates, these unstable behaviors correspond to Figs. A.1a, e and c, respectively [95, 528]. For a familiar example from nonlinear science, recall that the van der Pol oscillator, which we met in Sect. 6.6.2, is realized simply by connecting the output of a triode amplifier back to its input.

Although closed causal loops like that of Fig. 7.19 are not strange or even unusual in applied science, many puzzle over them, calling them vicious circles, and the otherwise thorough analysis by philosopher Jon Barwise and mathematician Lawrence Moss doesn't mention applications of their ideas to feedback amplifiers [62]. Yet since the late-1920s, closed causal loops are often employed in engineering practice – to stabilize amplifiers and control servomechanisms (not to mention your furnace and air conditioner) – and

---

[32]In condensed matter physics, this is called the *Curie principle* after Pierre Curie [203].

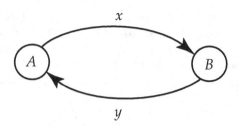

x

A B

y

Fig. 7.19. A closed causal loop

they have been widely employed in biology since the emergence of life, providing an analytic basis for Norbert Wiener's cybernetics (see Sect. 5.2.1). More generally, instabilities stemming from closed causal loops underly the phenomena of chaos (Chap. 2) and emergence, both in energy conserving systems (Chap. 3) and in highly dissipative systems (Chap. 4).

In the preceeding chapters, we have seen many examples of symmetry breaking and the emergence of new dynamic entities, including strange attractors, hydrodynamic solitons, candle flames, local modes in molecules and molecular crystals, shock waves, moduational instabilities in water waves and nonlinear optics, phase transitions in uniform condensed-matter systems, atmospheric vortices, solitons on biomolecules, Turing patterns, and nerve pulses on reaction-diffusion systems, among others. A central point of this book is that such emergent phenomena are new *things*, which should be given appropriate ontological status.

In the next chapter, we will see that reductionists tend to view symmetry breaking and emergence as leading to well-established phenomena – believing with Ecclesiastes that "there is no new thing under the sun" – whereas those of us who would challenge reductionism give emergent entities the same ontological status as the elements of the dynamics from which they emerge. Atoms are as real as protons, neutrons and electrons; molecules are as real as atoms; biomolecules are as real as molecules; and so on. A lynch mob is very real to the guest of honor.

In *Hydra*, we again recognize a mathematical connection between the initial growth of jets in the ink-drop coronet of Fig. 6.16 and the initial growth of tentacles in Fig. 7.11, confirming Thompson's intuition [961]. And as for Buridan's ass, any practical person (an undergraduate student in electrical engineering, for example) would not expect the beast to starve, for a small random motion toward one hay stack or the other would seed exponential growth in that direction. In each of these three cases, an instability appears on a system of lower symmetry, which breaks that symmetry by introducing a pattern that grows exponentially with time and eventually leads to the emergence of jets, tentacles or a sated ass.

Thus morphogenesis may be viewed as an ongoing process through which an evolving entity explores a succession of new stable structures of equal or higher symmetry. Can this viewpoint serve as a formulation for the evolution of Life?

7.2.6 Relational Biology

One of the first publications that approached morphogenics from broad theoretical perspectives is Nicholas Rashevsky's seminal paper "Topology and life: In search of general mathematical principles in biology and sociology," which opened vistas for research in mathematical biophysics in an area that he called *relational biology* [789, 790].

When Rashevsky launched his program in mathematical biophysics at the University of Chicago in the 1930s, he aimed to educate a new type of scientist who would study biological problems using the tools of mathematics and physics. Under this program, he launched nonlinear-diffusion studies of both morphogenesis and nerve impulse propagation, which were to be advanced respectively by Turing and by the team of Hodgkin and Huxley in 1952 [440, 790, 973]. As the examples presented in the present chapter show, Rashevsky's program continues to inspire and grows in importance as ever more precise empirical information on living systems accumulates. With the 1954 publication of "Topology and life", however, Rashevsky took a different tack. He redirected his research to studies of the global organizations of living creatures, seeking a principle that "connects the different physical phenomena involved and expresses the biological unity of the organism and of the organic world as a whole" [789].

While it has been ignored for reasons that Thomas Kuhn understood,[33] this work differs markedly from the detailed studies of individual problems – cell growth and division, nerve pulse excitation and propagation, cardiovascular dynamics, and so on – which had occupied Rashevsky and his colleagues during the 1930s and 1940s and were criticized as being too narrowly focused [8]. In his later efforts, he attempted to understand how complex single-celled creatures could have developed from simple prebiotic molecules and how complex organisms develop from relatively simple embryos, thus relating morphological development to the process of biological evolution.

An organized review of Rashevsky's papers on relational biology is available in the second volume of the third edition of his *Mathematical Biophysics* (published in 1960) [790], and an inspiration for his new departure was Thompson's *On Growth and Form*, which first appeared in 1917 [961]. While expressing confidence in physical science as a guide to unraveling the mysteries of life and describing ways that physical laws can determine biological forms, Thompson devoted several chapters of his opus to the geometrical similarities among related species. In conformity with the theory of evolution, he showed how topological (continuous) transformations can be constructed to relate the structures of mammalian leg bones, human faces,

---

[33]Some reasons for this ignorance are suggested in Sect. 5.2.2 and [8].

crustacea, hydroids, fish, reptile skulls, and mammalian skulls, among others (see Fig. 7.13).

Rashevsky's approach to this problem is based on the concept of a *directed graph* as in Fig. 8.2b. Interpreting the nodes of such graphs as representing functional features of single-celled organisms (contact with food, ingestion, digestion, excretion, locomotion, reproduction, etc.) and using the concept of a topological transformation, Rashevsky proceeds in two directions: he first derives the graph-theoretic implications of differentiating particular features of individual cells (one cell ingests, another digests, and a third excretes) to obtain a specialized multicelled organism, and he then considers how the complex graph of an individual cell might have evolved from a simpler (virus-like) protobiological molecule.

Although it has seemed overly theoretical to some [8], Rashevsky's topological network analysis remains relevant for several studies that arise in relational biology, including the following. First, as the appearance of new nodes is allowed under his formulation, topological network dynamics provide a natural theoretical basis for the concepts of emergence and hierarchical structures that are emphasized throughout this book.[34] Second, the general truth of Ernst Haeckel's widely debated observation that "ontogenesis is a brief and rapid recapitulation of phylogenesis" – which has been discussed in detail by Stephen Jay Gould [369] – can be understood in the context of topological network analysis, because similar transformations are required to go from egg and sperm to the adult of some species as were involved in evolution of that same species from some single-celled ancestor. Third, topological analysis supports the seminal work of Manfred Eigen and Peter Schuster on the emergence of life, according to which a three-level *hypercycle* (cycle of cycles of biochemical cycles) is deemed necessary in the structure of a prebiotic molecule (see Sect. 7.3.3 and [269]). Finally, topological network analyses promise to play roles in studies of cultures, from slime molds through ants and bees to cities and human culture [53, 74, 171, 471].

7.2.7 A Clash of Scientific Cultures?

At this point, a difference should be noticed between the efforts of mathematical biologists and those of developmental biologists to explain the growth of form. On the one hand, Lewontin regards Rashevsky's program at the University of Chicago as "leaving no lasting trace" upon its termination in the late 1960s (see Sect. 5.2.2) and concludes that Turing's reaction-diffusion model of morphogenesis, while "correct in its simple outline form", did not "play a significant part" in modern developmental biology (Sect. 7.2.4 and [556]). When an attempt was made to close down Rashevsky's program in the early

---

[34]In related work, several scientists are currently attempting to formulate relationships among levels of nonlinear dynamic hierarchies in ways that are suitable for mathematical analyses [43, 314, 717, 1007].

1950s, on the other hand, a group of his distinguished colleagues objected in a public letter, pointing out that this program was "the only one of its kind in the world" and "of great interest and importance in [...] biology, clinical medicine, mathematics, psychology, philosophy, and sociology" (Sect. 5.2.2 and [634]). In academe – where such passion is usually reserved for debates over how often the coffee pot should be scrubbed or whether to install a women's restroom on the third floor – this is serious stuff, but what was (or is) driving the spleen?

In her recent book *Making Sense of Life* [484], science historian Evelyn Fox Keller traces in some detail the background of this affair, describing it as a "cultural divide" similar to that observed between science and the humanities by Charles Percy Snow at about the same time [916], but more bitter. Keller sees, in other words, a struggle between two diverse groups of scientific researchers: the "visual culture of molecular embryology" and mathematical biologists, who are physical scientists and so think largely in the language of differential equations. As in any such conflict, there are several facets, including the following.

First, the two subcultures use the term "cause" in quite different ways. Under the Aristotelian definitions of Chap. 1, developmental biologists employ a *material* definition (the cause of a disease is a germ and the cause of a physiological trait is a gene), stemming from older concepts (the cause of his bad luck was a curse and she got her nose from her aunt). Guided by Galileo and Newton, physical scientists have focused their collective attention on *efficient* cause (the ball flies over the fence because it was struck by a bat and the airplane crashed because a wing strut broke), which stems from the lexicon of mechanics [136]. (Although this usage properly avoids the idea that a king can be cursed at his birth by an alignment of Jupiter and Saturn – we have noted in Chap. 1 that physical scientists do accept that a high tide is caused by an alignment of the Moon and the Sun.)

Roughly speaking, developmental biologists study the *what* of morphogenesis, whereas mathematical biologists study the *how*, with both groups largely unaware of the distinction. As we have seen, Turing was prepared to accept that there are some sorts of biomolecules that carry information about shape – call them morphogens or form producers or evocators, or what you will – while his interest was in the dynamical question: How do they get to the required locations in needed concentrations? His critics among the developmental biologists, on the other hand, are primarily concerned with knowing what enzymes are involved, for that is the information that can be related to a genetic code. Meinhardt has recently commented that morphogenesis is like the "legend of the Tower of Babel", which was not completed because the builders could not find a common language [368].

Second, there are some bad feelings between the two groups, which have arisen for several reasons. It is my observation that physicists taken en masse can seem arrogant even to other physical scientists, and there is little doubt that physical scientists believe themselves to be professionally superior to biologists. Physicists are convinced that their studies are most fundamental; thus they are doing the basic research in the context of which biology will someday be understood. As he lacked roots in the US community of physicists, Rashevsky may have been perceived by biologists as a safe target for retaliation; in any case, he bore the brunt of many attacks on his professional competence [8, 484, 485, 556], which in my view are wholly unjustified. In connection with the "hijacking" complaints about soliton researchers looking at DNA from Sect. 7.1.4 [321, 590], Meinhardt has recently commented that [368]: "experimentalists are not very enthusiastic if it turns out that a process was correctly predicted" – a harsh indictment of scientific ethics. All in all, many developmental biologists see the mathematical biologists as a barbarian horde who speak an alien tongue and are trying to colonize biology, whereas physical scientists often feel their pearls of wisdom are not appreciated by the biological hoi polloi.

Even when personal relations are not strained, third, the role of a biomathematician is often perceived differently by members of the two subcultures. Biologists see mathematicians helping with the organization of their data files, undertaking statistical analyses, and making those lovely three-dimensional, multicolored plots that tend to become cover art for journals. For the biomathematician, these are all trivial tasks, suitable for research students, perhaps, but not a basis for serious professional work. Following Stan Ulam's dictum from Sect. 5.2.2, on the other hand, an odd speculation can cause a biomathematician to begin proving a biologically irrelevant theorem, rather than struggling with interdisciplinary research. Mathematical biology, in other words, can fall through cracks in the laboratory floor if both parties don't concentrate on problems of central interest to biological science. Interestingly, the 1952 papers of Turing [973] and Hodgkin–Huxley [440] were both of fundamental biological importance and both involved cutting-edge mathematics, but neither was a cross-disciplinary collaboration. The former was by a mathematician working alone and the latter by two physiologists, but they do suggest that such collaboration is possible.

Finally, of course, there is the matter of money. With no inside knowledge of the attempt to close down Rashevsky's mathematical biology program in the early 1950s, I have nonetheless attended too many interdisciplinary faculty meetings over the past four decades not to suspect that a shortage of resources was a contributing factor in this sorry affair. In general, of course, such problems can be solved by making support available specifically for interdisciplinary programs, as is typically done in centers for nonlinear studies.

The above remarks are intended to be realistic but not discouraging. Although neuroscience is admittedly closer to physical science than developmental biology, I have had excellent personal and scientific experiences in collaborating with life scientists in the neuroscience programs at both the University of Wisconsin and the University of Arizona. Especially valuable –

both for ironing out language problems and for convincing physical scientists how complicated biology really is – have been joint supervision of theses and interdisciplinary courses, which mix faculty and research students from both subcultures.

7.3 Physical and Life Sciences

Empirical studies in the physical sciences over the past four centuries have gathered a wealth of information (on mechanical motion, diffusion of small particles, fluid dynamics, electrodynamics, atomic physics, and physical chemistry) showing how the world works, and this knowledge bears on the dynamics of living organisms in three ways. First, physical science allows us to understand how plants and animals do what they do. Second, this same empirical knowledge places limits on what individual organisms and collections of them can do, and these limits continue to play a central role in guiding the course of biological evolution. Finally, there are aspects of life sciences that are uniquely different from the physical sciences, and these differences are often overlooked by those who would develop science as a unified fabric of knowledge.

7.3.1 Mathematical Biology

Excellent material for university courses in mathematical biology and in mathematical physiology is now available in the books by mathematician James Murray [690] and by mathematicians James Keener and James Sneyd [483], which aim to teach neither mathematics nor biology but the border area between them where many discoveries are waiting to be made. Yet more recently Boye Ahlborn, a physicist, has developed a previously mentioned book for an interdisciplinary course centered on the "crack lines between physics and zoology" [10], not to mention my book on mathematical neuroscience, which stemmed from lecture notes for an interdisciplinary course in "neurophysics" presented at the University of Wisconsin during the mid-1970s and and has recently been brought up to date [865, 874].

For those who would disparage the efforts of Rashevsky to found mathematical biology in the 1930s, it is interesting to compare the content and approaches of these books with his *Mathematical Biophysics* [790]. Although current efforts are fortified by decades of theoretical and empirical investigations, the problems (ionic diffusion, cell structure and dynamics, mitosis, excitation and conduction of nerve pulses, electrodynamics of the central nervous system, morphology, and locomotion) remain the same, as do the various ways that physical scientists simplify biological reality in order to find exact solutions for biologically approximate models.

In addition to presenting the elements of thermodynamics in a manner that can be understood by biologists, the above authors describe several ways in which the laws of physics constrain the forms of living organisms, including the following. From *statics*, skeletal structures are related to the engineering of bridges, as moving bones and muscles are compared to simple machines; *dynamics* helps us see how we walk and ride a bicycle, as does *fluid dynamics* with the swimming of aquatic creatures and *aerodynamics* with the flight of insects, birds and bats, and these fluid studies also help to understand the global design and structure of the circulatory system and the lungs; a knowledge of *acoustics* is essential for understanding the auditory system, as is *optics* for the design of the eye (which gave Darwin such concern); *electromagnetics* helps to understand how fish communicate and birds navigate; *ionic diffusion theory* plus *electrostatics* tell much about how cells achieve homeostasis; and *electrophysiology* puts constraints on the sizes and shapes of nerve fibers – to mention but a few applications of physics and mathematics to biology. Students and teachers who would join this effort now have access to a substantial body of material – much of it closely related to nonlinear science – which forms the core of current research in mathematical biology, is truly interdisciplinary and would have pleased Rashevsky.

7.3.2 Collective Phenomena

The *synchronization* of pendulum clocks – first observed by an ailing Christian Huygens in 1665 (see Sect. 6.6.5) – is a general nonlinear phenomenon which occurs in several different contexts, including generators on electric power grids and the brain's neurons, as was pointed out by Norbert Wiener back in the 1950s [934, 1038]. To understand this phenomenon, note that the simplest model of coupled oscillators consists of two van der Pol equations with a small interaction, and if the free (uncoupled) frequencies of the two oscillators are sufficiently close together, they will "lock" or synchronize, having exactly the same common frequency [983]. It is well known to electrical engineers that synchronization is easier to achieve if the nonlinear parameter [ε in (6.9)] is larger [821], which is the case for biological oscillators. Because they are largely dissipative, in other words, biological oscillators are of the *blocking type* (large ε) rather than the *sinusoidal type* (small ε), so synchronization among them is to be expected.

A striking biological example of coupled oscillations is the synchronized flashing of Asian fireflies, which has been studied by field biologists for many years (Sect. 6.6.5 and [132, 133]). Although this phenomenon has been noted in travel books for centuries, it was greeted with "a strong aura of incredulity or even mysticism" when observations were first published in the scientific literature in the 1930s. Most agree that flashing by males is a means of sexual communication to the females, but the reason for *synchronized* flashing has not yet been determined to the satisfaction of the biological community. Could it be that the male flashings synchronize because it is a natural thing for them to do – like generators on an electric power grid – without there being any significance for survival?

Fig. 7.20. A fairy ring of mushrooms. (Courtesy of John Sorochan)

A related nonlinear phenomenon with which we are all familiar is the flocking of birds, schooling of fish, and swarming of bees, which is now readily modeled on a computers [257, 799]. In such nonlinear models, each element (bird or fish or bee) adjusts its position to be equidistant from its nearest neighbors and it adjusts its direction of motion and speed to be the average of the nearest neigbors. These simple rules allow a large number of creatures to move through space as a single entity – rather like a sociological soliton – with clear savings of energy and effort for the group. As there are eight nearest neighbors in three space dimensions, the number of birds in a flock must be large compared with about eight for this strategy to become effective; thus a dozen or so of migrating geese choose to travel in the well-known vee pattern.

From current nonlinear perspectives, reaction-diffusion systems (Sect. 4.3) exhibit a wide range of collective biological phenomena, including three-dimensional dynamics of cardiac tissue [1049, 1050], intercellular calcium waves during muscular contraction [926], pressure waves in the gastro-intestinal tract [483], slime-mold dynamics [925], and the fairy rings of mushrooms shown in Fig. 7.20 [48].

Often appearing on meadows and lawns after a rainy night, fairy rings were taken by the Celts as empirical evidence for mystical woodland creatures, and stepping inside a ring is still believed by some to bring bad luck. Groundskeepers fear them for another reason: they spoil the grass, whereas nonlinear scientists are delighted to see fairy rings and speculate about their underlying dynamics. Interestingly, the ring of mushrooms is merely external evidence of an underlying reality – a fungal root that is expanding outward under a reaction-diffusion dynamics, leaching nutrients from the outer edge of the circle as it goes. After a rainfall, the mushrooms sprout, giving visible evidence of the location of the fungus. Referring back to Fig. 4.5, the almost perfectly circular shape can be understood as a dynamic property of a reaction-diffusion system, under which an arc with a smaller radius of curvature travels more slowly and get eaten up by an arc with a larger radius of curvature. When two fairy rings collide, they leave behind a characteristic cusp, as is shown for the Belousov–Zhabotinsky reaction in Fig. 4.4. What did the ancient Celts make of that?

One of the most significant medical applications of reaction-diffusion waves is in understanding the three-dimensional dynamics of cardiac tissue, which has been carefully studied by Leon Glass, Valentin Krinsky, Art Winfree, and their colleagues [357, 358, 513, 1049, 1050]. This usage models the dynamics of *fibrillation* within cardiac tissue, wherein the heart muscle appears to quiver locally in rapid, irregular and unsynchronized activity rather than pump blood by a large-radius wave of synchronized contraction as it is intended to do. During fibrillation, in other words, a small-radius pattern similar to that in Fig. 4.6 becomes established in healthy cardiac tissue, making the medical problem dynamic rather than physiological. As no global contraction of the heart muscle occurs, blood ceases to be pumped through the circulatory system, and such ventricular fibrillation causes "sudden cardiac death", wherein a seemingly healthy person dies with little warning.

7.3.3 Population Dynamics

As we have seen in Sect. 1.2, an early biological application of mathematical analysis was by Pierre-François Verhulst to the dynamics of population growth, which led to one of the first solved problems of nonlinear science [778,993]. Verhulst's basic idea was to calculate the rate of growth dN/dt as a function of a population N by assuming proportionality at small values of N, which falls smoothly to zero when the population reaches some limiting value, as in (1.2), and then integrating the resulting ordinary differential equation. We have also seen in Sect. 2.4 that a discrete version of Verhulst's basic equation (2.3), which was studied by zoologist Robert May in 1976 [621], exhibits chaotic dynamics (see Fig. 2.3). (May's discrete version of Verhulst's equation is biologically motivated as a "pulsed-dynamics" model of a species that reproduces in the same season each year.) While Verhulst's approach works well for the data in Fig. 1.1 and allowed him to make a surprisingly accurate estimate of the present-day population of Belgium, biological populations are more complicated for several reasons, including the following [204]:

- As zoologist Warder Allee observed in 1931, most populations tend to decrease if they are too small; thus he proposed replacing (1.2) with [17]

$$\frac{\mathrm{d}N}{\mathrm{d}t} = \lambda N(N - \alpha N_0)\left(1 - \frac{N}{N_0}\right), \tag{7.8}$$

where $0 < \alpha \ll 1$. This is called the Allee effect. The additional factor of $(N - \alpha N_0)$ causes the population to fall to zero unless it exceeds a certain threshold fraction α of the final population N_0, and this factor also accounts for a growth rate that increases with population size.

- Because different age groups of a human population reproduce at different rates, one must know the age distribution of a population in order to accurately predict its rate of growth. Thus a country with a very young human population can expect much stronger future growth than a country with an older population.
- In addition to its distribution in age, the spatial distribution of a population is also an important factor in its dynamics. In 1937, this consideration independently inspired both statistician Ronald Fisher and mathematician Andrei Kolmogorov and his colleagues to publish studies which bring this feature into play. These analyses led to a reaction-diffusion system that is closely related to the equation guiding the fairy ring in Fig. 7.20 (Sect. 5.1.1 and [308, 505]).
- An important feature of population dynamics is the interaction between a predator species and its prey, which was independently studied by chemist Alfred Lotka in 1910 (motivated by problems in chemical kinetics) [577, 578] and by mathematician Vito Volterra in the early 1930s (motivated by variations in fish populations in the Adriatic Sea) [997]. In this case, the prey species feeds from some nutrient source that allows its population to grow in the absence of a predator, and the predator in turn feeds on the prey;[35] thus the system is modeled by the coupled nonlinear equations [204, 362, 690]

$$\begin{aligned}\frac{\mathrm{d}N_1}{\mathrm{d}t} &= \lambda_1 N_1\left(1 - \frac{N_1}{N_{10}}\right) - c_1 N_1 N_2, \\ \frac{\mathrm{d}N_2}{\mathrm{d}t} &= -\lambda_2 N_2 + c_2 N_1 N_2, \end{aligned} \tag{7.9}$$

where N_1 is the prey population and N_2 is the predator population. If $N_1 = 0$, N_2 falls to zero as $\exp(-\lambda_2 t)$, so the predators need the prey. If $N_2 = 0$, on the other hand, the prey population grows to its limiting value as in (1.1).

Under appropriate parametric conditions, predator and prey species can oscillate in the manner shown in Fig. 7.21 and are described as follows. Low predator population allows the prey population to grow rapidly toward its limiting value (N_{10}), but when the prey population becomes

---

[35]This situation is also reflected in laser dynamics, where the prey corresponds to population inversion – N in (6.12) of Sect. 6.7.1 – its nutrient source is the energy supplied by the pump (N_0), and the predator corresponds to the photon energy in the laser cavity (I).

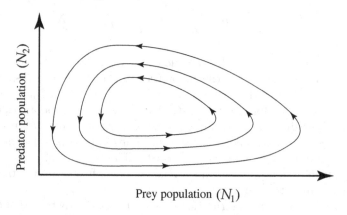

Fig. 7.21. A cycle in a phase space of predator and prey populations, indicating oscillatory behavior where *arrows* indicate the direction of time. (Redrawn from computations in [362])

large, the predator population begins to grow rapidly, which eventually causes the prey population to decrease, in turn decreasing the predators, and the cycle begins again. With the bilinear assumption that the rate of biomass transfer between prey and predator species is proportional to the product of their populations ($N_1 N_2$), the dynamic equations can be integrated to obtain closed trajectories in the predator–prey phase plane, corresponding to a family of exactly periodic solutions of a nonlinear ordinary differential equation [362].

• It is also of interest to consider populations that are competing for the same food supply, which is closely related to multimode operation of lasers (Sect. 6.7.1). In the early 1970s, I made a family of electronic models of such systems, which were essentially one-, two-, and three-dimensional arrays of van der Pol oscillators [860,862]. From both a quasilinear analysis and empirical observation, it was shown that the spatial dimension of the array must be two or more in order to have more than one oscillatory state stably present in the system. This is bad for the laser application because it leads to *multimoding*, which decreases the wavelength precision, but it is good for the biological application because it allows several species to stably compete for the same food supply in the same geographical region, thereby enhancing biological diversity.

All of this is fun for the nonlinear scientist, but real world population dynamics are far more complicated for several reasons, including the following. First, it is hardly ever the case that only two species interact in an isolated manner, as is assumed in Fig. 7.21. The more general case, which was considered in some detail at the beginning of the 1970s by physicist Elliott Montroll and his colleagues Narendra Goel and Samaresh Maitra, involves a large number of species – some predators, some prey, some both – so the nonlinear problem is defined in a phase space of dimension much greater than two. Thus linear analyses of behaviors near singular points (see Appendix A) becomes complicated, and the general solution must be obtained by numerical integration of coupled nonlinear equations [362]. Second, the parameters defining interactions among these species (rates of growth or decay, limiting populations, rates of biomass transfer, etc.) are often not well known. Finally, the biological system parameters may change with time in response to human interventions, unpredictable changes in climate, and so on. To explore these difficulties, the Arthur Rylah Institute for Environmental Research has recently published an empirical and numerical study of interactions among feral cats, foxes, native carnivores, and rabbits in Australia, drawing the following conclusions [809]:

> The preliminary simulation models explore the potential interactions between rabbits, foxes and feral cats, but they are based mainly on hypothesised relationships. The sensitivity of the model to small changes in rainfall suggests a more detailed understanding of the relationships is required. The tendency of the model to reach the lower defined minimum for all species when both predators are present also suggests a much better understanding of the relationships is required. Refinements to the model such as the utilisation of ratio-dependent functional responses may partly address issues of stability at lower resource densities. More specifically, there is a clear need to properly quantify the relationship between rabbits and the two predators. Numerical responses for the two predators should be determined in relation to both the abundance of rabbits (or juvenile rabbits) and simultaneously the abundance of alternative food sources. Based on diet studies both foxes and feral cats consume many prey species other than rabbits, but to our knowledge no quantitative information currently exists to build this into the predators' numerical responses. Ideally, numerical responses would be based on intake rates, but whether this is feasible is currently unknown. Foxes and feral cats have clearly defined breeding seasons, and future investigation of the relationships should focus on these characteristics. The pulsed dynamics resulting from breeding and non-breeding seasons will affect model behaviour and are likely to generate predictions that are substantially different to the continuous models we have used above.

Although the field data in this report show annual fluctuations of species' populations, it is not clear to what extent these stem from the oscillations shown in Fig. 7.21, chaotic variations (as in Fig. 2.1), temporal variations in the system parameters, or from some mixture of all three causes – not to mention complications arising from different spatial distributions of the populations. Evidently there is much empirical, numerical, and theoretical research to be done in sorting out this interesting and important area of nonlinear biology.

7.3.4 Immense Numbers

In school we all learn that the number of elements in a set is either *finite*, in which case we can make a list of them, or it is *infinite*, and we cannot make a list. This seems clear, but Walter Elsasser, a physicist with a strong interest in theoretical biology, showed that there is a third possibility: finite sets so large that the elements cannot be listed, which he termed *immense* [284–287]. How can this be?

The concept of an immense number can be understood by thinking about the difference between sets of *real* objects and sets of *possible* objects. Examples of sets of real objects include all Honda Civics produced in 2002 (121 328), the population of the US at 7:46 a.m. (EDT) on 17 October 2006 (300 000 000), and the number of baryons in the Universe ($\sim 10^{80}$) (see Sect. 6.9.4). These are all finite numbers that can be listed, and indeed they have been listed because they exist as sets of real objects.

Finite numbers that cannot be listed arise when we begin to think about possible sets, rather than existing sets of real objects. As an important biological example, consider the number of possible protein molecules. Each of these molecules comprises a chain of amino acids of which there are 20 varieties (determined by DNA codes), so the number of possible protein chains that are 200 amino acids long is

$$20^{200} = 10^{260} .$$

Because this number is far greater than the baryonic mass of the Universe, all possible protein molecules have not been made and never will be made. Also it would not be possible to list them all because there isn't enough paper or computer storage or brain cells or whatever to do so. Thus the number of possible proteins is immense.

Elsasser first proposed the concept of an immense number in the late 1950s to describe computational difficulties that arise in biology, but the same idea has been independently invented at least twice in the computer science community. Thus in 1938, Edward Kasner whimsically defined[36]

$$10^{100} = \text{a googol} ,$$

following a suggestion by his nine-year-old nephew, and thereafter an even larger finite number as [479]

$$10^{\text{googol}} = 10^{10^{100}} = \text{a googolplex} .$$

Using the ordinary notation of arithmetic, a googol is written as

$$10\,000\,000\,000\,000\,000\,000\,000\,000\,000\,000\,000\,000$$
$$000\,000\,000\,000\,000\,000\,000\,000\,000\,000$$
$$000\,000\,000\,000\,000\,000\,000\,000\,000\,000$$

---

[36]Not to be confused with *Google* (a search engine) or Nikolai Vasilevich *Gogol* (a Russian author).

and the claim (observation, really) is that this many objects of any sort cannot be assembled in our finite Universe. In 1976 Donald Knuth made the same distinction between *realistic* and *unrealistic* numbers, apparently unaware of the earlier definition of Kasner and Elsasser [501]. Interestingly, a googolplex cannot be written in the ordinary notation of arithmetic because there are not enough pencils and paper in the Universe to do so.

As Elsasser points out, one cannot select an exact boundary between large and immense numbers, but the value of 10^{100} is convenient, for three reasons. First, it is about equal to the atomic weight of the Universe times its age in milliseconds, so there is no way to make a list this long. Second, this number has been independently defined by computer scientists to indicate a value above which it is not possible to work numerically [198, 501]. Finally, it is a nice round number, easy to remember. Thus we will use Elsasser's terminology that finite numbers greater than a googol are immense, noting with Elsasser that because they "cannot be realized as objects or events in the real world", immense numbers "have no operational meaning" [287].

Like the concept of chaos which was considered in Chap. 2, immense numbers pop up everywhere once we begin thinking about them. As there are about 10^{120} possible games of chess, this number is immense, explaining why it is not possible to find a chess book that describes all games. In other words, there will always be interesting chess games that have never been played, whereas the number of tic-tac-toe games is finite and rather small, making it possible to play the game faultlessly after a bit of study. Like chess, Japanese *go* is estimated to have about 10^{174} possible games, while the number of possible bridge games is large but not quite immense.[37]

Often arising in the biological realms, immense numbers stem from what mathematicians call combinatorial explosions, in which a quantity grows at least exponentially with some index. Thus the number of one's antecedents increases combinatorially as 2^n (where n is a generational number), rapidly becoming much greater than the total human population. Taking four generations to a century, the number of our antecedents exceeds the population of the world less than eight centuries ago, making almost all of us descendents of William the Conqueror.

Throughout the eons of life on earth, to repeat, most of the immense number of possible protein molecules have never been constructed and never will be. Those particular proteins that are presently known and used by living creatures were selected in the course of evolution through a process of sifting and winnowing that was consistent with the laws of physics and chemistry but not determined by them. In the course of its evolution, interestingly, Life spent some two of its four billion years as a collection of one-celled organisms [607], which seems at first to have been a developmental hang-up, but perhaps this period of time was necessary for multicellular creatures to develop. Like

---

[37]The number of deals in bridge is "only" 5.36×10^{28}, but there are about 10^{47} different ways to play each deal, for a total of about 10^{75} different games.

an aspiring chess player who spends years learning the tricks of many but not all possible games before becoming a master, Life needed two billion years to evolve the genetic structures of many but not all useful proteins before proceeding on with the evolution of higher organisms. Although relatively simple by presently evolved standards, the *Hydra* of Fig. 7.11 is, after all, a very intricate dynamical system.

7.3.5 Homogeneous vs. Heterogeneous Sets

According to Elsasser, the concept of an immense number of possibilities has another important implication; it shows that the life sciences are qualitatively different from the physical sciences. This difference stems from the fact that measurements differ in the two realms.

As all helium atoms and benzene molecules and so on are identical, it is the practice in a laboratory of physical science to study *homogeneous sets* of identical objects. Thus the individual properties of atoms and molecules (mass, charge, dipole moment, etc.) can be determined to accuracies limited only by the nature of the macroscopic measurements. In thermodynamics where such properties as temperature and specific heat are defined as averages over many elements (atoms or molecules), good accuracy is also obtained because the numbers of elements measured is usually very large, eliminating problems with sampling error. Thus for a precision of one part in 10^5 in measuring the average kinetic energy (temperature), it is necessary to average over greater than 10^{10} elements, but this is easy to do as 10^{10} atoms of carbon weigh about 2×10^{-13} grams, which is far less than the amounts that physical chemists ordinarily manipulate.[38] These important features of physical science are so well established and widely accepted that they are scarcely noted in introductory chemistry texts.

In the life sciences, on the other hand, the measurement situation is very different for two reasons. First, the items being measured (*Escherichia coli*, *Hydra*, biological membranes, human hearts, human brains, chess games, and so on) are not identical. Often there is an immense number of possible members of a measurement class; thus a particular measurement may reflect a bias in the elements chosen to look at. Following work of biochemist Roger Williams [1042], Elsasser points out that the levels of various biochemicals in human blood differ by factors from 2 to 6, in gastric juice from 2 to 10, in human milk from 2 to 18, and bone density factors of 3 to 5. Second, the numbers of items that can be measured in the life sciences are usually limited to a few dozen, so sampling errors are often of the order of 15% or more [287].

Thus life scientists do not establish causal connections with the same degree of precision as in the physical sciences. This is an inherent feature of the life sciences, reflecting a nature that differs fundamentally from that of

---

[38] For an average measurement on N items, the root-mean-square sampling arror is $\pm 1/\sqrt{N}$.

the physical sciences. Physical scientists often belittle life sciences as lacking the mathematical precision of the "hard sciences" (physics and chemistry), and life scientists often enable this disparagement by attempting to make physiology and psychology "more like physics". Both are misguided. As life scientists have taken on a more challenging class of problems, they should not allow themselves to feel inferior to those who work with easier problems. Life scientists should not ape the physical sciences, but accept that they are not engaged in the same activity. More about these differences are presented in the following chapter.

7.3.6 Biological Hierarchies

Decrying the predations of minor poets upon their betters, Jonathan Swift has famously written:

> So, the naturalists observe, the flea
> Hath smaller fleas that on him prey;
> And these have smaller still to bite 'em;
> And so proceed, ad infinitum.

Appropriately, this is among the among the most parodied of English verse, often by scientists. To emphasize, for example, his observation that hierarchical systems can be extended both upward and downward, it was changed by mathematician Augustus De Morgan to:

> Great fleas have little fleas
> Upon their backs to bite 'em,
> And little fleas have lesser fleas,
> And so ad infinitum.
>
> And the great fleas themselves, in turn,
> Have greater fleas to go on,
> While these again have greater still,
> And greater still, and so on.

Joking aside, the idea is important. Much of science is organized in this manner, and in previous sections, Swift's doggerel has been applied to turbulence (Sect. 6.8.8) and to a possible structure of the Universe (Sect. 6.9.5). Here it suggests that the life sciences can be organized into a *biological hierarchy* which is roughly sketched as follows.

Biosphere

Species

Organisms

Organs

Cells

Processes of replication

Genetic transcription

Biochemical cycles

Biomolecules

Molecules

In the presentation of this formulation of the life sciences, four comments are appropriate. First, it is only the general nature of the hierarchy that is of interest here, not the details. One might include fewer or more levels in the diagram or account for branchings into (say) flora and fauna or various phyla. Although such refinements may be useful in particular discussions, the present aim is to study the general nature of a nonlinear hierarchy, so a relatively simple diagram is appropriate.

Second, the nonlinear dynamics at each level of description generate emergent structures, and nonlinear interactions among these structures provide a basis for the dynamics at the next higher level. As we have seen in Chaps. 3 and 4, the emergence of new dynamic entities stems from the presence of closed causal loops, in which positive feedback leads to exponential growth that is ultimately stabilized by nonlinear effects.

Third, closed causal loops may also provide a basis for the phenomenon of dynamical chaos, which was discussed in Chap. 2, and as in the fluid turbulence of Sect. 6.8.8, this chaos may span several levels of the biological hierarchy.

Finally, the number of possible entities that can emerge at each level is very large and often immense, implying that all possible biological hierarchies cannot be physically realized within our finite Universe.

Thus we see how this hierarchical structure embodies the key elements of nonlinear dynamics: emergence, chaos, and immense numbers of possible structures. Unlike simple Newtonian formulations, Life is organized as in the above diagram with causal influences running both up and down the levels. This is a far more difficult and important class of problems, of which the scientific community – not to mention the general public – is yet but dimly aware.

In my view, some of the bitterness described in Sect. 7.2.7 as a "clash of scientific cultures" stems from the fact that physical scientists and life scientists are engaged in quite different types of science, without being fully aware of those differences.

7.4 Neuroscience

It is almost a half century since the days of my doctoral research in electrical engineering, during which I became involved with neuroscience. This serendipitous interest arose because my project was to understand how electromagnetic waves propagate on Esaki (tunnel) junctions (see Sect. 6.6.6) with spatial dimensions greater than a wavelength, and as taught, I dutifully began by looking at linear normal mode solutions of Maxwell's equations. Associated experimental studies, however, soon showed that the waves were governed by a nonlinear reaction-diffusion (RD) equation that could not be approximated by any linear mathematical model, and is very biological in nature corresponding to the propagation of nerve pulses.

That this was a potentially fruitful line of study became clear from a number of observations, including the following:

1. In the context of nerve pulse generation and propagation, RD problems were of central interest to Nicholas Rashevsky, as I had learned in the early 1950s from reading the second edition of his *Mathematical Biophysics* (Sect. 5.2.2).

2. Balthasar van der Pol had recently generalized his famous oscillator equation (Sect. 6.6.2) to an RD model for nerve pulse propagation [984].

3. Related nonlinear wave problems on semiconducting electronic devices were being discussed in the Russian engineering journal *Radiofizika*, which in those post-Sputnik days was translated into English on a monthly basis under the ponderous title *Radioengineering and Electronic Physics*.

4. Alan Hodgkin and Andrew Huxley had recently published their masterful experimental, numerical and theoretical analyses of RD propagation of pulses on the squid giant axon (Sect. 4.1 and [440]), which was closely related to my research.

5. In 1962, Hewitt Crane proposed using an electronic analog of the nerve axon – he called it a *neuristor* – for information processing in a new class of computing machines (Sect. 4.2 and [199]).

6. Also in 1962, Jin-ichi Nagumo and his colleagues at the University of Tokyo descibed the same neuristor system that I was then studying in my laboratory as an electronic model for nerve pulses (Sect. 4.2 and [697, 857]). I was not then aware of Alan Turing's 1952 paper on pattern formation (Sect. 7.2.4 and [973]), but it was an exciting time nonetheless, with doable problems arising at almost every turn.

By the early 1960s and thereafter, the nature of pulse propagation on RD systems and the implications of such phenomena for understanding the brain's dynamics were of central interest to a growing group of applied scientists in Europe, Japan, Russia, and the United States, and it is now evident that this activity was a precursor to the nonlinear science revolution of the 1970s (Chap. 1). Although such studies are closely related to applied mathematics, oddly, that professional group did not engage with neuroscience problems until the 1970s, a decade after they became of wide interest to the international engineering community, two decades after the seminal papers by Hodgkin and Huxley and by Turing (Sects. 1.3, 4.2, and 7.2.4 and [440, 973]), and almost four decades after they had been proposed by Rashevsky as an important part of his program in mathematical biology [790]. Without going into immoderate detail, the aim of this section is to share some fruits of this effort that seem relevant to understanding the relationships between nonlinear science and biology.

7.4.1 Nerve Models

Quantitatively observed in the middle of the nineteenth century by Hermann Helmholtz and Julius Bernstein among others [80,123,420], nerve pulse propagation seemed at first to be cloaked in mystery. Is a nerve pulse alive? What are the differences between Luigi Galvani's animal electricity and Alessandro Volta's chemical electricity, not to mention Ben Franklin's atmospheric electricity? Was the "spark of Life" delivered by a bolt of lightning, as Mary Shelley assumed [899]? Can a living phenomenon have a physical model? Why is the speed of nerve pulse propagation some seven orders of magnitude less than that of light?

In the context of such questions, Ralph Lillie's "passive iron wire" nerve model, proposed in 1925, was an important advance in because it showed neuroscientists that a phenomenon with the properties of a nerve pulse need not be alive [563]. In this model, a piece of iron wire does not disolve in weak nitric acid because it rapidly becomes coated with a thin protective layer of iron oxide. When scratched, however, an uninsulated segment of the wire passes a loop of ionic current for a short time, breaking down the neighboring oxide layer, which then breaks down the layer of the next neighbor, leading to a propagating current loop much like that of a nerve pulse. Once it was established that a nerve-like pulse need not be alive, people noticed that a candle or a stick of incense are also nerve analogs, leading to the amusing speculation that Helmholtz sat at a candle-lit table struggling to imagine what sort of physical phenomenon could travel as slowly as a nerve pulse. He eventually concluded that the pulse must involve some movement of matter, overlooking the possibility that it might comprise a wave of activity.

In 1902 prior to Lillie's discovery, Bernstein had proposed his *membrane hypothesis* that the cell membrane somehow breaks down in the course of nerve pulse dynamics [81], but the physical nature of this breakdown was not made clear. In 1906 a physical chemist named Robert Luther demonstrated an autocatalytic chemical reaction, which he suggested might help to explain nerve pulse propagation (Sect. 4.1 and [583]), but the electrophysiologists would not learn of this for eight decades, far too late to help them formulate a theory of nerve pulse propagation.

Lillie enters the story in two other ways. First, Nagumo had an iron-wire net of two space dimensions in his laboratory during the 1960s, and it was here, Art Winfree told me, that he first saw the self-sustaining spiral waves that he would come to understand so well in the early 1970s (Sect. 4.3 and [1049,1050]). It was in trying to prove that spiral waves could not exist in a continuous medium that he discovered then in the Belousov–Zhabotinsky reaction. Second, it was Lillie who – together with Arthur Compton and Karl Lashley – invited Rashevsky to set up his mathematical biology program at the University of Chicago in the early 1930s (Sect. 5.2.2 and [816, 819]).

As is evident from the references to his work on nerve pulse excitation and propagation during the 1930s, Rashevsky's program was truly farsighted [790]. Not only did this group develop a credible theory for the *intensity–duration curves* that are typical of the threshold conditions for nerve pulse stimulation (higher-intensity inputs need to be maintained for shorter times to reach threshold), but they accounted for fluctuations in thresholds, formulated the *two-factor theory* of excitable processes (which would become a standard concept in several different contexts of neuroscience throughout the twentieth century), and introduced the *neural network* as a model of the neococortex. Additionally, they treated propagation on the *myelinated axons* of motor nerves, in which the activity jumps (*saltatory conduction*) from one active node to the next, achieving increased propagation velocity at much lower levels of energy consumption. This is not to claim that Rashevsky and his coworkers solved all of these problems to the degree that is considered satisfactory today, but they first formulated many of the key issues in physico-mathematical terms, showing the directions that future theories of nerve-pulse conduction would take. These contributions are emphasized because some claim – incorrectly, in my view – that Rashevsky's program has been of little or no long-term value in mathematical biology [484, 485, 556]. When I began my academic career in the early 1960s, his books provided a theoretical basis for my research program, and the interactions that I had with him during the 1960s were thoroughly professional and supportive. For this I remain grateful.

Presently, there are a number of nerve models, with complementing degrees of accuracy and ease of application. Some of these were discussed in Chap. 4, but a few words about all the presently known neural wave models are collected here.

The Hodgkin–Huxley (HH) System

A key feature of the HH model for a squid giant axon is their representation of how the transmembrane ionic current density J_{ion} depends on transmembrane voltage V as [196, 440]

$$J_{ion} = G_{Na}m^3h(V - V_{Na}) + G_K n^4(V - V_K) + G_L(V - V_L) , \qquad (7.10)$$

where the three components on the RHS represent respectively sodium ion current, potassium ion current, and "leakage" ion current (all that is neither sodium nor potassium) as in Fig. 7.3. In addition, m, h, and n are respectively sodium turn-on, sodium turn-off, and potassium turn-on variables, which obey first order kinetic equations. The basic reaction-diffusion equation is given in (4.2), where $j_{ion} = \pi d J_{ion}$; thus the HH system has five dynamical variables V, dV/dt, m, h, and n, plus nine constant parameters G_{Na}, V_{Na}, G_K, V_K, G_L, and V_L, in addition to the resistance per unit length r, and capacitance per unit length c, and the fiber diameter d, which are given in Table 4.1. Under a traveling-wave (TW) analysis, it is assumed that all five dynamical variables are functions of the TW variable $\xi \equiv x - vt$, where v is another parameter – the TW velocity – which can be adjusted under the analysis to obtain a finite solution using the phase-space techniques sketched in Appendix A [874].

On the upside, the HH formulation is a complete theory where all of the experimental parameters can and have been independently determined. This theory predicts the measurements shown in Fig. 4.1 plus a TW velocity of 18.8 m/s, which, compared with an experimental value of 21.2 m/s, implies an error of about 11%, as is to be expected from errors in measurements of the system parameters. From its status as a complete theory of nerve pulse dynamics, HH has become widely accepted as a general model for active waves in electrophysiology, used in cases where many of the system parameters cannot be measured but must be estimated (or guessed). On the downside, the many variables of the HH system increase the difficulties of numerical computations, although this is now less problematic as computing power available to the electrophysiologist has steadily increased over the past half century.

The Zeldovich–Frank-Kamenetskii (ZF) Equation

As described in detail in Sect. 4.1, the ZF equation can be used to model the sodium current on the leading edge of a HH pulse by a cubic polynomial, leading to (4.4), which has the exact solution given in (4.5) with TW velocity in (4.7). Recall that this equation and its exact solution were first derived for flame-front propagation in 1938 in a Russian publication which, unfortunately, Rashevsky didn't see [680, 874, 1083]. Apart from the advantages of having an exact solution and giving a fair estimate of TW speed on a squid giant axon, ZF has the disadvantage of not recovering. Like a candle, in other words, which goes from unburned to burned under the passage of a flame, the TW solution of ZF carries the system from one outer zero of the cubic polynomial to the other.

The Markin–Chizmadzhev (MC) Equation

Although not widely known among neuroscientists, one of the simplest means for representing recovery of a propagating nerve pulse was introduced by physicist Aleksandr Kompaneyets in the mid-1960s [507] and developed in detail by biophysicists Vladislav Markin and Yuri Chizmadzhev in 1967 [608]. This MC model assumes a diffusion equation as in (4.2)

$$\frac{1}{rc}\frac{\partial^2 V}{\partial x^2} - \frac{\partial V}{\partial t} = \frac{j_{mc}(t)}{c} \,, \tag{7.11}$$

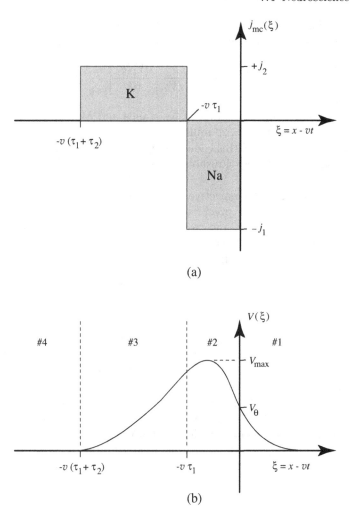

Fig. 7.22. (a) Ionic current in the MC model of a nerve axon as a function of the TW variable $\xi \equiv x - vt$. (b) Structure of the associated nerve pulse

but the RHS side is the prescribed function of position and time shown in Fig. 7.22a. Thus there is an inward flow of sodium ion current followed by an outward flow of potassium ion current, with the total charge transfer across the membrane equal to zero because $j_1\tau_1 = j_2\tau_2$. As the system is linear with constant coefficients in the four ranges (#1, #2, #3 and #4), the solution $V(\xi)$ can be written as general sums of exponential functions within these regions, matching magnitudes and slopes at the boundaries to obtain the pulse solution shown in Fig. 7.22b.

In this MC analysis, a critical parameter is the threshold voltage V_θ. When the leading-edge voltage reaches this value, it is assumed that the prescribed ionic current shown in Fig. 7.22a begins its course. The exponentially decaying solution in region #1 then implies that the TW speed is

$$v = -\frac{1}{rcV_\theta}\left[\frac{dV_1(\xi)}{d\xi}\right]_{\xi=0}, \tag{7.12}$$

where r and c are the values of resistance and capacitance per unit length given in Table 4.1, $V_\theta = 20$ millivolts, and the initial slope can be taken from Fig. 4.1. Interestingly, this equation, which relates TW speed to the shape of the leading edge below threshold was first derived by Hodgkin and Huxley in 1952 [440].

From a general perspective, the central point of (7.12) is that the exponentially decreasing precursor into region #1 *guides* the pulse (thereby determining its speed), and (retroactively) the main body of the pulse *generates* the precursor. This is an example of a *closed causal loop*, which can be represented by the following diagram:

Pulse generates precursor

$\downarrow$ $\uparrow$

Precursor guides pulse

The positive features of the MC model are twofold: it represents pulse recovery as in Fig. 7.22b, and it is relatively easy to solve, being linear within the four regions shown in this figure. Thus it is useful in numerical studies involving complicated axon geometries (branchings, etc.) where more realistic dynamical models like HH would be more difficult to analyze or require more computing time. The main drawback of MC is that the inward flow of sodium ions and subsequent outward flow of potassium ions does not arise naturally from the dynamics of the model (as in HH) but are a specified component of the theory (see Fig. 7.22).

The FitzHugh–Nagumo (FN) System

Finding the simplest way to have sodium and potassium ion currents emerge naturally from a dynamical theory motivates the FN model, which is described in Sect. 4.2. In this model, the inward sodium current is modeled by a cubic polynomial (as in the ZF model), and the outward flow of potassium ions is modeled by a *recovery variable – r* in (4.11). As this is a natural way to construct a neuristor, those of the early 1960s were physical examples of FN systems [697, 857].

Since the 1970s when the applied mathematics community became aware that reaction-diffusion (RD) processes are both theoretically interesting and practically important [440,697,973], the FN model has been a favorite context in which to discuss general properties and develop theorems [472, 639, 992]. From Fig. 4.3, it is seen that in addition to the stable TW pulse, there is an unstable pulse solution which is important in understanding the threshold conditions for launching a pulse on the nerve [874]. As the parameter ε in (4.11) is increased – which decreases the time delay for rise of the recovery variable – a critical value (ε_c) is reached, at which the faster stable pulse and the slower unstable pulse merge into one and above which TW pulse propagation is not possible. As the rates of ionic response in a squid nerve (and in its HH model) increase with temperature, ε in FN is called a *temperature parameter*, and the fact that the FN model has no TW pulse solutions above a critical value of this parameter corresponds to the experimental fact that a squid nerve will not carry pulses if its temperature is too high. This physical constraint is one of the reasons that temperature control is more important in animals than in plants.

As is discussed in Sect. 4.3, FN is a natural model for exploring the dynamics of RD systems in two or three space dimensions, merely by replacing d^2u/dx^2 in (4.11) by $d^2u/dx^2 + d^2u/dy^2$ or by $d^2u/dx^2 + d^2u/dy^2 + d^2u/dz^2$, which leads to the *nonlinear geometrical optics* developed by biophysicist Oleg Mornev [678].

The Morris–Lecar (ML) System

A key feature of the biologically realistic HH system is that TW pulse solutions are driven by an inrush of sodium ions, but this is not the case for many nerves, including those of the mammalian central nervous system [568]. Although detailed measurements of membrane properties are difficult to perform on such small nerves, a study of calcium-mediated membrane switching on the giant muscle fiber of the barnacle was reported by Catherine Morris and Harold Lecar in 1981, leading to a simple model for the transmembrane ionic current [681]. In this work, it was found that the initial inward flow of calcium ions did not experience deactivation, so their model for membrane ionic current does not include the turn-off (h) component of the HH system. Thus (7.10) is replaced with the expression [875]

$$J_{ml} = G_K n \left(V - V_K \right) + G_{Ca} m \left(V - V_{Ca} \right) + G_L (V - V_L) , \qquad (7.13)$$

where m and n are respectively calcium turn-on and potassium turn-on variables that obey first order kinetic equations.

Like HH, the ML system is a physically realistic model. It is one degree more complicated than FN because the initial inrush of calcium ions that drives the leading edge of the pulse is also modeled by a dynamic variable (m), so it has four dynamic variables compared with three for FN and five for HH. It is expected that the ML system will be of increasing interest in coming years, both for theoretical studies and as a realistic model of experiments in electrophysiology.

Myelinated Nerve Models

With a diameter of 476 microns, the HH standard axon is among the largest of such structures, and there is an evolutionary reason for this size. Larger axons have greater TW pulse speeds and the function of a squid's giant axon is to signal "escape!" to its anterior muscles, which then suddenly contract, and the squid darts away like a rocket, leaving behind a tasty cloud of black ink. As this ink is used as a pasta sauce in Naples, I speculate that its flavor evolved in order to further confuse predators.

From (4.8), recall that the TW pulse speed on an HH model axon is proportional to the square root of its diameter. Over a temperature range from $T = 15$ to $22°C$, my measurements on squid giant axons showed (to an experimental accuracy of about $±5\%$) that [893]

$$v = 20.3\sqrt{\frac{d}{476}}\left[1 + 0.038(T - 18.5)\right] , \qquad (7.14)$$

where v is in m/s and d is in microns. In order to double the speed of this vital signal, it follows that the axon diameter must grow by a factor of four, thereby increasing its area of cross section by a factor of sixteen. Thus it seems that the diameter of a squid giant axon has arrived at a value beyond which further increases would require the axon to take an inappropriate share of the animal's volume.

In the motor axons of vertebrates, this problem has been solved by using the myelinated nerve structure shown in Fig. 7.23a, in which a wave of activity jumps from one active node to the next, saving metabolic energy and greatly increasing the propagation speed for a given fiber diameter. In such myelinated nerves, it turns out that the speed of TW pulse propagation (v) is roughly proportional to the first power of the fiber diameter (d), as is shown in Fig. 7.23b from measurements on frogs and cats in the 1940s [456, 946].

More precisely, these data imply

$$v = 2d \qquad (7.15)$$

for frog axons at $24°C$, where v is in m/s and d is in microns. Similarly, for cat axons at $37.5°C$,

$$v = 5.6d . \qquad (7.16)$$

The linear dependence in these two equations stems from the fact that the internode spacing s is roughly proportional to the fiber diameter d [456, 946], and the signal speed is equal to the internode spacing divided by a relatively constant switching time. This linear dependence on d stems from an evolutionary optimization,[39] whereas the square root dependence in (7.14)

---

[39]From an evolutionary perspective, pulse speed increases with internode spacing s, until s becomes so large that the switching signal at one node cannot reliably

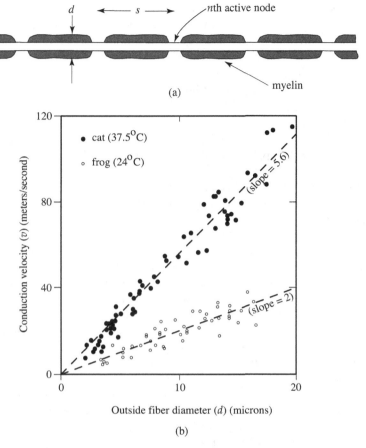

Fig. 7.23. (a) Structure of a myelinated motor axon (not to scale). (b) Measured values of TW pulse speed v vs. outside fiber diameter d for cats and frogs. (Data from [456] and [946])

can be deduced from a dimensional analysis of Luther's parameter factor in (4.9).

Comparing (7.14) with (7.15) and (7.16), it is clear that a bundle of several dozen vertebrate fibers have a smaller area of cross section than a squid giant axon and carry signals at greater speeds. Thus a vertebrate motor nerve is typically a bundle of the axonal fibers each carying a signal faster than does a squid giant axon, with the entire nerve bundle having about the same

---

reach the firing threshold of the next node. As the node resistance is proportional to $1/s$ and the internode resistance is proportional to s/d^2, the determining parameter is $(1/s)/(s/d^2) = (d/s)^2$, which will have a constant optimum value.

diameter as a squid axon.[40] With this remarkable evolutionary invention, vertebrates are able to send highly nuanced signals to their extremities – rather than the one-bit message of the squid – using much less metabolic energy per bit. Properly connected to the muscles of our arms and legs, these messages allow us to walk, run, dance, write, and play musical instruments – all far beyond the modest abilities of a squid.

Cortical Waves

Shortly after the papers appeared on nonlinear diffusion in biological pattern formation by Turing [973] and on the squid nerve pulse by Hodgkin and Huxley [440], a related paper by physicist Raymond Beurle was published on the propagation of an RD wave in the neocortex [84]. From the general perspectives of modern nonlinear science, these three papers – all published in highly-visible journals within a few years of each other – were read by quite different groups, underscoring the lack of communication in those years among scientists dealing with closely related mathematical structures.

In his seminal study, Beurle assumed first that the neural mass of the neocortex can be locally described by the fraction $F(x, t)$ of cells at position x that are firing at time t, and second that the probability of two cells at x and x' being interconnected is an exponentially decreasing function of the distance between them; thus, $p(x, x') \propto \exp(-|x - x'|/\sigma)$. In this theory, activity at a particular region of the cortex induces activity at neighboring regions, which leads to a *wave of information* propagating through the neural mass. In accord with the electrophysiology of the time, Beurle supposed all of the cortical neurons to be excitatory; thus the waves described by his theory correspond to the leading-edge formulations of Sect. 4.1, albeit with the activity averaged over many neurons.

One observation of Beurle's study is that a wave of information may involve the activity of only a small fraction of the local neurons, allowing waves of information to pass through each other with little interference. Thus many different messages may propagate throughout the neocortex, carrying information from one region to another. (That cortical information waves pass through one another leads some to confuse them with *solitons*, but the two phenomena are quite different: solitons conserve energy, whereas waves of information do not.) If the neural interconnections (synapses) are supposed to increase in strength upon exposure to the activity of a particular wave (a learning mechanism), Beurle described a means for holographic-like recall [874]. In this case, waves induced by external (sensory) stimulation become coupled to subsequent internal waves, leading to the possibility that a fragment of the original stimulation suffices to trigger the related internal response.

---

[40]The frog nerves studied experimentally by Galvani, Helmholtz and Bernstein (among many others) were actually nerve bundles.

Following Beurle's lead, biophysicist John Griffith developed a field theory of neural activity in the 1960s, where time and space dependencies are brought in through their lowest derivatives [375–377]. In this theory, the probability S of a neuron firing in the next time interval is a sigmoid function of the present firing rate F. Assuming $S(F) = F^2/(F^2 + \theta^2)$ with θ a *threshold parameter*, the first time derivative can be approximated as

$$\tau\frac{dF}{dt} \approx S(f) - F = -\frac{F^3 - F^2 + \theta^2 F}{F^2 + \theta^2} \ ,$$

where τ is a time delay for firing.

To introduce spatial dependence, Griffith reasoned from symmetry that the connectivity must be the same in both the $+x$ and the $-x$ directions; thus, the lowest space derivative is the second, implying the PDE

$$D\frac{\partial^2 F}{\partial x^2} - \frac{\partial F}{\partial t} = \frac{F^3 - F^2 + \theta^2 F}{\tau(F^2 + \theta^2)} \ , \tag{7.17}$$

where for a mean interconnection distance indicated by σ and a neural response time of τ, the diffusion constant is of order $D \sim \sigma^2/\tau$. Although (7.17) is similar to the ZF equation, its physical interpretation is different, because (4.4) describes a voltage on the leading edge of a nerve pulse, whereas (7.17) represents a wave of activity propagating through the neocortex.

To bring *recovery* into the picture, biomathematicians Hugh Wilson and Jack Cowan took advantage of the fact that some of the neocortical neurons were later found to be inhibitory; thus, they developed a theory in two dependent field variables [1044, 1045]:

- $E(x,t)$: the fraction of excitatory neurons that are firing as a function of x and t, and
- $I(x,t)$: the corresponding fraction of inhibitory neurons,

which are assumed to interact as [1043]

$$\tau\frac{dE}{dt} = S_E\left(\int \Big[w_{EE}(x,x')E(x') - w_{IE}(x,x')I(x')\Big]dx' + P(x,t)\right) - E \ ,$$
$$\tau\frac{dI}{dt} = S_I\left(\int \Big[w_{EI}(x,x')E(x') - w_{II}(x,x')I(x')\Big]dx' + Q(x,t)\right) - I \ .$$
$$\tag{7.18}$$

In these coupled integro-differential equations:

1. the nonlinearity of neural response is introduced through the sigmoid functions

$$S_E(y) \equiv \frac{100y^2}{\theta_E^2 + y^2} \quad \text{and} \quad S_I(y) \equiv \frac{100y^2}{\theta_I^2 + y^2} \ ,$$

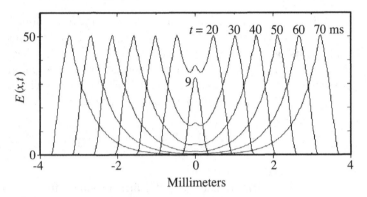

Fig. 7.24. Outgoing activity-wave solutions of (7.18) generated by a brief pulse of excitation near the origin

2. diffusion stems from the interconnection probabilities between neurons at x and x',
3. $w_{ij}(x, x') = b_{ij} \exp\left(-|x - x'|/\sigma_{ij}\right)$,
4. external (sensory) inputs to the excitatory and inhibitory cells are represented by $P(x, t)$ and $Q(x.t)$, respectively.

Although it is difficult to fix the many parameters of such a model, Wilson suggests the following values as reasonable "guesstimates" for the human neocortex: $\sigma_{EE} = 40$ μm, $\sigma_{EI} = \sigma_{IE} = 60$ μm, $\sigma_{II} = 30$ μm, $\theta_E = 20$, and $\theta_I = 40$, and he has made available several *Matlab* codes for exploring the resulting dynamics [1043]. Those familiar with *Matlab* are encouraged to play with these codes and explore a spectrum of behaviors, including stationary patterns of activity, transients, localized oscillations, and waves of activity, an example of which is shown in Fig. 7.24.

In his book, Wilson offers many examples of such dynamics, discussing ways in which neural field theories similar to (7.18) can model a variety of mental phenomena, including phase transitions, hallucinations, and epileptic seizures [1043].

7.4.2 The Multiplex Neuron

A typical neuron in the human central nervous system has a single *axonal tree*, which carries pulse signals away from the neuron and several *dendritic trees* which carry signals into the neuron (see Fig. 7.27). As is described in Sect. 7.2.3, pulse propagation away from the neuron is most efficient if the axonal branching is governed by Rall's law, for which the geometric ratio (GR) [defined by (7.7) in relation to Fig. 7.16] is unity. But what of the dendritic branchings?

In fact, Rall's law was originally proposed for dendrites back in the 1950s when it was universally assumed that dendrites are passive structures, merely

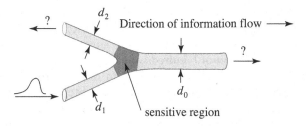

Fig. 7.25. Geometrical structure of a branching dendrite. (The sensitive region is where an all-or-none switching decision can be made by the branch)

gathering synaptic input signals and presenting a linear weighted sum of them to the *initial segment* of the outgoing axon, which in turn would compare this sum with a threshold voltage. If this weighted sum is greater than the neural threshold, according to the theory, the neuron will fire and send a pulse out onto its axonal tree. Under this assumption, an entire dendritic tree was viewed as an equivalent *passive cylinder*, each branch of which obeys Rall's law. Apart from its being the simplest picture of dendrites, a subjective reason for supposing passive dendrites is that it simplifies the flow of causality, as there is only one switching element in a neuron: the initial segment of the outgoing axon, which is adjacent to the cell body.

Although this simple picture is still assumed by some neuroscientists and many philosophers of neuroscience, empirical evidence has been accumulating over the past three decades that dendritic trees carry active pulses into the cell body.[41] Thus on a dendritic tree branching as shown in Fig. 7.25, one might ask whether a pulse coming into daughter #1 will manage to propagate to the parent and the other daughter, for which the relevant geometric ratio is

$$\text{GR} = \frac{d_0^{3/2} + d_2^{3/2}}{d_1^{3/2}} . \tag{7.19}$$

For dendritic branches that obey Rall's law ($d_1^{3/2} + d_2^{3/2} = d_0^{3/2}$) and are symmetrical ($d_1 = d_2$), this GR evidently equals 3, but it is not unusual to find dendritic GRs between 2 and 5 [874].

If the GR of a dendritic branch is large enough to block a pulse incoming on *one* of the daughters, then the simultaneous appearance of pulses on *both* daughters will be able to proceed to the parent, and the branch will act as a Boolean AND circuit. In other words, a pulse on daughter #1 *and* a pulse on daughter #2 will cause a pulse to appear at the parent, but just one of these pulses will not. In the early 1970s, it was noted independently by

---

[41] This evidence includes observations on Purkinje cells of the cerebellum [568, 569], and on pyramidal cells in the hippocampus [26, 79, 427] and in the neocortex [649, 936], among others [874, 937].

several researchers that this effect could provide a basis for *dendritic logic*, vastly increasing the computing power of a neuron [490, 874, 888, 1014]. If this possibility is accepted, then the neuron no longer looks like a single all-or-none switch but a collection of hundreds of such switches located at the dendritic branches. In modern computer jargon, in other words, a neuron would be more like a chip than a single gate. Neurologist Stephen Waxman has termed a neuron with such extensive dendritic computing the *multiplex neuron* because it can handle several lines of causality at the same time [1014].[42]

To decide whether the multiplex neuron is a realistic possibility, one can proceed either theoretically (calculating GRs at which blocking occurs for the above-mentioned nerve models), numerically (by computing the propagation of nerve pulses through model branches with different GRs by integrating relevant PDEs), or empirically (by comparing the GRs observed for dendritic branches with theoretical and numerical estimates, and by studying pulse propagation through biological branches). Let us consider some results of these investigations [874].

Theoretical Calculations of Critical GRs

One approach to finding the critical GR for an AND gate is to analyze a branch using the MC model of an HH nerve. This has been done by Markin, Chizmadzhev and Vassili Pastushenko, who find blocking for GR > 5.2, defined as in (7.19) [743–745]. More recently, Mornev has shown that this problem can be exactly solved for the ZF equation, finding blockage for GRs greater than a critical value of 5.9 [677, 874]. It should be emphasized that these calculations are for simplified models of the HH system.

Numerical Estimates

Numerical calculations on the full HH system indicate that the critical value of GR above which a single pulse will not pass through a branch is 12.7, which is significantly larger than the above results for simple models of the HH system [19, 490, 741] and larger than observed GRs of dendritic branching. A corresponding numerical computation for the ML system gives a significantly smaller value of 3.4 [19].

In an extensive series of such HH computations – which were especially impressive for the early 1970s – Boris Khodorov and his colleagues found that the critical GR for the HH system is significantly lowered for groups of pulses [490]. With an incoming pair of pulses, for example, these Russian researchers found, the first goes through a branch for GRs up to 12.7, but the second becomes blocked at much smaller values of GR. The reason for this is that the branch has not yet fully recovered from the passage of the first pulse when the second pulse arrives.

---

[42]In communications engineering, the term "multiplexing" means to send several messages simultaneously over the same channel.

Table 7.2. The GR range for some typical dendrites calculated from observations of branching exponents using (7.20). Apical dendrites consist of a single tree, whereas basal dendrites comprise several trees. Branching exponents are from [36]

Cell type	Branching exponents (Δ)	GR range
Purkinje	2.36 ± 1.2	2.3–3.5
Stellate	2.24 ± 1.12	2.4–3.5
Granule	2.58 ± 1.8	2.3–4.8
Motoneuron	1.69 ± 0.48	2.6–3.4
Pyramidal (apical)	1.99 ± 0.79	2.5–3.4
Pyramidal (basal)	2.28 ± 0.89	2.4–3.1

Empirical Studies

To get an idea of the GRs to be expected in real dendrites, consider Table 7.2, where branching exponents Δ for (7.6) (Leonardo's law) are recorded for a variety of mammalian dendrites [36]. Assuming that the two daughter branch diameters are equal ($d_1 = d_2$) implies a ratio of parent diameter to either one of the daughter diameters of $d_0/d_1 = d_0/d_2 = 2^{1/\Delta}$; thus the corresponding critical geometric ratio is

$$\mathrm{GR}_{\mathrm{crit}} = d_0^{3/2} + d_2^{3/2} d_1^{3/2} = 2^{3/(2\Delta)} + 1 . \tag{7.20}$$

For GRs above $\mathrm{GR}_{\mathrm{crit}}$, a single pulse will be blocked at the branch, so pulses on both daughters are needed to fire the parent: an AND branch. For GRs below $\mathrm{GR}_{\mathrm{crit}}$, a single pulse on one daughter *or* on the other daughter can fire the parent; thus it is an OR branch. (Note that the values of Δ in Table 7.2 differ greatly from the value of 3/2 assumed under Rall's law.)

In the last column of Table 7.2 are recorded values of GR calculated from (7.20), and these values suggest a range for which blockage might or might not occur, thereby allowing both AND branches and OR branches. As the active ion in these dendrites is calcium rather than sodium, the GR values should be compared with the critical GR of 3.4, which was numerically computed from the ML system [19]. If the branches in Table 7.2 have GRs greater than 3.4, they will be AND branches, while lesser values will be OR branches; thus one would like to see this value about in the middle of the ranges of the third column, which it is not. One reason for this discrepancy is that the daughters were assumed to be of equal diameter is calculating the third column; if this condition is dropped, the range of GR values in the third column becomes larger. Another reason is that the value of 3.4 for a critical GR in the ML system may be too large because it was calculated for a fully developed pulse, whereas in real dendritic dynamics, a pulse would be growing from its threshold level upon leaving a previous branch. Finally, as Khodorov has shown, closely spaced pulses are more easily blocked [490].

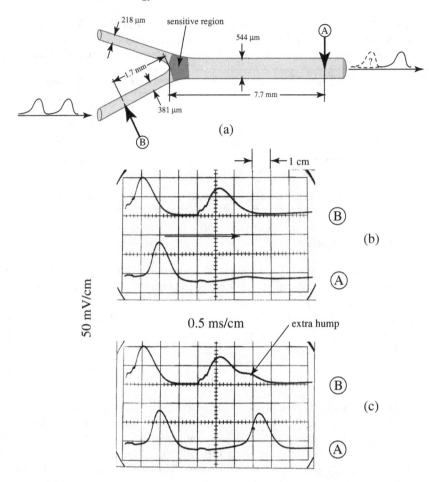

Fig. 7.26. All-or-none switching action in the branching region of a squid giant axon at 20.3°C [894]. (**a**) Geometry of the preparation, showing the point of upstream recording of a pair of incoming pulses at B and the point of downstream recording at A (not to scale). (**b**) Blocking of the second pulse. (**c**) Passage of the second pulse

For branchings of squid giant axons (which are well modeled by the HH system), I have done experiments on the passage of double pulses through branching regions, an example of which is shown in Fig. 7.26 [874, 894]. As the incoming pairs of pulses are launched upstream from electrode B, the relevant GR for this experiment is

$$\frac{5.44^{3/2} + 2.18^{3/2}}{3.81^{3/2}} = 2.14 \ .$$

In the upper oscilloscope tracings, two pulses passing electrode B are seen, but the time interval between these two pulses is very slightly less in Fig. 7.26b than in Fig. 7.26c. (This slight decrease in pulse spacing was adjusted by a knob on the pulse generator.) The lower oscilloscope recordings show the observations at electrode A, where it is seen that the second pulse is blocked for the slightly lower value of pulse spacing. Based on the HH system, these observations are in accord with Khodorov's numerical computations [490], and they demonstrate that the sensitive region of a branch (see Fig. 7.26a) acts like an all-or-none switch. (That the switching actually occurs at the branching region has been verified by showing that the time the switching of the second pulse to travel back to electrode B corresponds to the location of the extra hump in Fig. 7.26c.)

From presently available knowledge, therefore, it seems that a typical cortical neuron is at least as complicated as the information processor suggested by the cartoon of a multiplex neuron in Fig. 7.27, where the dendrites are no longer viewed as linear passive structures but nonlinear computing systems with the sensitive regions acting as AND or OR Boolean functions and a NOT function is provided by inhibitory synapses.

As they seem to have all the features needed for a general information processing system, it is interesting to consider what a large collection of such neurons can do.

7.4.3 The McCulloch–Pitts Model of the Brain

Of all who joined Nicholas Rashevsky's program in mathematical biology during the 1930s, perhaps the most interesting figure – and certainly the most tragic – was young Walter Pitts. Self-taught and blessed with remarkable intelligence, he had run away from an unsupporting home at the age of fifteen and settled in Chicago, reading mathematics in libraries and participating in the activities of Rashevsky's group [911]: the only academic department where he ever felt at home [7]. It was at one of Rashevsky's seminars that Pitts met Warren McCulloch, who had a decisive influence on the course of his life.

A generation older than Pitts, McCulloch had recently come to Chicago as a professor of psychiatry at the University of Illinois, and although their socio-economic and academic backgrounds differed greatly, they shared a common goal – to understand the logic of the human brain. With a medical degree and a decade of experience treating damaged and malfunctioning brains, McCulloch was in search of the *psychon* or *simplest psychic act*, while Pitts had taught himself enough about the mathematics of logic (Boolean algebra) to be able to interact intelligently with leaders in the field [911]. Together they made an ideal team, each strong where the other was weak, and the central idea of their first work together was to map the logic of the brain onto the all-or-none behavior of its constituent neurons, using Boolean algebra as a

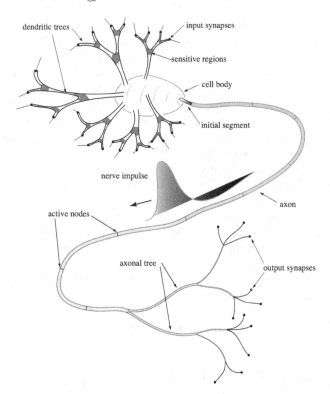

dendritic trees

input synapses

sensitive regions

cell body

initial segment

nerve impulse

axon

active nodes

axonal tree

output synapses

Fig. 7.27. A cartoon of a multiplex neuron

tool. Rashevsky was supportive of this collaboration, as it was in a line of his research activities since coming to Chicago in the early 1930s [7, 790].

In the paper which they wrote after a year of intense effort [633], McCulloch and Pitts assumed a simple model for an individual neuron. Briefly, they supposed that the dendritic trees (see Fig. 7.27) gather a linear weighted sum of the incoming synaptic signals and compare this sum with a threshold level at the base (initial segment) of the axonal tree. If the weighted sum of input signals exceeds the threshold, then a pulse is launched on the main trunk of the axonal tree. Once launched, that pulse travels out to the first axonal branching, where two pulses are generated that proceed down both secondary fibers. This process continues until all synapses at the tips of the tree have received pulses, with none being lost at axonal branchings. Formally, these assumptions can be expressed as the input–output relationship

$$V_j(t + \tau) = H\left[\sum_{k=1}^{N} \alpha_{jk} V_k(t) - \theta_j\right] , \tag{7.21}$$

where $H(\bullet)$ is the *Heaviside step function* with properties

$$H(x) = \begin{cases} 1 & \text{for} \quad x \geq 0\,, \\ 0 & \text{for} \quad x < 0\,. \end{cases}$$

In this formulation, a model brain is represented by N of such *MP neurons* that are joined by an $N \times N$ interconnection matrix

$$A = [\alpha_{jk}]\,, \tag{7.22}$$

with each element indicating how the firing of the kth neuron influences the tendency of the jth neuron to fire.

There are at least two reasons why McCulloch and Pitts chose such a simple model for their basic neuron. First, not as much was known about the dynamic properties of real neurons in 1943 as now, and it seemed prudent to avoid speculation about more exotic possibilities for neuronal behavior. Second, and perhaps more important, these authors were interested in using their model neuron as a basis for a theory of how the brain works, and a simpler model for the neuron eased this more ambitious task.

Globally, they divided neural systems into "nets with circles" and "nets without circles", the former being more difficult to analyze because of the presence of closed causal loops and the latter involving logical implications that move from input terminals to output terminals through the system.[43] In their analysis of network properties, they discussed only nets without circles. Since the 1950s, several variations of the MP system have been suggested for computer-based models of a brain, with diverse means for adjusting the "synaptic" interconnection weights (α_{jk}) during the course of neural activity [146, 380, 449, 820]. Under the names *linear threshold unit* (LTU), *threshold logic unit* (TLU), and *perceptron*, the MP concepts are widely employed by the engineering community as a basis for designing machines that can learn to recognize and classify patterns [662, 719, 1035].[44] As was subsequently emphasized by mathematician John von Neumann, McCulloch and Pitts had essentially invented the modern digital computer.

Shortly after the 1943 paper appeared, Pitts was introduced to MIT's Norbert Wiener, who was sufficiently impressed by this twenty-year old autodidact to take him on as a doctoral student – the two becoming "like father and son, in contrast to the more collegial relationship Pitts shared with McCulloch" [911]. As the importance of cybernetics became widely recognized in the early 1950s (see Sect. 5.2.1), the president of MIT decided to bring the entire McCulloch group to the Research laboratory of Electronics, expecting

---

[43]Unaware of the work on feedback in electrical engineering in the 1920s and 1930s, McCulloch had been worried that closed neural loops, which are evident from histological studies of brain, could lead to logical contradictions (A equals NOT A, for example). By 1943 he had learned that there is always a time delay of a signal around a loop, so this is not a problem [28].

[44]The proliferation of names for the same concept reflects attempts to establish proprietary positions.

that Wiener would work closely with McCulloch, who – together with von Neumann – was by then one of the leading proponents of cybernetics [189].

Just at the time of this move, and perhaps because of it, Wiener became deeply and inexplicably angry with McCulloch, refusing to have any further contact with him and also dropping Pitts. The reason for this strange turn of events is not clear. Michael Arbib who was working as a graduate student with both scientists in the early 1960s speculates that it was because Wiener could not share center stage with McCulloch [28], while historians Flo Conway and Jim Siegelman place blame on the machinations of Wiener's jealous and unstable wife [189], but whatever the cause, the effect on young Pitts was devasting [911]. He destroyed his research notes and sank into despond, becoming progressively dependent on alcohol and more exotic chemicals until his death in 1969 at the age of 46, just before the dawn of the nonlinear science revolution. Who can know what this brilliant young man might have produced had he been blessed with a loving family and a kind and supportive mentor?

7.4.4 Hebb's Cell Assembly

Not long after the MP paper appeared in 1943, an entirely different formulation of the brain's dynamics was proposed by a Canadian psychologist named Donald Hebb [412]. Although he was aware of the mathematical studies of the brain by Rashevsky, Pitts, McCulloch, von Neumann, and others, Hebb felt that these researchers had "been obliged to simplify the psychological problem almost out of existence". During the 1940s, Hebb had become impressed with several sorts of evidence that cast doubt on current mathematical assumptions and suggested that more subtle theoretical perspectives were needed to explain psychological observations [412]. In contrast to the Chicago group, therefore, he began with the facts of psychology and those of neurology and attempted to "bridge the long gap" between them.

Among such facts is the surprising robustness of the brain's dynamics, an example of which is provided by a Vermont railroad workman named Phineas P. Gage, who in 1848 survived having a tamping iron (about a meter long and an inch in diameter) go through his brain [589]. Explaining such robustness is problematic for the MP model; thus there is a story that during the 1940s von Neumann called McCulloch in the middle of the night to say [28]: "I have just finished a bottle of creme de menthe. The thresholds of all my neurons are shot to hell. How is it I can still think?" Equally forthright, Hebb put the question thus [410, 412]: "How can it be possible for a man to have an IQ of 160 or higher, after a prefrontal lobe has been removed, or for a woman [...] to have an IQ of 115, a better score than two-thirds of the normal population could make, after losing the entire right half of the cortex?"

In response to this problem, Hebb proposed that the brain's neurons typically act as members of functional groups, which he called *cell assemblies* and defined as follows:

Any frequently repeated, particular stimulation will lead to the slow development of a "cell-assembly", a diffuse structure comprising cells [...] capable of acting briefly as a closed system, delivering facilitation to other such systems and usually having a specific motor facilitation. A series of such events constitutes a "phase sequence" – the thought process. Each assembly may be aroused by a preceding assembly, by a sensory event, or – normally – by both. The central facilitation from one of these activities on the next is the prototype of "attention".

His first publication on the cell assembly stemmed from observations of chimpanzees raised in a laboratory where, from birth, every stimulus was under experimental control. Such animals, Hebb noted, exhibited spontaneous fear upon seeing a clay model of a chimpanzee's head [411]. The chimps in question had never witnessed decapitation, yet some of them "screamed, defecated, fled from their outer cages to the inner rooms where they were not within sight of the clay model; those that remained within sight stood at the back of the cage, their gaze fixed on the model held in my hand" [413, 414, 498]. Such responses are clearly not reflexes; nor can they be explained as conditioned responses to stimuli, for there was no prior example in the animals' repertory of responses. Moreover, they earned no behavioral rewards by acting in such a manner. But the reactions of the chimps do make sense as disruptions of highly developed and meaningful internal configurations of neural activity, according to which the chimps somehow recognized the clay head as a mutilated representation of beings like themselves.

Another contribution to the birth of his theory was Hebb's rereading of Marius von Senden's *Space and Sight* [1005], which was originally published in Germany in 1932. In this work, von Senden gathered records on 65 patients who had been born blind due to cataracts up to the year 1912. At ages varying from 3 to 46 years, the cataracts were surgically removed, and a variety of reporters had observed the patients as they went about handling the sudden and often maddeningly novel influx of light. One of the few generalizations over these cases, von Senden noted, was that the process of learning to see "is an enterprise fraught with innumerable difficulties, and that the common idea that the patient must necessarily be delighted with the gifts of light and colour bequeathed to him by the operation is wholly remote from the facts." Not every patient rejoiced upon being forced to make sense of incoming light that was all but incomprehensible, and some found the effort of learning to see to be so difficult that they simply gave up. That such observations are not artifacts of the surgery or uniquely human was fortuitously established through observations on a pair of young chimpanzees that had been reared in the dark by a colleague of Hebb [805]. After being brought out into the light, these animals showed no emotional reactions to their new experiences. They seemed unaware of the stimulation of light and did not try to explore visual objects by touch. Hebb conjectured that the chimps showed no visual response because they had not yet formed the neural assemblies needed for visual perception.

In addition, Hebb pointed out that the learning curve for an individual subject in a behavioral experiment is not the smoothly increasing function shown in psychology textbooks. This difference arises because the textbook curves are averages over many learning experiments, whereas the observations in a particular experiment are influenced by whether the subject is paying attention to the task. Thus the factor of *attention* (otherwise called attitude, expectancy, hypothesis, intention, vector, need, perseveration, or preoccupation), Hebb felt, must somehow be included in any satisfactory theory of learning.

More particularly, cell assemblies were assumed to have the following properties. First, each complex assembly comprises a "three-dimensional fishnet" of many thousands of interconnected cells sparsely distributed over much of the brain. Second, the interconnections among the cells of a particular assembly grow slowly in numbers and strength as a person matures in response to both external stimuli and internal dynamics and are therefore tailored to particular experiences. Third, a mechanism suggested for the growth of neuronal interconnections postulated the strengthening of synaptic contacts through use.[45] Fourth, upon ignition – effected through some combination of external stimuli and the partial activities of other assemblies – a particular assembly remains briefly active, yielding in a second or so to partial exhaustion of its constituent neurons. Fifth, during the period of time that an assembly is active, the attention of the brain is focused upon the concepts embodied in that assembly. Finally, as one assembly ceases its activity, another ignites, then another, and so on, in a temporal series of events called the *phase sequence*, which is experienced by each of us as a train of thought.

As a simple example of assembly formation, consider how an infant might learn to perceive the triangle T shown in Fig. 7.28a. The constituent sensations of the vertices are first supposed to be centered on the retina by eye movement and mapped onto the primary visual area (V1) of the optical lobes of the neocortex (located in the back of your head). Corresponding subassemblies E, F, and G then develop in the secondary visual area through nontopological connections with area V1. The process of examining the triangle involves elementary phase sequences in which E, F, and G are sequentially ignited. Gradually, these subassemblies fuse together into a common assembly for perception of the triangle T. With further development of the assembly T

---

[45] That this feature has become widely known among neural network mavens as a Hebbian synapse amused Hebb because it was one of the few aspects of the theory that he did *not* consider to be original [658]. In fact a real neuron has several means for altering its behavior, including changes in the geometry of dendritic spines or branching, variations in the distributions of ionic channels over the dendritic and axonal membranes, development of dendro-dendritic interactions, changes in amplification levels of decremental conduction, and so on [874].

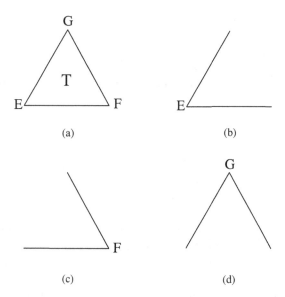

Fig. 7.28. Diagrams related to the process of learning to see a triangle

– which reduces its threshold for ignition upon strengthening of the internal connections among the subassemblies for E, F, and G – a glance at one corner with a few peripheral cues serves to ignite the entire assembly representing T. At this point in the learning process, T is established as a second-order cell assembly for perception of a triangle, including E, F, and G among its constituent subassemblies.

Is there empirical evidence supporting Hebb's theory? Upon formulating the cell-assembly theory for brain dynamics, Hebb and other psychologists began the demanding process of empirical evaluation that is central to science. By the mid-1970s, their collective efforts had produced the following results.

Robustness

There is a social analogy for the cell-assembly concept in which the brain is likened to a community, the neurons to its individual citizens, and cell assemblies to cultural subgroups: families, sport clubs, political parties, circles of close friends, and so on. From this perspective, the remarkable robustness of Phineas Gage's brain to severe physical damage can be understood. If a motorcycle club gets into a fight and loses several of its members, the strength of the club is not permanently reduced because new members can be added. Similarly, a damaged cell assembly can recruit additional neurons to participate in its activities. (Such recruitment of new assembly members may occur during rehabilitation from a stroke, a lobotomy, back injury, or other forms of neurological damage.) Furthermore, because the cells of an assembly may be widely dispersed over much of the brain, partial destruction of the brain does not completely destroy any of the assemblies. Thus, the cell-assembly theory offers the same sort of robustness under physical damage as neurophysiologist Karl Pribram's hologram [771], but it is more credible because a regular structure that can reinforce scattered waves of neural activity is not required.

Learning a New Language

As a graduate student in the "post-Sputnik" days of the late 1950s, I had the experience of learning to read Russian, having no prior knowledge of the language whatsoever. This effort proceeded in stages, beginning with the task of recognizing Cyrillic letters and associating these new shapes with novel sounds. Upon mastering the alphabet, it became possible to learn words comprising these letters, and with enough words, sentences and then paragraphs could eventually be understood. Thus it is my empirical observation that language learning proceeds in a step-by-step process, during which it seems that a hierarchically organized memory of the language is slowly constructed. Interestingly, full perception of a letter or word involves the melding of visual, auditory, and motor components, which underscores the concept of subassemblies being distributed widely over the brain. This general idea of hierarchical learning and memory has been rather carefully formulated in the early 1990s by cyberneticist Valentino Braitenberg and neurologist Friedemann Pulvermüller [120]. Although the acquisition of our basic skills lies buried in the forgotten past, most learning seems layered, with each stage necessarily mastered before it becomes possible to move on to the next. In the context of Hebb's theory, these stages involve the formation of subassemblies from which assemblies of higher order will subsequently emerge.

Ambiguous Perceptions

No discussion of the brain can neglect the mention of ambiguous figures, which have fascinated Gestalt psychologists for generations, and my favorite example – the Necker cube – is shown in Fig. 7.29. In attempting to "bridge the long gap between the facts of neurology and those of psychology", Hebb has offered an explanation for the properties of such perceptions [412]. Gestalt phenomena are thus understood in a visceral manner by supposing that an assembly is associated with the perception of each orientation. Upon regarding Fig. 7.29, for example, I sense something switching inside my head every few seconds as the orientations change. From the cases of people learning to see that were cited by von Senden [1005], it is clear that the ability to perceive an object in three spatial dimensions is learned, and the Necker cube is particularly interesting because perceptions of its two possible orientations would seem to be of equal likelihood.

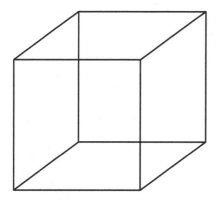

Fig. 7.29. The Necker cube

Stabilized Images

In Hebb's view, some of the strongest evidence in support of his theory was obtained from *stabilized-image* experiments, which were carried out at McGill University in the early 1960s [413, 414, 658, 774, 775]. The experimental setup is sketched in Fig. 7.30, where a simple geometric figure (e.g., a triangle or a square) is projected as a fixed image onto the retina of the subject. In other words, the image will project onto the same cells of the retina even though the eyeball moves. The subjects, in turn, are asked to lie down, relax and simply report what they see, and because this is an introspective experiment, typical results are displayed in a thought balloon.

At first, subjects report seeing the entire figure, but after a few moments the figures change. Habituation effects (perhaps electrochemical changes in the stimulated retinal neurons) cause entire parts of the figures to disappear or to fall out of perception, and it is the manner in which perceptions of the figures alter that is of particular interest. Subjects reported that the component lines or angles (i.e., subassemblies) of a triangle and a square would suddenly jump in and out of perception all at once. These observations are as expected from Hebb's original formulation of the cell-assembly theory and from the learning sequence for a triangle indicated in Fig. 7.28 because the constituent subassemblies are themselves perceptual units. Consequently, stabilized-image experiments confirm an important prediction of the cell-assembly theory.

Learning Environments for Animals

According to the cell-assembly theory, adult thought processes involve continuous interactions among assemblies, which in turn become organized by sensory stimulation and internal interactions during the learning period of a young organism. How does adult behavior depend on opportunities for percept formation during development? Experiments show that rats reared in a rich perceptual environment – a "Coney Island for rats" – are notably

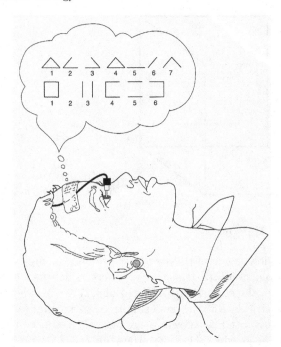

Fig. 7.30. Sketch of contact lens and optical apparatus mounted on the eyeball of a reclining observer. The wire is connected to a small lamp that illuminates the target. The thought balloon shows sample sequences of patterns perceived by the subject with images that are stabilized on the retina by the apparatus. In the upper row the target is a triangle, and in the lower row, it is a square. (After a photograph in Pritchard [775])

more intelligent as adults than those raised in restricted environments, providing yet another confirmation of the theory [658, 780]. As is anticipated from the cell-assembly theory, this positive influence of perceptual stimulation occurred only during youthful development. Extra stimulation of adult rats is far less effective in increasing their intelligence because the assembly structures are then well established.

Similar experiments with Scottish terriers showed even more striking differences, again as expected from Hebb's theory [963]. This is because the fraction of the neocortex that is not under the influence of sensory inputs – the *associative cortex* – is larger for a dog than a rat. Thus, the internal organization of the dog's brain should play a greater role in its behavior. Terriers reared in single cages, where they could not see or touch other dogs, had abnormal personalities and could neither be trained nor bred. Other studies showed that dogs reared in such restricted environments did not respond to pain, as if they had been lobotomized [654].

Sensory Deprivation of Humans

In his original formulation of the cell-assembly theory [412], Hebb speculated that perceptual isolation would cause emotional problems because the phase sequence needs the guidance of meaningful sensory stimulation to remain organized in an intelligible manner and not merely meander through the space of assemblies. To test this aspect of the theory, experiments on perceptual isolation were performed by Heron and his colleagues in the 1950s [426, 658]. In these studies, the subjects were college students who were paid to do nothing. Each subject lay quietly on a comfortable bed wearing soft arm cuffs and translucent goggles, hearing only a constant buzzing sound for several days. During breaks for meals and the toilet, the subjects continued to wear their goggles, so they averaged about 22 hours a day in total isolation.

Many subjects who took part in the experiment intending to plan future work or prepare for examinations were disappointed. According to Hebb [413], the main results were that a subject's ability to solve problems in his or her head declined rapidly after the first day as it became increasingly difficult to maintain coherent thought, and for some it even became difficult to day-dream. After about the third day, hallucinations became increasingly complex. One student said that his mind seemed to be hovering over his body like a ball of cotton wool. Another reported that he seemed to have two bodies but did not know which was really his. Such scientific observations are in accord with a variety of anecdotal reports from truck drivers, shipwreck survivors, solitary sailors, long-distance automobile drivers, and the like – extended periods of monotony breed hallucinations.[46]

After the perceptual isolation experiments were concluded, subjects experienced difficulties with visual perception lasting for several hours and were found to have a significant slowing of their electroencephalograms or brain waves. They also seemed more vulnerable to propaganda. Although the specific results of these experiments were not predicted by the cell-assembly theory, the disorganizing effect of sensory deprivation on coherent thought had been anticipated.

Structure of the Neocortex

While presenting a plausible theory for the dynamics of a brain, Hebb's classic book contains but one lapse into mathematical notation: he discusses in some detail the ratio

$$\frac{A}{S} \equiv \frac{\text{total associative cortex}}{\text{total sensory cortex}}$$

---

[46]Reporting on his famous solo flight across the Atlantic Ocean, for example, US aviator Charles Lindbergh noted "vapor-like shapes crowding the fuselage, speaking with human voices, giving me advice and important messages" [564].

for various mammalian species [412]. This ratio relates the area of the neo-cortex that is not directly tied to sensory inputs – the *associative* (A) regions – to the area of the *sensory* (S) regions, which are under direct environmental control from eyes, ears, and senses of touch and smell. If this ratio is zero, all of the cortex is under sensory control, and necessary conditions for behaviorist psychology are satisfied. On the other hand, larger values of the ratio imply increasing opportunities for the cortex to construct abstract cell assemblies with dynamics beyond direct control of the senses.

In general, Hebb pointed out, this A/S ratio increases as one moves through mammalian species from rat to dog to primate to human, in general agreement with two aspects of brains' behaviors. First, as most would agree, the character of a human's inner life is significantly more intricate than that of a chimp, which in turn is more than for a dog or a rat. Second, the time required for *primary learning* (until adulthood is reached) increases with the A/S ratio. Human infants are essentially helpless and remain so for several years as they slowly build the myriad assemblies upon which the complexities of their lives will eventually be based.

Experiential Responses

Finally consider the hallucinations induced by neurosurgeon Wilder Penfield and his colleagues through electrical stimulation of the neocortex during the 1950s [752]. The purpose of such stimulation was to locate the origin of epileptic activity in order to remove the offending portion of the cortex. In the course of this study, the records of 1 288 brain operations for focal epilepsy were examined. Gentle electrical stimulation was applied to the temporal lobes of 520 patients, of whom 40 reported experiential responses.[47]

M.M. was a typical case. A woman of 26, she had her first epileptic seizure at the age of five. When she was in college, the pattern included hallucinations, both visual and auditory, coming in "flashes" that she felt she had experienced before. One of these "had to do with her cousin's house or the trip there – a trip she had not made for ten to fifteen years but used to make often as a child. She is in a motor car which had stopped before a railway crossing. The details are vivid. She can see the swinging light at the crossing. The train is going by – it is pulled by a locomotive passing from the left to right and she sees coal smoke coming out of the engine and flowing back over the train. On her right there is a big chemical plant and she remembers smelling the odor of the chemical plant."

During the operation, her skull was opened and the right temporal region was explored to locate the epileptic region. Figure 7.31 shows the exposed temporal lobe with numbered tickets marking the sites where positive responses were evoked. Penfield termed these responses *experiential* because

---

[47]Stimulating currents between 50 and 500 microamperes were used in pulses of 2 to 5 milliseconds at frequencies of 40 to 100 hertz.

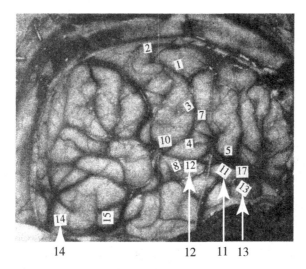

Fig. 7.31. The right temporal lobe of M.M. with numbers that indicate points of positive response. (Redrawn from Penfield and Perot [752])

the patient felt as if she was reliving the experience – not merely remembering it – even as she remained aware that she was lying in the operating room and talking to the doctor. The following experiential responses were recorded upon electrical stimulation at the numbered locations [752]:

11. She said, "I heard something familiar, I do not know what it was."
11. Repeated without warning. "Yes, sir, I think I heard a mother calling her little boy somewhere. It seemed to be something that happened years ago." When asked if she knew who it was she said, "Somebody in the neighborhood where I live." When asked she said it seemed as though she was somewhere close enough to hear.
11. Repeated 18 minutes later. "Yes, I heard the same familiar sounds, it seems to be a woman calling. The same lady. That was not in the neighbourhood. It seemed to be at the lumber yard."
13. "Yes, I heard voices down along the river somewhere – a man's voice and a woman's voice calling." When asked how she could tell it was down along the river, she said, "I think I saw the river." When asked what river, she said, "I do not know, It seems to be one I was visiting when I was a child."
13. Repeated without warning. "Yes, I hear voices, it is late at night, around the carnival somewhere – some sort of a travelling circus. When asked what she saw, she said, "I just saw lots of big wagons that they use to haul animals in."
12. Stimulation without warning. She said, "I seemed to hear little voices then. The voices of people calling from building to building

somewhere. I do not know where it is but it seems very familiar to me. I cannot see the buildings now, but they seemed to be run-down buildings."

14. "I heard voices. My whole body seemed to be moving back and forth, particularly my head."

14. Repeated. "I heard voices."

The 40 cases surveyed by Penfield and Perot show that this particular example of experiential response is not at all unique. In accord with Hebb's cell-assembly theory, they conclude that, in the human brain, there is a remarkable record of the stream of each individual's awareness or consciousness and that stimulation of certain areas of the neocortex – lying on the temporal lobe between the auditory and visual sensory areas – causes previous experience to return to the mind of a conscious person.

Thus by the middle of the 1970s there was a considerable body of empirical evidence consistent with or directly supporting Hebb's theory of cell assemblies as an organizing principle of the human brain [865]. The theory itself had also undergone modification because in Hebb's original formulation only excitatory interactions among neurons were assumed in accord with neuroscientific knowledge in the 1940s [412]. In 1957, the theory was reformulated by Hebb's colleague Peter Milner into a Mark II version which includes inhibitory interactions, and this is the version that is currently meant by "cell-assembly theory" [657] and was confirmed by early numerical experiments during the 1950s [810].

In the early 1980s, a widely promoted effort in the physics community was sparked by papers of John Hopfield showing that collective neural states are related to mathematically similar collective effects in condensed-matter physics called spin glasses [449, 450]. This Hopfield model of the brain's dynamics is based on MP neurons together with the assumption that the elements of the interconnection matrix (7.22) can vary with learning and that the matrix itself is symmetrical, which is not physiologically realistic. Borrowing nonlinear analytical techniques from physics, an energy (Lyapunov) functional is constructed which is bounded from below and decreasing under the dynamics, thus proving the existence of stable oscillatory states, which are interpreted as states of mental attention. By the 1990s this line of analysis had been taken up by many condensed-matter physicists with interest in the brain dynamics, showing among other things that the maximum number of patterns (stable states) that can be stored by N model neurons is about $p \sim 0.14N$ [21, 432]. As the number of neurons in the brain is usually estimated to be about 10^{10}, this implies the number of stable brain states to be about 10^9, which is also the number of seconds in 30 years – a reasonable value. Although the physicists working in this area invariably cite Hebb's book, it is merely to support their assumption that the elements of (7.22) change under learning, which indicates that they hadn't actually read the book or been aware of the substantial body of psychological literature supporting Hebb's cell-assembly theory.

In fact, the Hopfield model is a special case of Hebb's cell-assembly theory [874], for which the number of stored patterns was studied back in the 1960s by Charles Legéndy, an engineer who conservatively estimated that the maximum number of complex Hebbian patterns that can be stored is

$$p \sim \left(\frac{N}{ny}\right)^2 , \tag{7.23}$$

where n is the number of neurons in a subassembly and y is the number of subassemblies in an assembly [550–552]. For $N = 10^{10}$ and $ny = 10^5$, the number of brain states according to (7.23) is 10^{10} – also a reasonable value.

Although a physicist, Hermann Haken does not share the insularity of the Hopfield model community, as his synergetic approach (see Sect. 5.2.6) has been widely applied throughout the physical, biological and social sciences [391, 393, 394]. In the context of brain modeling, one of his order parameters can be taken as the firing rate (F) of a cell assembly [392], and harking back to the Verhulst model of Chap. 1, it is convenient to write the growth law for a single assembly as [874]

$$\frac{dF}{dt} = F(1 - F) ,$$

where time is measured in units of the synaptic delay. (Thus the same growth equation describes the firing rate of a cell assembly, the population of Belgium, and the onset of research in nonlinear science.) So motivated, let us model the dynamics of two identical neural assemblies with inhibitory interactions by the coupled ODE system (see Appendix A)

$$\begin{aligned}
\frac{dF_1}{dt} &= F_1(1 - F_1) - \alpha F_2 , \\
\frac{dF_2}{dt} &= F_2(1 - F_2) - \alpha F_1 ,
\end{aligned} \tag{7.24}$$

where $0 \le F_1 \le 1$ and $0 \le F_2 \le 1$ because F_1 and F_2 represent the fraction of neurons in each assembly that are firing. When positive, the parameter α introduces an inhibitory interaction between the two assemblies because the $-\alpha F_2$ term in the first equation reduces dF_1/dt and similarly for the second equation.

To see how these equations model the role that inhibition plays in the formation of cell assemblies, note that if we let α be negative, only excitatory interactions among the neurons are allowed. In this case, as is seen from Fig. 7.32a, all points on the (F_1, F_2) phase plane move to (1,1), and no individual assemblies are permitted to ignite. In other words, all neurons always end up firing at their maximum rates. Milner's Mark II version of the cell

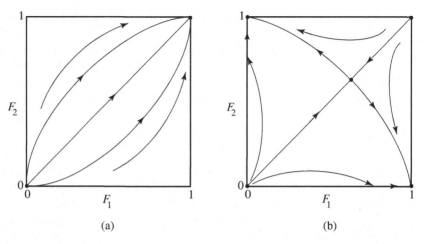

Fig. 7.32. Phase-plane trajectories for a cell-assembly model. (**a**) A phase-plane plot from (7.24) with $\alpha < 0$ (only excitatory interactions). (**b**) A similar plot for $\alpha > 1/3$ (excitatory and inhibitory interactions)

assembly theory overcomes this difficulty by allowing $\alpha > 0$, whereupon one finds a singular point at

$$F_1 = F_2 = 1 - \alpha ,$$

where the time derivatives are zero. For $0 < \alpha < 1/3$, this singular point is stable, but for $\alpha > 1/3$, it becomes *unstable,* as shown in Fig. 7.32b. Stable states of the system are then at either $(F_1, F_2) = (1, 0)$ or $(0, 1)$; thus, with sufficiently large inhibition, (7.24) suggest that assemblies can be individually ignited.

Now revisit Fig. 7.29 and experience how your perception switches back and forth between the two orientations of the Necker cube. Although it is easy to see the cube in either orientation, note that you cannot perceive both orientations at the same time. (How rapidly can you switch between perceptions of the two orientations? Might the speed of these transitions be taken as a measure of how well your brain is working?) Evidently, (7.24) with $\alpha > 1/3$ and the corresponding phase-plane diagram shown in Fig. 7.32b model the switching on and off of assemblies that correspond to the dynamics of those in your head as you regard the Necker cube.

From an engineering perspective, the interactive dynamics of two assemblies are like a flip-flop circuit widely used in the design of information storage and processing systems – the van der Pol oscillator of Sect. 6.6.2 with large ε, for example. With a cell assembly, however, the bit of information being switched on or off is not the voltage level of a transistor but an intricate psychological perception embodied in the connections among hundreds of subassemblies scattered about the brain that have developed in response to the lifelong experiences of the organism. Although this has been a "bottom-up" discussion of the brain's dynamics, it suggests the utility of "top-down" approaches. Regarding assembly firing rates as *order parameters* for higher level representations of the brain's dynamics, for example, Haken has been able to model a variety of psychological experiments [391, 392].

To represent more than two assemblies, (7.24) can be generalized to

$$\frac{dF_1}{dt} = +F_1(1 - F_1) - \alpha F_2 - \alpha F_3 - \cdots - \alpha F_n ,$$

$$\frac{dF_2}{dt} = -\alpha F_1 + F_2(1 - F_2) - \alpha F_3 - \cdots - \alpha F_n , \qquad (7.25)$$

$$\vdots$$

$$\frac{dF_n}{dt} = -\alpha F_1 - \alpha F_2 - \alpha F_3 - \cdots + F_n(1 - F_n) ,$$

where $0 \leq F_j \leq 1$ for $j = 1, 2, \ldots, n$. In this n-assembly model, interestingly, all of the previous analysis (for $n = 2$) can be carried through; thus there is a singular point for positive α (the inhibitory case) at

$$F_1 = F_2 = \ldots = F_n = 1 - (n - 1)\alpha , \qquad (7.26)$$

which is stable (unstable) for $\alpha < (>) \; \alpha_c = 1/(2n - 1)$. Below this critical value α_c of inhibition, all of the assemblies can become simultaneously active, which is not of interest from a neurological perspective. Above the critical level, on the other hand, (7.25) has a singular point defined by (7.26) which becomes unstable with a switching time τ_{sw} given by

$$\tau_{sw} = \frac{1}{(2n - 1)\alpha - 1} , \qquad (7.27)$$

counterintuitively implying that the rate for a neural system to change from one perception to another *increases* with inhibition (α).

Although obtained from a very simple model, (7.27) is in accord with Hebb's suggestion that we humans are more intelligent than our fellow mammals in part because we can switch our attention more quickly from one assembly to another, allowing more rapid trains of thought [413, 414, 475]. Following a related suggestion by William James in his *Principles of Psychology* that attention be paid to transitions between one stable mental state and another [466], physicist Harald Atmanspacher and his colleagues have recently been studying correlations between Necker-cube perception reversals and event-related potentials (ERPs) measured as EEGs, finding well defined ERPs that have been tentatively related to a "toy model" of a cell-assembly phase space that is similar to (7.25) [509]. In contrast to clinical observations that rates of perception reversal for a Necker cube are less (i.e., slower) for subjects with brain lesions than for a normal control group [1006], it has been observed that a certain type of meditation practised by Tibetan

Buddhist monks ("one-point" but not "compassion" meditation) can prolong periods of perceptual rivalry [159]. Evidently, cell-assembly dynamics modeling in psychological research is of practical and theoretical interest, with much important work remaining to be done [392].

By the middle of the 1970s, the Mark II version of Hebb's cell assembly enjoyed firm empirical support [865] which has become even more solid over the past decade for two reasons. First, Anders Lansner, Erik Fransén and their computer science colleagues have been taking advantage of the great increases in computing power to conduct numerical studies of Hebb's system with physiologically realistic neuron models [324–327, 531, 532, 532–534]. These studies confirm most of Hebb's original speculations, including after activity of about 400 ms for an excited assembly, reasonable ignition thresholds, and pattern completion for partially stimulated asemblies. Second, the cell-assembly concept is becoming increasingly useful for understanding observations on laboratory preparations, including mollusk [1058], locust [540,541,1020], moth [181], rat [220, 349, 708, 710], ferret and mouse [631]. All of these results and more have led neurobiologists Miguel Nicolelis, Erika Fanselow, and Asif Ghazanfar to comment in 1997 as follows [709]:

> What we are witnessing in modern neurophysiology is increasing empirical support for Hebb's views on the neural basis of behavior. While there is much more to be learned about the nature of distributed processing in the nervous system, it is safe to say that the observations made in the last 5 years are likely to change the focus of systems neuroscience from the single neuron to neural ensembles. Fundamental to this shift will be the development of powerful analytical tools that allow the characterization of encoding algorithms employed by distinct neural populations. Currently, this is an area of research that is rapidly evolving.

7.4.5 Cognitive Hierarchies

If we are prepared to accept that the ten billion neurons of a human neocortex act in functional groups along the lines of the previous section, it follows that the global organization of the brain is hierarchical. To see this, note that the fundamental dynamical feature of a neuron is its threshold for incoming pulses, below which the neuron rests and above which it fires, sending pulses out to other neurons. A moment's consideration shows that a primary assembly of such neurons has the same basic property; if a sufficient number of its constituent neurons are active (firing) then the entire assembly will become active and send signals of that fact to the assemblies with which it is connected. For a secondary assembly, comprising primary assemblies, the same picture emerges; if a sufficient fraction of its constituent primary assemblies are active, the secondary assembly will become active – and so on for various higher-level assemblies. These considerations lead directly to a *cognitive hierarchy* which corresponds to the biological hierarchy that was presented in Sect. 7.3.6 and can be sketched as follows [866]:

Human culture
Psychology
Phase sequences
Complex assemblies

⋮

Assemblies of assemblies of assemblies
Assemblies of assemblies
Assemblies of neurons
Neurons
Nerve impulses
Nerve membranes
Membrane proteins
Molecules

Although this diagram differs from the biological hierarchy in some important ways, the comments in Sect. 7.3.6 still apply. In particular, the details of the diagram are not of primary concern (others might organize the levels differently), and – more important – each cognitive level has its own nonlinear dynamics, involving closed causal loops of positive feedback, out of which can possibly emerge a very large and often immense number (see Sect. 7.3.4) of interacting entities. As in the biological hierarchy, a necessarily small subset of these many possibilities does in fact emerge, providing a basis for the nonlinear dynamics of the next higher level. Perhaps the most significant difference between the biological and cognitive hierarchies lies in the internal levels

Complex assemblies

⋮

Assemblies of assemblies of assemblies
Assemblies of assemblies
Assemblies of neurons

which cannot be directly observed but are induced from indirect evidence using arguments presented in the previous section. For the biological hierarchy, on the other hand, the emergent entities at every level are seen using electron or optical microscopes, X-rays, or direct observation.

Thus the concepts of nonlinear science are woven into the fabric of the cognitive hierarchy, just as for the biological hierarchy. The protein solitons described in Sects. 7.1.2 and 7.1.3 lead to an understanding of the nonlinear dynamics of active membranes as discussed in Sect. 7.2.2. The ways that active membranes support nerve pulses have been considered in Sects. 4.1, 4.2, and 7.4.1, and the nonlinear dynamics of neurons were presented in Sect. 7.4.2. Finally, the nonlinear organization of neurons into complex cell assemblies was described in the previous section, and the relation of complex assemblies to psychology is the main subject of Hebb's classic book [412] (see also the important studies in the late 1960s by psychologist Abraham Maslow [615, 616]).

Although the emergence of human culture from psychology lies beyond the scope of this book, Philip Chase has recently reviewed the evidence that our culture arose near the end of the Middle Pleistocene epoch (about 200 000 years ago [171]), using a definition of emergence that is directly drawn from nonlinear science and basing his conclusions on datings of cave paintings, decorated bones, shell beads, other artistic objects, and ritual practices for burial. Contrasting forms of present-day human culture were perceptively compared in 1934 by Ruth Benedict in her book *Patterns of Culture* by looking at three quite different groups: the gentle Zuñi of New Mexico, the paranoid Dobu of southeastern New Guinea, and the destructively competitive Kwakiutl of Vancouver Island [74].

Since first reading it in a freshman class on "world cultures", I have felt close to Benedict's classic because it awoke my interest in the phenomenon of emergence, more than a decade before I had conceived of nonlinear science [873]. An introduction to her book by Franz Boas, Benedict's mentor-cum-colleague, includes a poem by Goethe which was translated for us by an Austrian student as:

One who would know a living thing,
Tries first to drive its spirit out.
Then with the pieces in his hand,
He lacks its unifying bond.

In the following chapter, we will seek this "unifying bond" in living systems.

8 Reductionism and Life

> There is no record of a successful mathematical theory which would treat the *integrated* activities of the organism as a whole. [...] And yet this integrated activity of the organism is probably the most essential manifestation of life.
>
> Nicholas Rashevsky

As was emphasized in the first chapter of this book, the nexus between nonlinear science and philosophy can be traced back to Aristotle's four categories of causality [29], because nonlinearity mixes strands of causal implication. Although fully aware of Newtonian physics, modern thinkers often overlook the implications of nonlinear science, tacitly believing that chains of logical inference can always be traced from whatever happens back to the sources of what made it so and thereby supporting the devotion to explanatory reductionism that pervades our Western culture. Here we will see that the reductive perspective ignores emergent and chaotic phenomena that arise in much of physical science and most of the biological and social sciences. In the course of their current attempts to understand the nature of Life, it is hoped that philosophers of science and the literate public will begin to see how complicated reality is, and thus to recognize the implications of the nonlinear-science revolution.

8.1 Newton's Legacy

Since Aristotle's mechanics were overturned by Isaac Newton's radical reformulation in the seventeenth century, the reductive program has been successful in prising out explanations for natural phenomena in many fields, in addition to reformulating our collective concept of Nature itself. As a part-time alchemist, Newton did not restrict himself to one way of pursuing knowledge [359], but his name has become attached to the reductive program, which is now widely accepted by the scientific community as the fundamental way to pose and answer questions in a search for truth. How does it go?

8.1.1 The Reductive Program

As scientists are aware, an effective approach to understanding natural phenomena proceeds in three steps:

- **Analysis.** Assuming some higher-level phenomenon is to be explained, separate the underlying system into disjoint elements, the behaviors of which are to be individually investigated.
- **Theoretical Formulation.** Guided by empirical studies, imagination and luck, obtain a formulation (or model) of how the components might be related and interact.
- **Synthesis.** In the context of this model, derive the higher-level phenomenon of interest and quantitatively compare its theoretical behavior with empirical observations.

Studies of natural phenomena that have successfully used this reductive approach include: *planetary motion* (based on the concepts of mass and gravity, related by Newton's laws of motion), *hydrodynamics* (based on the concepts of fluid density and Newton's laws, related by the Navier–Stokes equation), *electromagnetic radiation* (based on the concepts of electric charge, electric fields and magnetic fields, related by Maxwell's electromagnetic equations), *atomic and molecular structures* (based on the concepts of mass and electric charge, related by Schrödinger's equation for quantum probability amplitudes), *nerve impulse propagation* (based on the concepts of voltage, membrane permeability and ionic current, related by the Hodgkin–Huxley equations for the dynamics of current flow through a voltage sensitive membrane), the *structures of protons and neutrons* (based on the concepts of leptons and quarks and strong, weak, and electromagnetic forces, related by the Standard Model), and *the evolution of our Universe* (based on the concepts of mass and curved space, related by the Einstein–Hilbert gravitational equations) – an impressive list indeed.

Generalizing from such examples, many scientists agree with physicist Steven Weinberg and biologist Jacques Monod that all natural phenomena can be understood "in principle" by reducing them to fundamental laws of physics [669, 1023, 1024, 1026, 1027].[1] Others deny such *explanatory reductionism*, maintaining that there are natural phenomena that cannot be described in terms of lower-level entities – Life and human consciousness being outstanding examples [23, 191, 623, 624].[2] In an extreme form, the denial

---

[1] Weinberg was awarded the 1979 Nobel Prize in Physics for unifying weak and electromagnetic forces.

[2] In the late 1980s and early 1990s, the polemics between these two camps were sharpened in the United States by a political struggle over an expensive Superconducting Super Collider (SSC), which would test high-energy predictions of the Standard Model. Funding for the SSC was terminated by the House of Representatives in 1993 after two billion dollars had been spent and fourteen miles of tunnel had been dug.

of explanatory reductionism is called *substance dualism*: the view of René Descartes that important aspects of the biological and cognitive realms do not have a physical basis. A more moderate position is *property dualism* (also called *functionalism* [1080]), which accepts a physical basis for all biological and cognitive phenomena but asserts nonetheless that some emergent aspects of these sciences cannot be explained in terms of atomic or molecular dynamics [313, 623, 1080]. How can we decide between such divergent views of Nature?

8.1.2 Supervenience and Physicalism

To statements of belief there is not a scientific response, but if we are able to agree on the physical basis of Life and Mind, the scope of the discussion narrows. Let's assume, therefore, that all biological, cognitive and social phenomena *supervene* on the physical in the following sense (as philosopher Jaegwon Kim has put it in the context of cognitive science [491, 492]):

> Any two things that are exact physical duplicates are exact psychological duplicates as well.

Related to Ernst Mayr's *constitutive reductionism* [623, 624], this position is called *physicalism* by philosophers [1080], and among biologists and neuroscientists it is now widely accepted for the phenomena of Life and Mind, whereas Henri Bergson's vitalism [78] is almost universally rejected, not to say despised. Thus two questions arise:

- Does explanatory reductionism follow from physicalism?
- Does physicalism allow property dualism?

Over the past two decades, these questions have been considered by Kim, who reluctantly concludes that physicalism does indeed imply explanatory reductionism and sits uneasily with property dualism [491, 492]. Let us review his arguments with reference to Fig. 8.1.

This figure represents higher-level mental phenomena (M_1 and M_2) that supervene on lower-level physical descriptions (P_1 and P_2), where supervenience is indicated by the vertical dashed lines. In other words, there cannot be a difference in M_1 without a corresponding difference in P_1, with the same relationship between M_2 and P_2 [1080].

Now suppose that empirical studies in psychology have established *causality* between M_1 and M_2 (indicated by the horizontal arrow in Fig. 8.1), under which an observation of M_1 always leads to a subsequent observation of M_2. Under the assumption of physicalism, P_1 (P_2) must be present to provide a basis for M_1 (M_2); thus we could as well say that P_1 causes P_2, which is a formulation of the upper-level causality in terms of the corresponding lower-level properties. In other words, one can reduce the causal relation between M_1 and M_2 to a corresponding causal relation between P_1 and P_2,

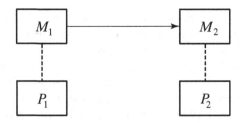

Fig. 8.1. The causal interaction of higher-level mental phenomena (M_1 and M_2) that supervene on lower-level physical properties (P_1 and P_2)

thereby supporting explanatory reductionism and undercutting property dualism. There is no claim that such an explanatory reduction is conveniently formulated or has been achieved, but merely that it is possible "in principle".

8.1.3 Practical Considerations

Beyond Weinberg's claim and Kim's logic, there is a practical argument supporting the reductive view. Even if explanatory reductionism were not to hold for all aspects of biological or mental organization, it is still a prudent strategy for biologists and cognitive scientists to take as a working hypothesis, because the riddles of one generation often become standard knowledge of the next. Recalling from earlier chapters that physicist Hermann Helmholtz could not come up with an explanation for his measurements of a small speed of nerve impulse propagation [864] and that physiologist Edward Reid's empirical observations of active ion transport across cell membranes were ignored for seven decades by biologists fearing the taint of Bergsonian vitalism [795, 796, 854], who would claim that a mystery of today couldn't be reductively explained by a a fresh-eyed young thinker of tomorrow? Thus the dualist (substance or property) is ever in danger of giving up too soon, and one might say that it is the *duty* of a scientist to search for reductive explanations for empirical observations.

Evidently, explanatory reductionism grounded in physicalism is a serious philosophical position. Those who disagree, as I do, must offer substantial objections.

8.2 Objections to Reductionism

Although many physicists (especially elementary-particle physicists who seek a "theory of everything") are reductionists [1026, 1027], condensed-matter physicists (those who study global phenomena in aggregates of atoms and molecules) often challenge such claims. Thus condensed-matter physicist Philip Anderson has asserted that [23]:[3]

> [...] the reductionist hypothesis does not by any means imply a "constructionist" one: The ability to reduce everything to simple fundamental laws does not imply the ability to start from those laws and reconstruct the universe. In fact the more the elementary-particle physicists tell us about the nature of the fundamental laws, the less relevance they seem to have to the very real problems of the rest of science, much less to those of society. The constructionist hypothesis breaks down when confronted with the twin difficulties of scale and complexity.

What is it about "scale and complexity" that creates difficulties for the constructionist hypothesis?

8.2.1 Googols of Possibilities

As was discussed in Sect. 7.3.4, biological computations are often problematic because the numbers of possible emergent structures at each level of Life's hierarchy are of the order of a *googol* (10^{100}) or more. Thus biological science differs fundamentally from physical science, which deals with *homogeneous* sets having identical elements. A physical chemist, for example, has the luxury of performing as many experiments as are needed to establish scientific *laws* governing the interactions among (say) atoms of carbon and hydrogen as they form identical molecules of benzene. In life sciences, on the other hand, the number of possible members in interesting sets is typically *immense*, so experiments are necessarily performed on *heterogeneous subsets* of the classes of interest (see Sect. 7.3.5). Because the elements of heterogeneous subsets are never exactly the same, it follows that experiments on them cannot be precisely repeated; therefore causal regularities cannot be determined with the same degree of certainty in the biological, cognitive and social sciences as in the physical sciences.

In other words, life scientists establish *rules* rather than laws for biological and social dynamics, and your doctor can only know the probability that a certain pill will effect a cure. Thus the horizontal arrow from M_1 to M_2 in Fig. 8.1 might be drawn fuzzy or labeled with an estimate of its reliability, in order to indicate a deviation from strict causality.

8.2.2 Convoluted Causality

Contrary to the assumption of reductionist scientists, causality can be and usually is a very complicated (intricate) phenomenon. Nonlinear dynamics

---

[3] Anderson was awarded the 1977 Nobel Prize in Physics for his work on disordered systems.

offer many examples of sensitive dependence on initial conditions, leading to the "fortuitous phenomena" discovered by mathematician Henri Poincaré (see Sect. 2.2 and the epigraph to Chap. 2) and dubbed the butterfly effect by meteorologist Edward Lorenz (Sect. 1.3), but such effects have long been informally recognized. Political scientists speak of tipping points, and among computer engineers and neuroscientists, the corresponding idea of a threshold level at the input of an information processor (below and above which different outcomes transpire) is an essential concept [886].

As a problem long of interest to me, consider the dynamics of dendrites, which carry signals into neurons from sensors and other neurons (see Sect. 7.4 and [863,874,894,937]). Until the mid-1980s, it was widely assumed that these highly branching structures computed linear weighted sums of their inputs, largely because this perspective helped neuroscientists follow their strands of theoretical causality. Real dendrites, on the other hand, are now known to be highly nonlinear, offering many additional tipping points to the dynamics of every neuron.

How are such twisted skeins of causality to be sorted out? Whether one is concerned with establishing dynamic laws in the physical sciences or seeking rules in the life sciences, the notion of causality requires careful consideration [136]. As was noted in Chap. 1, Aristotle identified four types of cause: material, formal, efficient, and final [29], and for those familiar with the jargon of applied mathematics, the following paraphrasing of his definitions may be helpful:

- At a particular level of the biological hierarchy, a *material cause* would be a time or space average over a dynamic variable at a lower level of description, which would appear as a slowly varying *parameter* at the level of interest. (Examples are the density of atoms in a chemical reaction or their temperature.)

- Again, at a particular level of the biological hierarchy, a *formal cause* might arise from the more slowly varying values of a dynamic variable at a higher level of description, which enters as a *boundary conditions* at the level of interest. (The walls of a chemist's test tube, the drum with a marching column of soldiers, or the temperature inversion inducing the formation of a tornado are examples.)

- An *efficient cause* is represented by a *stimulation–response* relationship, which is usually formulated as a differential equation with a dependent variable that responds to a forcing term. We physical scientists spend our formative years solving such problems, with the parameters (material causes) and boundary conditions (formal causes) specified. Apart from Galileo's influence [136], this widely-shared early experience may explain why most of us automatically assume that every natural event can be described in terms of efficient causes.

- A *final cause* requires understanding the *intention* of some cognitive entity. In modern philosophical jargon, this is the "problem of action" which has recently been analyzed in some detail by philosopher Alicia Juarrero from the perspectives of information theory and nonlinear dynamical systems [475]. Although the reality of intentions has been strongly rejected by Monod [669], Juarrero has shown how intention can emerge in a higher-level state space that includes both internal mental states (the cell assemblies of Sect. 7.4.4) and external physical and cultural phenomena.

As Aristotle well knew, several such causes are often involved in a single outcome, comprising a web rather than a chain; thus we should expect that parameter values, boundary conditions, forcing functions, and cognitive intentions will all combine to influence a given dynamical process. Clearly, causality is more convoluted than reductionists would have us believe. But there is more.

8.2.3 Nonlinear Causality

In applied mathematics, the term "nonlinear" is usually defined in the context of relationships between efficient causes and their effects. Suppose that a series of experiments on a certain system has shown that cause C_1 gives rise to effect E_1; thus

$$C_1 \rightarrow E_1 \;,$$

and similarly

$$C_2 \rightarrow E_2$$

expresses the relationship between cause C_2 and effect E_2. From the discussion in Chap. 1, this relation is *linear* if

$$C_1 + C_2 \rightarrow E_{12} = E_1 + E_2 \;. \tag{8.1}$$

If, on the other hand, E_{12} is *not* equal to $E_1 + E_2$, the effect is said to be a *nonlinear* response to the causes. Equation (8.1) indicates that for a linear system any efficient cause can be arbitrarily divided into components $(C_1, C_2, \dots, C_n)$, whereupon the effect will be correspondingly divided into $(E_1, E_2, \dots, E_n)$. Although convenient for analysis,[4] this property is not found in many areas of physical science (as we have seen in Chap. 6) and rarely in the life sciences (see Chap. 7).

Far more common is the nonlinear situation, where the combined effect from two causes is not equal to the sum of their individual effects, and the whole is not equal to the sum of its parts. Nonlinearity is inconvenient for the academic analyst who wishes to publish many papers and advance his career, because multiple causes interact among themselves, allowing possibilities for more outcomes, obscuring relations between cause and effect, and confounding the constructionist. For just this reason, however, nonlinearity plays key roles in the course of biological dynamics.

---

[4]Providing a basis for Fourier analysis and Green function methods [875] and for the quantum theory sketched in Appendix B.

8.2.4 Time's Arrow

Our concept of causality is closely connected with our sense of time; thus, the statement "C causes E" implies (among other things) that E does not occur before C [136] – yet the properties of time may depend on the level of description [328, 329], particularly in biological systems [1049, 1050]. For example, the dynamics underlying molecular vibrations are based on Newton's laws of motion, in which time is *bidirectional*. In other words, the direction of time in Newton's formulation can be changed without altering the qualitative behavior of the system. (The Solar System could as well run backward as forward, as could the pendulum of a clock.) For systems like a nerve axon that consume energy, on the other hand, time has an "arrow"; thus a change in time's direction makes an unstable nerve impulse stable and vice versa [772]. In Fig. 4.3 under time reversal, for example, the dashed branch would represent stable impulses traveling more slowly than the unstable impulses represented by the solid branch – a striking qualitative difference.

In appealing to Fig. 8.1, therefore, the reductionist must recognize that the nature of the time used in formulating the causal relationship between P_1 and P_2 may differ from that relating M_1 and M_2.

8.2.5 Downward Causation

Explanatory reductionists focus their attention on efficient causality, which acts upward through the biological hierarchy [669, 1026], possibly explaining a tendency in current science to ignore other types of causality [136]. Importantly, material, formal and final causes can also act *downward*, because variables at the upper levels of a hierarchy can place constraints (boundary conditions, for example) on the dynamics at lower levels and higher-level intentions can supply material for lower-level activities (think of a runner deciding to take glucose tablets during a marathon). To sort things out, theoretical biologist Claus Emmeche and his colleagues have recently defined three types of downward causation [22, 288]:

- **Strong Downward Causation (SDC).** Under SDC, it is supposed that upper-level phenomena act as efficient agents in the dynamics of lower levels. In other words, upper-level organisms can influence or modify the physical and chemical laws governing their molecular constituents, for example by changing the attraction between electric charges. Although New Agers may disagree,[5] there is presently no empirical evidence for the downward action of efficient causation, so SDC is almost universally rejected by life scientists.
- **Weak Downward Causation (WDC).** Rejecting vitalism, WDC assumes that the molecules comprising an organism are governed by some

---

[5]See *What the Bleep!?* at http://www.noetic.org/links/bleep_guide.cfm for a New Age perspective on SDC.

nonlinear dynamics in a phase space, having *attractors* which include the living organism (see Appendix A). Each of these attractors is endowed with a corresponding *basin of attraction*, within which the dynamics are stable. Under WDC, an external higher level stimulation might move lower level phase-space variables from one basin of attraction to another, causing a global change in the dynamics of the organism – from fight to flee or feed, for examples. Another example of WDC is as a theoretical explanation for *punctuated evolution*, whereupon a new species appears suddenly on an evolutionary time scale as the phase point describing a species moves into a new basin of attraction [367, 370, 481].

Because many examples of such nonlinear systems have been studied both experimentally and theoretically, there is little doubt about the scientific credibility of this means for downward causation. Thus biologists Stuart Kauffman [481] and Brian Goodwin [367] have used a suggestion of mathematician Alan Turing [973] to present detailed discussions of ways that WDC can influence the morphological development and behavior patterns of living organisms.

- **Medium Downward Causation (MDC).** Accepting WDC and again rejecting vitalism, proponents of MDC go further in noting that higher-level dynamics (e.g., the emergence of a higher-level structure) can modify the parameters of an organism's lower-level phase space through the downward actions of formal and final causes. Examples of MDC include modifications of DNA codes caused by interactions among species under Dawinian evolution and the changes in interneuronal coupling (synaptic strengths) in our brains upon learning.

Evidently both MDC and WDC must be considered in any serious account of Life's dynamics.

8.2.6 Open Systems

In contrast with most formulations of classical physics, in which energy is conserved, living organisms are *open systems*, requiring an ongoing input of energy and matter (sunlight or food plus oxygen) to maintain their metabolic activities. The *steady state*[6] of an open system, which was defined and discussed in 1950 by biologist Ludwig von Bertalanffy, corresponds to a stable singular point in an underlying phase space (see Appendix A). Von Bertalanffy considered the corresponding thermodynamics and also defined the important concept of *equifinality* in which different initial conditions can lead to the same final state (Chap. 4 and [1001]). From the phase-space perspective of Appendix A, equifinality is expected because a solution trajectory can enter the same basin of attraction from several different directions, all of which eventually arrive at the same stable singular point.

---

[6]The corresponding German term *Fliessgleichgewicht* was coined by von Bertalanffy.

Among the many biological applications of these concepts mentioned by von Bertalanffy are biophysicist Nicholas Rashevsky's theoretical cell model [790], morphogenesis (Sect. 7.2), and nerve impulses (Sects. 4.2 and 7.4 and [874]). Although stationary states of open systems seem strange to physicists, who think in terms of energy-conserving fields, they are well known to electrical engineers, working as they do with nonlinear systems comprising batteries (sources of energy) and resistors (sinks of energy) in addition to energy-storage elements (inductors and capacitors) [861].

A familiar example of a steady-state solution in an open system is provided by the flame of a candle [300]. Starting with the size and composition of a candle, it is possible to compute the (downward) propagation velocity of the flame (v) by equating the rate at which energy is dissipated by the flame (through radiation of light and conduction of heat) to the rate at which energy is released from the wax of the candle [875]. This analysis establishes a rule for finding where the flame will be located at a particular time; thus corresponding to

$$M_1 \rightarrow M_2 \ ,$$

in Fig. 8.1, such a rule is the following: if the flame is at position x_1 at time t_1, then it will be at position

$$x_2 = x_1 + v(t_2 - t_1) \tag{8.2}$$

at time $t_2 > t_1$, where v is the velocity of the flame. Because a candle and its flame comprise an open system, however, a corresponding relation

$$P_1 \rightarrow P_2$$

cannot be written in terms of the positions and velocities of the constituent atoms and molecules – not even "in principle". Why not? Because the molecules comprising the physical substrate of the flame are *continually changing* [87]. Although its velocity is the same, the flame's heated molecules of air and wax vapor at time t_2 are different from those at time t_1. Thus, knowledge of the detailed positions and speeds of the molecules present in the flame at time t_1 tells us nothing about the speeds and positions of the constituent molecules at time t_2. What remains constant is the flame itself – a higher-level *process* which can be computed from local averages over its constituents.[7]

As noted in the epigraph to Chap. 4, a candle flame is a useful example of an open system that is also a closed causal loop. The causal loop occurs as heat from the flame "feeds back".

---

[7]It might be asserted that "in principle" one could compute the dynamics of all the matter and all the radiation of the universe, but this would require an "omniscient computer", which is similar to (and perhaps equivalent to) the Calvinist notion of God. True or not, this speculation tells us nothing about reductionism.

8.2.7 Closed Causal Loops and Open Networks

In his influential analysis of reductionism, Kim asks (with Aristotle): "How is it possible for the whole to causally affect its constituent parts on which its very existence and nature depend?" [493]. Causal circularity between higher and lower levels is deemed unacceptable by reductionists because it violates the following *causal-power actuality principle*:

> For an object, x, to exercise, at time t, the causal/determinative powers it has in virtue of having property P, x must already possess P at t. When x is being caused to acquire P at t, it does not already possess P at t and is not capable of exercising the causal/determinative powers inherent in P.

There are two objections to this claim, one theoretical and the other empirical. Theoretically, Kim incorrectly supposes that an emergent structure somehow pops into existence at time t, which would indeed be surprising. Typically, however, an emergent entity (or coherent structure) begins from an infinitesimal seed (noise) that appears at a lower level of description and develops through a process of exponential growth (instability). Eventually, this growth is limited by nonlinear effects (see Fig. 1.1), and a stable entity is established (the collective research activity in nonlinear science, for example). Think of lighting a bonfire. Upon being barely lit, a small but viable flame grows rapidly before settling down to its natural size for the available fuel.

Thus, using Kim's notation, both x and P should be viewed as functions of time (t), which may be related by nonlinear ordinary differential equations as

$$\frac{dx}{dt} = F(x, P) , \qquad \frac{dP}{dt} = G(x, P) ,$$

where F and G are general nonlinear functions of both x and P. Importantly, the emergent structure is not represented by $x(t)$ and $P(t)$ (which are functions of time and can be infinitesimally small), but by x_0 and P_0, satisfying

$$0 = F(x_0, P_0) , \qquad 0 = G(x_0, P_0) .$$

Assuming that x_0 and P_0 are an asymptotically stable solution of this system,

$$x(t) \longrightarrow x_0 , \qquad P(t) \longrightarrow P_0 ,$$

as $t \to \infty$, they establish a dynamic balance between downward and upward causations of the emergent structure (see Appendix A). Thus, Kim's causal-power actuality principle is a theoretical artifact stemming from his static analysis of a dynamic situation. In simpler terms, we saw in Sect. 7.4.3 that

Kim's concern about circular causal loops was shared by Warren McCulloch until he learned that there is always a time delay around a closed loop [28].

Empirically, we have seen in Sect. 6.6.3 that there are many examples of closed causal loops in the feedback mechanisms of mechanical and electrical systems, and the fact that philosophers have been bothered by this concept underscores the balkanization of knowledge prior to the nonlinear-science revolution. In the course of his attempts to realize mathematician David Hilbert's dream of puting all mathematics on an axiomatic basis, for example, philosopher Bertrand Russell called closed loops of logical implication "vicious circles" and devised various intellectual contortions to deal with this perceived problem, including his *theory of types* [62].[8] As mathematician Kurt Gödel soon showed, however, Hilbert's procrustean program was doomed to failure and Russell was wasting his considerable intellectual energy [696].

Going back to inventor and engineer James Watt in the eighteenth century, for example, *negative* feedback (NFB) has been used to govern the speed of engines. Since the 1920s, NFB loops stabilize the performance of electronic amplifiers, making long distance telephone communications possible, and closed causal loops of NFB play key roles in control systems and in mathematician Norbert Wiener's science of cybernetics (see Sect. 5.2.1 and [1036]).[9] In biological contexts, NFB plays a role in allosteric regulation of enzyme activity (Sect. 7.1.3 and [1059]), and biochemists Manfred Eigen and Peter Schuster suggested over two decades ago that closed causal loops of *positive* feedback (PFB) around at least three hierarchical levels of dynamic description were necessary for the emergence of living organisms from the Hadean oceans (Sects. 5.2.2 and 7.3.6 and [269]). Finally, Erich Harth has proposed "creative loops" of positive feedback as being essential elements in the dynamics of the human brain [391].

To realize feedback in engineering applications, a signal from one terminal (say the output) is brought back to the input, as shown in Fig. 8.2a. Here **A** causes **B**, which in turn causes **A**, closing a causal loop and conflating the concepts of cause and effect, to the distress of logically oriented philosophers. If the net gain around the loop exceeds unity, the feedback is positive, and exponential growth of the dynamic variables occurs, whereupon nonlinear effects will eventually limit the growth as a new entity emerges.

As noted above, a familiar example of a closed causal loop with exponential growth is the flame of a candle that has just barely been lit. The heat from the small flame (**A**) begins to melt wax (**B**), which is then drawn

---

[8] Among the many examples of such problematic loops, I am amused by Groucho Marx's statement: "I wouldn't want join a club that would have a bum like me as a member."

[9] A system with NFB must be carefully designed to avoid unwanted oscillations, which are called singing in amplifiers, hunting in control systems, and Parkinson's disease (*paralysis agitans*) in humans.

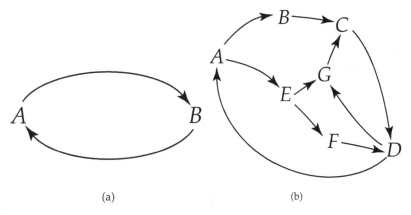

Fig. 8.2. Feedback diagrams in which the arrows indicate the actions of causality. (a) A simple loop. (b) A simple cartoon of the organization of a complex network

into the flame, providing the fuel to make it larger. As the flame becomes larger, yet more heat is produced so more fuel is released, and so on. In other words, **A** causes **B** to increase, which in turn causes **A** to increase, both growing together until they reach the levels of a fully developed flame, the size of which is determined by nonlinear (saturation) effects. The resulting emergent entity is a flame, and the system is evidently open, because energy is being produced at **B** and dissipated at **A**. Thus instead of being logically problematic, as they are for the reductionist, closed causal loops are essential elements of nonlinear science.

Nonlinear science offers many examples of PFB and the subsequent emergence of coherent structures. In the physical sciences, structures that emerge from PFB loops include tornadoes, tsunamis, optical solitons, rogue waves, ball lightning (I suppose), and Jupiter's Great Red Spot, among many others. Biological examples include the nerve impulse, cellular reproduction, flocks of birds and schools of fishes, and the development of new species, in addition to the emergence of Life itself. Psychologically, there are emotions (love, hate, rage, fear, joy, sadness; and so on), recognition of external objects by sound, sight or touch; and the birth of an idea; while in the social sciences, one finds lynch mobs, natural languages, the founding of a new town or city, and the emergence of human culture [74,171] – to name but a few examples that come to mind [878]. The exponential growth stemming from closed loops of PFB is also an essential element of low-dimensional chaos, as we saw in Chap. 2.

In biological and cognitive systems (see Sects. 7.3.5 and 7.4.5), downward causations (WDC and MDC) lead to opportunities for more intricate closed causal networks, which can involve many different levels of both the biological and cognitive hierarchies as is suggested in Fig. 8.2b. In this diagram, the node **A** might represent the production of energy within an organism, which induces a muscular contraction **B**, leading the organism to a source of food **C**, which is ingested **D**, helping to restore the original energy expended by **A**. Additionally, **A** might energize a thought process **E**, which recalls a positive memory of taste **F**, further encouraging ingestion **D**. The thought **E** might also induce the generation of a digestive enzyme **G** which also makes the source of food seem more attractive. Finally, ingestion **D** might further induce generation of the enzyme **G**. In this simple example – which is intended only as a cartoon – the network comprises the following closed loops of causation: **ABCD**, **CDG**, **AEFD**, and **AEGCD**, where the letters correspond to entities at various levels of both the biological and the cognitive hierarchies. With sufficient gain around some of these closed loops, the entire network shown in Fig. 8.2b becomes established as a new entity – one's first experience of eating a Neapolitan pizza, say, or the birth of a new organism.

Building on this example, PFB networks can lead to the emergence of entities with seemingly unbounded complexity, relating physiological, mental, intentional and motor levels. Analysis of such networks is not a trivial matter because the time and space scales for models of living creatures differ by many orders of magnitude as one goes from the biochemical levels of a single cell to the dynamics of a whole organism, and – as molecular biologist David Goodsell has shown in a remarkable little book called *The Machinery of Life* [366] – a single cell is a very complicated dynamical entity. Thus the modeler wonders what can be said about the organization of a living system from a theoretical perspective.

8.3 Theories of Life

While much of modern biology is rightly directed toward understanding causal connections between phenotypical (adult) characteristics and DNA codes, common sense suggests that such reductive analyses ought not to comprise it all. Some appropriate fraction of the total research effort should be devoted to understanding the organizational aspects of living creatures, asking questions like: Are biological organisms mechanistic, as Descartes claimed? Is that real duck swimming in the pond over there but an improved version of Jacques de Vaucanson's eighteenth-century *canard digérant* (digesting duck) (see Fig. 8.3) which quacked, ate, drank, swam, flapped its wings and defecated?[10] Can we build living machines or must they grow? Is a living organism something more than a mechanism? If so, how does it differ? Can studies of nonlinear science help biologists see the difference?

---

[10]Descartes never saw Vaucanson's mechanical duck as it was constructed decades after he died, but he had been youthfully impressed by hydraulic automata at the Palace of Versaille [230].

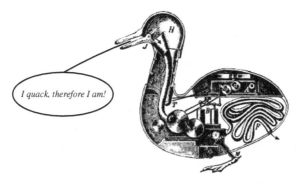

Fig. 8.3. Jacques de Vaucanson's *canard digérant* [932]. (Having the ability to defecate, this mechanical model is in amusing accord with recent observations that "biology is inherently messy" [485])

8.3.1 Artificial Life vs. Autopoiesis

Motivated by such questions, the term "artificial life" (ALife) was coined during a workshop organized by computer scientist Chris Langton, sponsored by the Los Alamos Center for Nonlinear Studies, and held in Los Alamos during September of 1987 [97,530]. An important cause of this "new science" was the recent growth in computing power (see Fig. 1.3), which became widely evident in the mid-1980s, stimulating the imaginations of computer mavens.[11] In simplest terms, the aim of ALife is to study biological phenomena by seeing how (or if) Life can be simulated on computers; thus this development is a sophisticated outgrowth of mathematician John Conway's relatively simple Game of Life which has fascinated scientists and novices alike since the 1970s [397]. Generally, of course, ALife aims for more advanced computer realizations of previous mechanical models, like the *canard digérant* shown in Fig. 8.3 and Wiener's amusing little robot from the 1950s (Sect. 5.2.1). As with Turing's famous definition of computer intelligence [972], the logic is that if entities are given more and more of the properties of Life, they will eventually become alive.

Among the biological behaviors that have been simulated under ALife are flocking, exploring environments, reproduction, competing for limited resources, and evolution by mutation and selection, but more generally in Langton's view, ALife studies the *synthesis* of life-like behavior rather than the *analysis* of biological organisms and considers "life-as-it-could-be" rather than "life-as-we-know-it". To these ends, ALife includes proponents of both "weak" and "strong" varieties, with the former group assuming that their field can provide tools to assist the biologists in formulating and testing their

---

[11]In July of 2006, interestingly, the term "artificial life" recalled over four million pages on Google.

theories, and the latter asserting that new life forms will be created within their computers.

Coined in the early 1970s by Chilean biologists Humberto Maturana and Francisco Varela and taken from Greek words meaning "self-producing" or "self-creating", *autopoietic* refers to the ability of a living organism to set its own agenda, as opposed to the term *allopoietic* which means "being produced" [990]. Thus these scientists were thinking of a living cell as an autonomous dynamical system, which is organized as a network of processes (see Fig. 8.2) that continually regenerates its constituent relations [366]. In the words of Maturana and Varela [619]:

> An autopoietic machine is a machine organised (defined as a unity) as a network of processes of production (transformation and destruction) of components that produces the components which: (i) through their interactions and transformations continuously regenerate and realise the network of processes (relations) that produced them; and (ii) constitute it (the machine) as a concrete unity in the space in which they (the components) exist by specifying the topological domain of its realisation as such a network.

Thus autopoiesis is a top-down approach toward understanding living systems, proceeding – appropriately, it seems to me – from Life's empirical properties to its theoretical formulation, as opposed to the bottom-up approach of ALife. By concentrating their attention on Life's organization, Maturana and Varela have helped to move biological research in an important direction.

8.3.2 Relational Biology

Two key publications that also approach the phenomenon of Life from organizational perspectives are Rashevsky's seminal paper "Topology and life: In search of general mathematical principles in biology and sociology" (see Sect. 7.2.6 and [789, 790]) and a recent book by Robert Rosen entitled *Life Itself: A Comprehensive Inquiry Into the Nature, Origin, and Fabrication of Life*. Together with some related essays [818], Rosen's book presents an account of ideas that began to ferment during his graduate studies on "relational biology" in Rashevsky's mathematical biophysics program (Sect. 5.2.2), importantly showing that not all natural systems can be simulated [814, 815, 817, 818].

As it is primarily concerned with the organization of living systems, relational biology has an interesting duality with respect to reductionism, which can be oversimplified as follows. Whereas the reductionist would "throw away the organization and keep the underlying matter", the relational biologist would "throw away the matter and keep the underlying organization" [817]. Of course the reductionist does not entirely ignore the organization as he expects to recapture it upon understanding the dynamics of the material

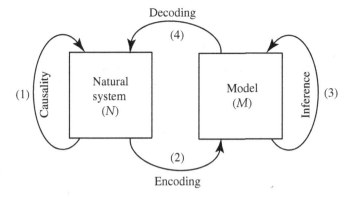

Fig. 8.4. Rosen's diagram of a scientific model

constituents, and the relational biologist does not ignore matter as she expects to realize an interesting organization in *some* material context, but their starting perspectives and emphases are different.

Under reductionism, organization is merely *epiphenomenal*, whereas from the relational perspective as Rosen puts it: "the organization of a system has become the main object of study. In essence, organization has become a *thing*, with its own formal images or models, its own attributes, and its own modes of analysis." Thus relational biology is in accord with both the spirit and the practice of modern nonlinear science, where mathematically similar dynamics are realized in quite different material contexts, and emergent entities (strange attractors, solitons, nerve impulses, and so on) are viewed as real dynamic objects and given appropriate ontological status.

8.3.3 Mechanisms

As scientific models are often conflated with the underlying reality they strive to represent, Rosen considered these two aspects of an analysis separately, using Fig. 8.4 to emphasize the dichotomous relationship between a natural system (N) and a theory (M) that has been constructed to model some features of it [817]. The temporal course of causality (1) in the natural system is mimicked more or less accurately by logical or recursive implications (3) in the model. To go from the natural world to initial conditions for the model requires the taking of data (Tycho Brahe's measurements of positions of the planets, for example, or recordings of spectral peaks by a physical chemist), which Rosen calls encoding (2) to the model, and to compare predictions of the model with measurements of events in the natural world requires a process of decoding (4) from the model (Johannes Kepler's calculations of elliptical planetary orbits, say, or predictions of molecular resonances).

Constructing models of natural systems is problematic (how do we know what's *really* out there?), but Rosen grounded his work by studying mathematical models of formal systems. Thus the natural system N in Fig. 8.4 is replaced by a set S, so his entire formulation remains within the realms of mathematics, wherein he defines two general types of models:

1. Analytic models, which are constructed from direct (Cartesian) products of measurements and are tied to the notion of a semantic description.[12] Thus each component of an analytic model on a set S takes the form

$$M(S) = \prod_\alpha f_\alpha(S) \,,$$

a direct product over measurements (f_α) on all the elements of S.

2. Synthetic models, which are constructed from direct sums of disjoint subsets of S and are tied to the notion of a syntactic description.[13] In formulating a synthetic model, the set S is resolved into a direct sum of subsets $\{u_\alpha\}$, having corresponding elements u_α on which measurements $\varphi_\alpha(u_\alpha)$ are defined, with each φ_α an arbitrary function of u_α. Then a component for a synthetic model is defined as

$$P(s) = \sum_\alpha r_\alpha \varphi_\alpha(u_\alpha) \,,$$

a direct sum over measurements (φ_α) on the elements of u_α (where the r_α are numbers). As the $\{u_\alpha\}$ are the "disjoint elements" assumed under "analysis" in Sect. 8.1.1, synthetic models are evidently reductive. For examples, the elements of a subset u_α might represent atoms (while the subsets represent molecules) or DNA base pairs (and the subsets represent codons) or neurons (with the subsets representing primary cell assemblies).

Although analytic and synthetic models are often assumed to be equivalent, Rosen shows they are not necessarily so, claiming this to be "the central question of all theoretical science". He shows that synthetic models can always be formulated as analytic models, but from his proof that direct products are more inclusive than direct sums,[14] it follows that analytic models cannot in general be formulated as synthetic models. This result is closely related to previous demonstrations that mathematical systems cannot in general be reduced to axioms (à la mathematician Kurt Gödel [696]) and that meanings

---

[12]The direct product of two sets X and Y, denoted $X \times Y$, is the set of all ordered pairs whose first component is a member of X and whose second component is a member of Y. Descriptions are "semantic" because meanings are not excluded from the dynamics at high levels.

[13]The direct sum of two disjoint sets X and Y, denoted $X + Y$, is the smallest set containing both X and Y. The formulation is "systactic" because meanings are not defined at low levels.

[14]To construct this proof, Rosen shows that synthetic models have largest and smallest forms, whereas analytic models do not.

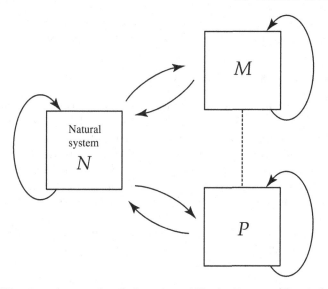

Fig. 8.5. A more detailed version of Kim's diagram (Fig. 8.1)

cannot in general be reduced to grammatical rules in the context of linguistics [818, 895].

Viewing the above discussion of Kim's analysis from Rosen's perspective, Fig. 8.1 can be redrawn as in Fig. 8.5, where M is an analytic model (because each component may depend on all aspects of the system) and P is a synthetic model (because it is constructed on disjoint subsets of the fundamental elements of S). As M is more general than P, Kim's main conclusion – that physicalism requires explanatory reductionism – is incorrect.

Rosen distinguishes between *modeling*, which at its best is a creative activity at high levels of description, and *simulation*, which is based on a low (reductive) level of description and is *algorithmic*, meaning that it can be done by a computer. For example, the Zeldovich–Frank-Kamenetskii (ZF) equation (see Sect. 4.1 and [874, 1083])

$$D\frac{\partial^2 u}{\partial x^2} - \frac{\partial u}{\partial t} = \frac{u(u-a)(u-1)}{\tau} , \qquad (8.3)$$

is a model for both flame-front propagation and nerve-impulse propagation, as is (8.2). Thus with appropriate values of the parameters D, τ and a, (8.3) describes the temperature of a candle flame as a function of position x and time t, which in turn fixes the velocity v of its propagation in (8.2). Simulation of a candle flame, on the other hand, requires a procedure that computes a solution of fluid equations for the molecules of wax being drawn into the flame, taking account of chemical reactions and the release of energy under combustion. Those systems – like the candle – for which all models – like (8.2) and (8.3) – can be simulated are called *mechanisms* or *simple systems*, a classification that includes most of the physical systems described in Chap. 6 and many of the biological systems in Chap. 7.

8.3.4 Complex Systems and Chaotic Emergence

Importantly, Rosen has done for models of the natural world what Gödel did for his science. What was this? In 1931, Gödel startled his colleagues by proving that mathematical systems of modest complexity (number theory, for example) cannot be put on an axiomatic basis, as there will always be true theorems that cannot be proven under a finite set of axioms [696].[15] In other words, there is more to a mathematical system than is captured by its axioms. Let's see how this unanticipated result translates into natural science.

Rosen called a system that cannot be simulated a *complex system*. Thus analytic models of complex systems cannot in general be reduced to synthetic models – the defining property of a simple system or mechanism. Physical examples of complex systems include fluid turbulence (shown in Figs. 2.5, 6.12 and 6.13 and discussed in Sect. 6.8.7) and the hierarchy of emergent universes (see Fig. 6.19 in Sect. 6.9.5). Thus although turbulence can be modeled by the Burgers equation

$$D\frac{\partial^2 u}{\partial x^2} - \frac{\partial u}{\partial t} = (c + \beta u)\frac{\partial u}{\partial x}$$

(with appropriate choices for the parameters D, c and β), its simulation is not possible, as applied mathematician Horace Lamb noted with amusing despair. Why not? The sizes and shapes of applied mathematician Lewis Richardson's big whirls, little whirls, and lesser whirls in a turbulent fluid cannot be determined by integration of the Navier–Stokes equation, because the locations and times at which these myriad new entities begin to emerge from their infinitesimal seeds are not under numerical control. In other words, the emergence of each whirl is not numerically stable, leading to uncontrolled errors (errors that grow exponentially with time) that arise when attempting to compute the turbulent internal geometry of the system. In the context of mental self-organization, some philosophers call such an uncontrolled process the *chaotic emergence* of a complex system [703, 704].

Similarly, the hierarchy of universes suggested in Fig. 6.19 is modeled by the discrete Mandelbrot map

$$z_{n+1} = z_n^2 + c \,,$$

---

[15] An example of such an unproven theorem in number theory may be Christian Goldbach's mid-eighteenth conjecture that any even number can be expressed as the sum of two prime numbers.

where z is a complex variable and c is a complex number.[16] (Recall from Sect. 6.9.5 that Mandelbrot's iteration starts with $z_0 = 0$, and c is included in the set if $|z_n|$ remains bounded as $n \to \infty$.) Although it has not been shown in detail, I expect that a corresponding integration of the Einstein–Hilbert equation (6.17) cannot be carried out because the geometries of the big black holes, little holes, and lesser holes would not be under numerical control, leading again to chaotic emergence of a complex system.

Let us next consider some examples of biological systems that have been mentioned in Chap. 7 and in this chapter, asking which are complex and which are merely mechanisms.

Biological Evolution

The Darwinian development of life forms over the past four billion years on Earth is thought to be chaotically emergent because – as evolutionary biologist Stephen Jay Gould has convincingly argued – the course of *punctuated evolution* would differ greatly on a second run [370, 625]. In other words, the times and locations of the emergence of new species (appearing as abruptly beginning "punctuations" in the fossil record) stem from small seeds (two individuals, say), implying a loss of computational control. It follows that simulations based on lower-level (reductive) descriptions are not feasible because the process is chaotically emergent; thus analytic models cannot be expressed as synthetic models, and the reductive program cannot be carried through.

The Birth of Life

In addition to the punctuated course of evolution (a biological question), there is the biochemical puzzle of how life emerged in the Hadean oceans, some four billion years ago. Even if the details of this momentous occurrence were known to us, it might involve one of many chemical instabilities. Also even computing the state of the seas – like determining next month's weather – would not be possible. Thus reductively determining how and when the pieces of the protobiological molecule came together and began to evolve into living matter is not possible.

Cognitive Hierarchies

The development of cognitive structures in a human brain is a chaotically emergent process (see Sect. 7.4.4 and [703, 704]), because identical twins do

---

[16]Here the standard terminology is unfortunate. The *numbers* z and c are said to be complex, indicating that they are constructed from a real and an imaginary component (i.e., they are two-dimensional vectors). This is unrelated to whether or not the *system* is complex in Rosen's sense.

not develop the same patterns of thought [873]. According to psychologist Donald Hebb, internal hierarchies of cell assembly structures emerge in each person's head during the course of his or her existence, and these necessarily differ as they become organized in response to the happenstance of experience (Sect. 7.4.5 and [412–414]). For similar reasons, chaotic emergence is expected to characterize the development of natural languages and the growth human cultures [74, 171].

Artificial Life

In general, ALife systems can be either mechanisms or complex systems, depending on the nature of the computations involved [97, 530]. Some ALife models, like Conway's Game of Life [397], are stably simulated on computers; thus they are mechanisms for which analytic models can be reduced to synthetic models, and they merely mimic Life [818]. Other ALife models may be numerically unstable in the sense that two successive computations do not give the same result, indicating that such models are not mechanisms, but this is only a necessary (not a sufficient) condition for being alive. In either case, however, the weak ALife program still promises contributions to biological research, which may be its strongest asset.

Protein Folding

A complicated dynamical system that seems simulatable is a folding protein, upon which biochemists continue to put great effort due to the technical importance of this problem. The crux of protein folding is not numerical instabilities stemming from emergence or chaos, because in principle the computational task is merely to find the configuration (shape) having a minimum of free energy. Unfortunately, however, a typical protein has a very large (nearly immense [287]) number of nearly degenerate conformational states (states of almost equal energy) [330], which makes the true energy minimum difficult to define and determine. This problem may eventually be overcome with further increases of computing power in machines such as IBM's Blue Gene[17] or Folding@Home – a cooperative system that employs the spare computing power of many participants.[18]

Human Morphologies

The forms of living organisms are largely determined by their DNA codes, but humans with identical genomes begin to deviate upon aging, because mind–body interactions – stemming from downward acting intentions [475] – induce unpredictable variations in eating patterns, physical activity, use of

---

[17]http://en.wikipedia.org/wiki/Blue_Gene
[18]http://folding.stanford.edu/

recreational drugs, and so on. In support of this claim, a recent study of 30 male and 50 female monozygotic (MZ) twins by geneticist Mario Fraga and his colleagues in Spain, Sweden, Denmark, Britain, and the United States shows "widespread 'epigenetic drift' associated with aging" involving acetylation of histones H3 and H4 (Sect. 7.1.5 and [318, 614]). In other words, changes in gene *expression* were found to occur with aging, which in turn led to differences in the onset of common diseases. For those of my age, a vivid if anecdotal example of epigenetic drift is provided by the saga of Canada's Dionne quintuplets – five winsome MZ girls born in 1934 who were the subjects of many articles and photo essays during the 1930s and 1940s. To take but one stark measure of their aging, Emilie died in 1954 of an epileptic seizure, Marie in 1970 of a stroke, Yvonne in 2001 of cancer, while Annette and Cecile are still alive as I write in 2006.

Dynamics of Life

Although the general form of an organism is determined by its genome (the offspring of a duck doesn't look like a dog), there remains the problem of knowing how it moves and interacts with its environment. As Rosen has demonstrated [817, 818], living organisms are complex systems rather than mechanisms, so analytic models cannot be reduced to synthetic models, and the dynamics of an individual being's behavior cannot be simulated. While Jacques de Vaucanson's *canard digérant* (see Fig. 8.3) may have swum like a duck and quacked like a duck, in other words, it did not simulate the behavior of a *particular* duck.

Whatever the ultimate judgments on the simulatability of these examples, there is now a precise – albeit more restrictive – meaning for the term "complex system" than those that have been loosely employed in academic and general discourse.[19] Under this new definition, a complex system can be modeled (of course), but it cannot be simulated from its reductive elements because it combines loops of PFB with downward causation – just the properties that lead to chaos and uncontrolled emergence and in turn are the components of chaotic emergence.

Why should we care about this? Why does it help us to know that biology is complex rather than mechanistic? Perhaps the broadest answer to this question is that one's thinking becomes released from the shackles of reductionism, in which it is difficult to understand how the constituent atoms (or molecules or DNA codons or neurons or whatever) can have intentions [475]. Thus the concept of a complex system does the work of Descartes's substance dualism without rejecting physicalism.

On reductionism, interestingly, Rosen concludes as follows [817]:

---

[19]For example, a complex system is defined on Wikipedia as "a system of many parts which are coupled in a nonlinear fashion".

The sequence of state transitions of a system of particles, governed by Newton's laws of motion, starting from some initial configuration, is then the analog of a *theorem* in a formalism, generated from an initial proposition (hypothesis) under the unfluence of the production rules. And just as [proponents of] formalization in mathematics believed that everything could be formalized *without loss*, so all truth could be recaptured in terms of syntax alone, so particle mechanics came to believe that every material behavior could be, and should be, and indeed must be, reduced to purely syntactical sequences of configurations in an underlying system of particles.

Hence the power of the belief in reductionism, the scientific equivalent of the formalist faith in syntax. Though of course Newtonian mechanics has had to be supplemented and generalized repeatedly, the basic faith in syntax has not changed; indeed, it has been bolstered and made more credible by these very improvements. And there has been no Gödel in physics to challenge that credibility directly. But there is biology.

And there has been a Rosen and there is also nonlinear science.

8.3.5 What Is Life?

This question was famously asked by physicist Erwin Schrödinger in a widely-read little book based on a series of public lectures that he presented at Dublin's Trinity College in February of 1943 [853]. Following an earlier suggestion of biophysicist Max Delbrück,[20] Schrödinger stated for the first time in clear physical terms that genetic information is embodied in a code of molecular valence bonds, starting physicist Francis Crick and biochemist James Watson on their search for the structure of DNA [674, 853]. For this reason, Schrödinger's book is often cited as support for reductive formulations of biology, but therein he also asked the question: "Is life based on the laws of physics?" which he answered as follows:

> From all we have learned about the structure of living matter, we must be prepared to find it working in a manner that cannot be reduced to the ordinary laws of physics. And that is not on the ground that there is any "new force" or what not, directing the behaviour of the single atoms within the living organism, but because the construction is different from anything we have yet tested in the physical laboratory.

As Schrödinger knew biology in addition to being one of the great physicists of the twentieth century [674], this claim offers little comfort to reductive life scientists, but they seldom mention it. The rest of us, however, may ask

---

[20]Delbrück was awarded the 1969 Nobel Prize in medicine for his work on viruses.

where to find these phenomena that "cannot be reduced to the ordinary laws of physics", and the answer is: within the realms of nonlinear science. During his wartime tenure as founding director of the Dublin Institute for Advanced Studies, interestingly, Schrödinger tried to move the physics community in the direction of nonlinear science – through the organization of conferences and his personal research on nonlinear electromagnetism as a basis for elementary particle theories (see Sect. 6.1.1) and on Einstein's gravitational theory (Sect. 6.9) – but he was decades ahead of his time.[21]

Dedicated to Schrödinger's nonreductive perspective is a recent book with the same title by evolutionary biologist Lynn Margulis and science writer Dorion Sagan which emphasizes the organizational aspects of living creatures [607]. Following Rashevsky's relational biology, finally, this question (What is Life?) motivated Rosen throughout his decades-long effort to understand and formulate organizational principles for biology [819]. Merely *asking* the question is significant because this act assumes Life to be a noun rather than an adjective (as in: Is that thing alive?). By treating "the *integrated* activities of the organism as a whole" [789], the relational analysis of Rashevsky and Rosen locates Life within organizational stucture and focuses attention on a fundamental fallacy of modern biology [817, 818].

Following beliefs of the physics community, many biologists continue to see their field as a special case of reductive physics for which the details are yet to be worked out – thereby avoiding any taint of vitalism. But Schrödinger suggests that reductive physics is the special case, neglecting much that falls within the realms of both nonlinear science and biology. Although not identical fields, nonlinear science and biology extensively overlap (as we have seen in the previous chapter), and both are more general than reductive physics because they are not restricted to simple systems (mechanisms, which can be reductively simulated) but include the unbounded and unpredictable phenomena of chaos, emergence, and chaotic emergence as they arise from interactions among the hierarchical levels of complex systems.

Generalizing the concept of emergence from the nonlinear dynamics at a single level of description to include organizational networks – comprising both downward and upward causations among many levels of a being's biological, cognitive, and cultural hierarchies as in Fig. 8.2b – leads to a very complicated organizational *thing*, the noun that we call *Life*. Far greater than the sum of its parts (Death), Life is not a mechanism, like some refined version of de Vaucanson's duck (Fig. 8.3). Nested within a nonlinear complexity that is unrivaled in the known Universe, Life combines all aspects of a creature's dynamics: its physiological, conceptual, motivational, and motor activities in an unsimulatable autopoietic network.

Nonlinear science is not yet close to describing this phenomenon, but here, in my view, is where to seek answers to the remaining riddles of biology.

---

[21]Walter Moore's biography gives an excellent survey of Schrödinger's efforts to promote nonlinear science during his tenure in Dublin [674].

9 Epilogue

> So much has been done, exclaimed the soul of
> Frankenstein – more, far more, will I achieve: tread-
> ing in the steps already marked, I will pioneer a
> new way, explore unknown powers, and unfold to
> the world the deepest mysteries of creation.
>
> Mary Wollstonecraft Shelley

In the summer of 1816, young Mary Wollstonecraft Godwin had fled England with her lover, Percy Bysshe Shelley, and their infant son for Geneva,
where they mixed with Lord Byron, his mistress (Mary's step-sister), and
John Polidori (his personal physician). As the weather was unpleasant, these
five passed the days at first by reading German ghost stories and then challenged themselves to write tales of terror. Byron responded with a fragment
and Polidori came through with a short story that founded modern vampire
fiction [769], but the laurel surely went to Shelley née Godwin for her novel
Frankenstein, or The Modern Prometheus [899], the plot of which involves
the scientific creation of Life – a question that continues to haunt us almost
two centuries later.

At the beginning of the nineteenth century, the time was ripe for Shelley's
tale, as Luigi Galvani had recently discovered "animal electricity", whereby
a frog's leg muscle is caused to twitch through the action of an agent that
might or might not be the same as Alessandro Volta's chemical electricity or
Ben Franklin's atmospheric electricity [864]. Did lightning from the heavens,
some wondered, provide the spark of Life for inert matter? By 1850, young
Hermann Helmholtz had used Galvani's frog preparation to show that animal electricity is indeed a unique phenomenon [420], but the surprisingly
small speed that he measured was not explained for a century, until the phenomenon of nerve-pulse propagation was recognised as an important example
of reaction diffusion within the more general area of nonlinear dynamics.

Growing up in a small New England town during the 1930s and 1940s, I
learned of Victor Frankenstein and his desolate creation through the movies –

a staple of entertainment in those pre-television days. Although fanciful versions of Mary Shelley's original tale had occasionally appeared in stage plays throughout the nineteenth century, her copyright-free concept was enthusiastically taken up by Hollywood's Universal Studios, which produced a flock of films that liberally mixed Frankenstein and his monster with vampires, other strange creatures, and those infernal hordes of villagers with their wild eyes and drooling torches. Thus emerged a genre that was eventually spoofed by *Young Frankenstein*, a film in which Victor's grandson (played by Gene Wilder in perhaps his greatest screen role) had the benefit of a modern neuroscience education. For an eleven-year-old walking home alone in the dark, however, Mel Brook's classic sendup was three decades in the future, and there was nothing amusing about *Frankenstein Meets the Wolf Man*, the dread of which was compounded by the possibility that one or both of this pair might be quietly following *me*. Such are the images that nest in our brains.

Looking back, it seems oddly significant that I graduated from the university in the same year that physiologists Alan Hodgkin and Andrew Huxley resolved the riddle of nerve-pulse propagation, combining experimental physiology with imaginative mathematics and heroic calculations in a tour de force that helped to set the stage for the nonlinear-science revolution of the 1970s [440]. Upon joining the academic community, my primary aim was to understand the nature of pulse conduction, which is so central to the movement of our muscles, the reception of signals from our eyes and ears, the formation of neocortical memories, and the beating of our hearts – all stuff that Victor Frankenstein needed to unfold for us "the deepest mysteries of creation". Motivated by this profound relation to biological dynamics, my professional energies have been devoted to nonlinear science for more than four decades.

So what is the secret of Life? Although rooted in nature, living beings are organized as *immensely complex dynamic hierarchies*, where "immense" is used in the technical sense to denote a finite number of possibilities that is too large to list [287] and "complex" implies a class of natural systems that cannot be reductively modeled [817, 818]. Biological hierarchies achieve their immense complexities through processes of *chaotic emergence*, a phrase that was coined by philosophers to describe mental self-organization [703, 704], and can be applied to Darwinian evolution, the growth of biological forms, and their daily dynamics, in addition to some physical examples (fluid turbulence and the hierarchical expansion of our physical Universe). Because they tend to regard biologically oriented physical scientists as academic claim jumpers who willfully ignore the messy features of Life [556], many biologists remain unaware of these ideas, openly or tacitly assuming reductive perspectives that have been discredited by research in nonlinear science.

Guided by the concepts of immensely complex dynamic hierarchies and of chaotic emergence, however, a fully nonlinear formulation of Life steers between the Scylla of Newtonian physics and the Charybdis of Cartesian dualism, driving a wooden stake through the heart of reductionism while suggesting that there may be something to Henri Bergson's vitalism after all.

A Phase Space

Current use of the term *phase space* in nonlinear science is conveniently exemplified by (2.1) – the Lorenz equations – and the corresponding sketch of Fig. 2.1. This is a *three-dimensional* phase space because there are three first order ODEs involving three *dependent variables* (x, y and z) governing the state of the system at time t (the *independent variable*), which can be indicated as a *phase point*

$$P = (x, y, z) ,$$

in this space. The functions on the right-hand sides (RHS) of (2.1) then give the components of a "velocity" vector

$$V = \left(\frac{\mathrm{d}x}{\mathrm{d}t}, \frac{\mathrm{d}y}{\mathrm{d}t}, \frac{\mathrm{d}z}{\mathrm{d}t} \right) ,$$

which determines how fast the phase point is moving in the phase space and in what direction. From a geometrical perspective, (2.1) show how V is computed from a position vector; thus

$$P \xrightarrow{\ (2.1)\ } V .$$

Integrating (2.1) then indicates where the phase point will be – or what the state of the system will be – at some future time, thus mapping out a *solution trajectory* or *orbit* in the phase space.

Equations (2.1) are called *autonomous* because the RHS functions depend only on the position of the phase point, and not, for example, on the time t (or on the positions of the celestial bodies, as the astrologers would suggest). Assuming that the RHS functions are uniquely defined, then trajectories cannot cross, for this would be a contradiction.

Chaotic trajectories in a phase space of two dimensions, called a *phase plane*. In analyzing dynamics on a phase plane, we begin with two equations of the form

$$\frac{\mathrm{d}x}{\mathrm{d}t} = F(x, y) , \qquad \frac{\mathrm{d}y}{\mathrm{d}t} = G(x, y) ,$$

and the first step is to find *singular points* (SPs) where the velocity vector V is zero. Singular points (x_0, y_0) evidently satisfy $F(x_0, y_0) = 0$ and $G(x_0, y_0) = 0$, which can be determined either analytically or numerically.

Having found a singular point, it is interesting to ask how the solution trajectory behaves in the vicinity of this SP. This can be answered by noting that as $V = 0$ at a SP, it must have small magnitude nearby; thus the governing equations can be written as

$$\frac{dx}{dt} = ax + by + \text{higher order terms},$$

$$\frac{dy}{dt} = cx + dy + \text{higher order terms},$$

where $a \equiv \partial F/\partial x|_{x_0,y_0}$, $b \equiv \partial F/\partial y|_{x_0,y_0}$, $c \equiv \partial G/\partial x|_{x_0,y_0}$ and $d \equiv \partial G/\partial y|_{x_0,y_0}$. As the "higher-order terms" involve products or powers of x and y, they can be neglected in a region sufficiently close to the SP, where the trajectory is governed by the *linear* matrix equation

$$\begin{pmatrix} dx/dt \\ dy/dt \end{pmatrix} \doteq \begin{pmatrix} a & b \\ c & d \end{pmatrix} \begin{pmatrix} x \\ y \end{pmatrix} = M \begin{pmatrix} x \\ y \end{pmatrix}$$

and $\doteq$ indicates that the higher-order (nonlinear) terms have not been included. Sufficiently close to the SP, then, a trajectory evolves as

$$\begin{pmatrix} x \\ y \end{pmatrix} \doteq e^{\lambda t} \begin{pmatrix} x(0) \\ y(0) \end{pmatrix},$$

where λ is a solution of the determinantal equation

$$\det(M - \lambda I) = \det \begin{pmatrix} a - \lambda & b \\ c & d - \lambda \end{pmatrix} = \lambda^2 - (a + d)\lambda + ad - bc = 0.$$

As this is a quadratic equation, there are two solutions, which we can label λ_1 and λ_2. In a curious mixture of German and English, it is customary to say that λ_1 and λ_2 are the *eigenvalues* of the matrix M, and the solutions of

$$\begin{pmatrix} a - \lambda_1 & b \\ c & d - \lambda_1 \end{pmatrix} \begin{pmatrix} x_1 \\ y_1 \end{pmatrix} = 0, \qquad \begin{pmatrix} a - \lambda_2 & b \\ c & d - \lambda_2 \end{pmatrix} \begin{pmatrix} x_2 \\ y_2 \end{pmatrix} = 0$$

are the corresponding *eigenvectors*. [The original English usage was to call λ_1 and λ_2 characteristic values and (x_1, y_1) and (x_2, y_2) the corresponding characteristic vectors but the bastard terms have carried the day.]

Let us now examine the generic behaviors of solution trajectories near singular points of a phase plane. Such a solution is of the general form

$$\begin{pmatrix} x \\ y \end{pmatrix} = A \begin{pmatrix} x_1 \\ y_1 \end{pmatrix} + B \begin{pmatrix} x_2 \\ y_2 \end{pmatrix},$$

where A and B are constants chosen so that x and y are real. (Thus if λ_1 and λ_2 are real then A and B are real, but if λ_1 and λ_2 are a complex-conjugate pair, then A and B are also a complex-conjugate pair.) The six possibilities are sketched in Fig. A.1, where axes have been scaled and the origin of the (x, y) plane has been moved to the SP for visual convenience.

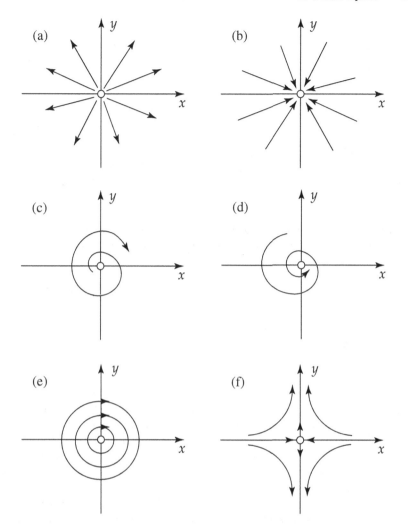

Fig. A.1. The six generic singularities in the phase plane. (**a**) Outward node. (**b**) Inward node. (**c**) Outward spiral. (**d**) Inward spiral. (**e**) Center. (**f**) Saddle

(a) Outward node. This is the case when both λ_1 and λ_2 are positive real numbers.
(b) Inward node. In this case, both λ_1 and λ_2 are negative real numbers.
(c) Outward spiral. Here λ_1 and λ_2 are a complex-conjugate pair with positive real parts.
(d) Inward spiral. In this case, λ_1 and λ_2 are a complex-conjugate pair with negative real parts.
(e) Center. In this final case, both λ_1 and λ_2 are purely imaginary numbers.

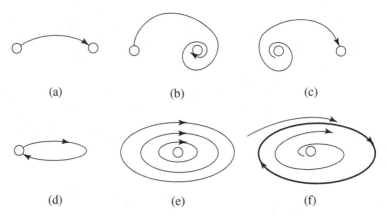

Fig. A.2. Finite nonlinear trajectories in the phase plane. (**a**), (**b**) and (**c**) show heteroclinic orbits. (**d**) A homoclinic orbit. (**e**) Nested cycles. (**f**) A limit cycle

(f) Saddle. In this case, both λ_1 and λ_2 are real, but one is positive and the other is negative.

As for global structures of nonlinear solutions on a phase plane, the requirement that trajectories cannot cross is a strong constraint, with the possibilities for finite nonlinear trajectories limited to those shown in Fig. A.2.

The first three examples in Figs. A.2a, b and c are *heteroclinic orbits* in which the phase point moves from one SP to another as time progresses from $-\infty$ toward $+\infty$. Figure A.2d is an example of a *homoclinic orbit*, in which the phase point leaves from a saddle point at $t = -\infty$ and returns to the same saddle as $t \to +\infty$. As discussed in Chap. 4, the examples of Figs. A.2a and d are particularly important for understanding the nature of impulse propagation on reaction-diffusion systems. The nested cycles of Fig. A.2e are topologically equivalent to the center in Fig. A.1e, but the example in Fig. A.2f is more interesting for technical applications. This is a *limit cycle*, in which periodic behavior is approached by trajectories of either larger or smaller amplitude, and the dynamic behavior corresponds to that of an electronic oscillator – a radio transmitter, for example. An inward node, and inward spiral, or a limit cycle will be embedded in a *basin of attraction* throughout which the same qualitative behavior obtains.

These examples capture the essential features of nonlinear dynamics on the phase plane, explaining why Newton and Euler were able to completely solve the two-body problem of planetary dynamics. Although there seem at first to be 12 dependent variables in this problem (the positions of both bodies in three-dimensional space and the coresponding components of their velocities), only the relative motions of the two bodies are relevant, reducing the number of dependent variables to 6. This number is further reduced to 2, because total energy is conserved along with angular momentum in three directions; thus the two-body problem can be studied on a two-dimensional surface which is topologically equivalent to the phase plane.

If we increase the number of phase-space dimensions to 3, as in the Lorenz system of (2.1) and Fig. 2.1, the possible dynamical behaviors change dramatically because the prohibition on trajectories crossing on a surface no longer restricts the dynamics. From examination of the RHS functions of (2.1), it is evident that there is always one SP at

$$(x, y, z) = (0, 0, 0) .$$

For $r > 1$, there are two additional SPs at

$$(x, y, z) = \left(\pm\sqrt{b(r-1)}, \pm\sqrt{b(r-1)}, r-1 \right) ,$$

where it was noted in Chap. 2 that r represents a positive temperature difference driving the system. We can proceed as in the two-dimensional example by investigating the linear behaviors through analysis of the matrix

$$M = \begin{pmatrix} -\sigma & \sigma & 0 \\ r & -1 & 0 \\ 0 & 0 & -b \end{pmatrix} ,$$

which has only real eigenvalues:

$$\lambda_{1,2} = -(\sigma+1)/2 \pm \sqrt{(\sigma-1)^2/2 + r} , \qquad \lambda_3 = -b ,$$

for r positive.

Thus there may be heteroclinic and homoclinic solution trajectories, as in Figs. A.2a and e, but knowing this tells us nothing about the possibility of finding the *strange attractor* shown in Fig. 2.1, which was discovered numerically by Lorenz [575]. Although the possibility of strange attractors had been known to Poincaré in the nineteenth century, teachers and students of physics and engineering remained largely unaware that such phenomena existed into the 1970s.

As we saw in Chap. 2, another interesting example of low-dimensional chaos is generated by the Hénon–Heiles Hamiltonian of (2.4), which was proposed in 1964 as a simple model of planetary dynamics. This energy functional is conserved under the dynamics, and it generates first-order ODEs in four dependent variables (x, y, $p_x \equiv \dot{x}$, and $p_y \equiv \dot{y}$) through the *Hamiltonian equations* [364]:

$$\frac{dx}{dt} = +\frac{\partial H}{\partial p_x} = p_x ,$$

$$\frac{dy}{dt} = +\frac{\partial H}{\partial p_y} = p_y ,$$

$$\frac{d\dot{x}}{dt} = -\frac{\partial H}{\partial x} = -\omega_1 x - 2\varepsilon xy ,$$

$$\frac{d\dot{y}}{dt} = -\frac{\partial H}{\partial y} = -\omega_2 y + \varepsilon(y^2 - x^2) ,$$

where p_x and p_y are *momenta* associated respectively with x and y. While these equations seem to define a four-dimensional phase space, conservation of energy requires solution trajectories to lie on surfaces defined by

$$H(p_x, p_y, x, y) = E \,,$$

where the total energy E remains constant over time. Thus one of the dependent variables can always be expressed as a function of the other three, reducing the essential phase-space dimension to 3 as is indicated in Fig. 2.4.

The first two of the four Hamiltonian equations merely repeat the definitions of $\dot{x}$ and $\dot{y}$, whereas the second two can be regarded as a pair of second order equations:

$$\frac{d^2 x}{dt^2} + \omega_1 x = -2\varepsilon xy \,, \qquad \frac{d^2 y}{dt^2} + \omega_2 y = \varepsilon(y^2 - x^2) \,.$$

These two equations define oscillatory behavior on a torus as is sketched in Fig. 2.4a. For $\varepsilon = 0$, all trajectories lie on such tori, and for $\varepsilon \ll 1$, the KAM theorem tells us that most of them do. For $\varepsilon \approx 1$, chaotic regions begin to open up in the phase space, as is indicated in the Poincaré section of Fig. 2.4b.

Although the term "phase space" is now widely used in the above sense [399, 494], its etymological origin is unclear to many because "phase" has diverse meanings. In addition to the standard definition of "a distinct stage of development", physical chemists speak of ice, liquid and vapor as the phases of water, astronomers observe the phases of Venus and the Moon, and engineers use phase to measure the degree to which a periodic function has progressed through a cycle – for example, a $90°$ phase change indicates passage through a quarter of a cycle. Perhaps this confusion is why electrical engineers in the United States coined the term "state space" in the early 1960s for a phase space, corresponding to the German term *Zustandsraum*.

The concept of a $6n$-dimensional space (with axes comprising the positions of n particles in ordinary three-dimensional space and their corresponding velocities or momenta) goes back to Joseph Lagrange and was developed by William Rowan Hamilton and Henri Poincaré (among others) in the latter decades of the nineteenth century, but Ludwig Boltzmann wrote only of the "differential equations of motion" in 1887 [107]. The term "phase" for a particular "configuration and velocity" of a set of interacting particles was employed by James Clerk Maxwell in 1879 (the year of his death) [620] and then picked up by Josiah Williard Gibbs [172]. Thus Gibbs wrote in the preface to his 1902 classic *Elementary Principles in Statistical Mechanics* that his studies were directed to [354]:

> [...] the phases (or conditions with respect to configuration and velocity) which succeed one another in a given system in the course of time.

This definition comes close to "a distinct stage of development", and the term "phase" may have been in Gibbs's mind through his earlier analyses of thermodynamic states [351, 352] and of heterogeneous systems [353].

Maxwell's terminology was originally coined to apply to the space of the $6n$-dimensional Hamiltonian equations governing the energy-conserving dynamics of n interacting particles, but present usage of the term "phase space" includes the solution space for any set of first-order nonlinear ODEs – for example the Lorenz system defined in (2.1).

B Quantum Theory

The birthing of quantum theory (QT) is a checkered tale that began in 1901 with Max Planck's derivation of the black-body radiation law, which gives the relative intensity of light emitted from a hot object as a function of its wavelength. In this derivation, Planck was compelled – against his intuition – to propose that the energy of an atomic oscillator with frequency ω is not continuously variable but restricted to discrete steps of size $\hbar\omega$, where

$$\hbar = 1.052 \times 10^{-34} \text{ joule seconds}$$

is a universal constant (now called *Planck's constant*) [763]. As one of the seminal papers of his 1905 "annus mirabilis", Albert Einstein extended Planck's idea to explain the photoelectric effect (in which incident light causes electrons to be emitted from a metal if it has a sufficiently high frequency), by suggesting that light is particle-like, with a light particle (called a photon) of frequency ω having an energy $E = \hbar\omega$ [273].[1] In 1913, Niels Bohr proposed a model for the hydrogen atom in which a relatively light electron rotates about a heavier protonic nucleus in fixed orbits with angular momenta quantized in units of $\hbar$. Also light is emitted or absorbed at frequencies given by $\Delta E/\hbar$, where ΔE is the difference in energies of two electronic states, allowing calculation of the absorption and emission spectra of hydrogen [102]. Louis de Broglie – in his 1924 doctoral thesis – proposed that atomic particles with momentum p should be characterized by a wavelength $\lambda = 2\pi\hbar/p$, as was soon confirmed at the Bell Telephone Laboratories by electron diffraction experiments of Clinton Davisson and Lester Germer [216].

Thus by the mid-1920s it was known that an atomic particle of energy E and momentum p is characterized by a frequency ω and wavenumber k given by

$$\omega = \frac{2\pi}{T} = \frac{E}{\hbar} \text{ and } k = \frac{2\pi}{\lambda} = \frac{p}{\hbar},$$

where T is a temporal period. In the autumn of 1925, Erwin Schrödinger began looking into this matter and soon concluded that there should exist a wave equation to describe the dynamics of an atomic particle. Although

---

[1]It was for this idea that Einstein was awarded the Nobel Prize in 1918.

he seldom allowed work to interfere with vacations, Schrödinger devoted his 1925–26 winter holidays to this problem, thereby continuing "a twelve-month period of sustained creative activity that is without parallel in the history of science" [674]. He soon constructed the linear wave equation

$$i\hbar\frac{\partial\Psi(\boldsymbol{x},t)}{\partial t} = V(\boldsymbol{x},t)\Psi(\boldsymbol{x},t) - \frac{\hbar^2}{2m}\nabla^2\Psi(\boldsymbol{x},t) \,, \tag{B.1}$$

in which $\boldsymbol{x} \equiv x\boldsymbol{i} + y\boldsymbol{j} + z\boldsymbol{k}$ is the location of the particle in three dimensional space, m is its mass, $V(\mathrm{x})$ is its potential energy as a function of position, and $\nabla^2 \equiv \partial^2/\partial x^2 + \partial^2/\partial y^2 + \partial^2/\partial z^2$.

Now known as the Schrödinger equation (SE), this is an important physical formulation because it gives the discrete energies of quantum states as eigenvalues of a linear PDE. Two results of this program, which Schrödinger immediately obtained, are as follows [909]. (i) Letting $\boldsymbol{x} = x$ (i.e., assuming a one-dimensional problem) and setting $V(x) = Kx^2/2$ (the potential energy of a linear spring) allows one to derive energy levels as

$$E_n = \hbar\omega(n + 1/2) \,, \tag{B.2}$$

which (except for the $1/2$) is the relation that Planck was unhappily forced to assume in 1901 in his derivation of the black-body radiation formula. (ii) By writing the operator ∇^2 in spherical coordinates, the eigenvalues of (B.1) correspond exactly to the experimentally observed energy levels of atomic hydrogen. Furthermore, Schrödinger soon showed that his formulation is logically equivalent to a far more complicated matrix mechanics that had been proposed for QT by Werner Heisenberg in 1925 [416].

According to his biographer, Walter Moore, the first of a series of papers that Schrödinger wrote on this equation [850] was cited more than 100 000 times by 1960 and "has been universally recognized as one of the greatest achievements of twentieth-century physics" [674]. To help the reader appreciate the nature and implications of the Schrödinger equation, some brief comments on salient features of QT are listed.

Solutions of the SE

- **Free Particle.** If $V = 0$ in (B.1) – corresponding to a free classical particle – a general solution is

$$\Psi(\boldsymbol{x},t) = \int F(\boldsymbol{k})\exp\left[i(\boldsymbol{k}\cdot\boldsymbol{x} - \omega t)\right]\mathrm{d}\boldsymbol{k} \tag{B.3}$$

where $\boldsymbol{k}$ is a three-dimensional wavenumber vector, $F(\boldsymbol{k})$ is the spatial Fourier transform of the initial conditions (at $t = 0$) and $\omega = \hbar|\boldsymbol{k}|^2/2m$ is a corresponding *dispersion relation*.

- **Trapped Particle.** If V in (B.1) is a potential well (sufficiently negative in a localized region to trap a classical particle) and is independent of time, the time dependence in a component of the wave function appears merely as an exponential factor. Thus elementary solutions have the form $\psi(\boldsymbol{x})\exp(-\mathrm{i}Et/\hbar)$, where $\psi(\boldsymbol{x})$ obeys the *eigenvalue equation*

$$\nabla^2\psi(\boldsymbol{x}) + \frac{2m}{\hbar^2}\left[E - V(\boldsymbol{x})\right]\psi(\boldsymbol{x}) = 0\ , \tag{B.4}$$

with bounded solutions for particular values of E. If one denotes these *energy eigenvalues* by E_n and the corresponding *eigenfunctions* of (B.4) by $\psi_n(\boldsymbol{x})$, a solution of (B.1) is

$$\Psi(\boldsymbol{x},t) = \sum_n a_n\psi_n(\boldsymbol{x})\exp(-\mathrm{i}E_n t/\hbar)\ , \tag{B.5}$$

where the a_n are complex constants that can be chosen to fit some initial conditions (at $t = 0$, say). The solution of (B.1) can be written in this component form because it is a *linear* PDE, in contrast to the nonlinear Schrödinger equation that was introduced in Chap. 3. It was by carrying through this calculation for a linear spring potential and for the $1/r$ potential of electrostatic attraction between proton and electron in a hydrogen atom that Schrödinger obtained agreement with the Planck formula and the spectrum of atomic hydrogen, convincing him that (B.1) is physically relevant.

- **More General Cases.** Depending on the value of the energy (E), solutions for a time-independent potential (V) will be propagating (radiating) – as in (B.3) – in some regions of space and and trapped in other regions – as in (B.5). If V depends weakly on time, perturbation theory can be used to calculate transitions from one bound state to another, corresponding to emission or absorption of radiation. If V depends strongly upon time, the classical motion may be chaotic, leading to a complicated quantum spectrum, which is described in Chap. 6.

What Is Ψ?

Although (B.1) evidently has physical relevance, it was not immediately clear what aspect of reality is represented by the wave function Ψ because this quantity is a complex number (comprising both a real and an imaginary part). Is the real part of physical significance? Or should (B.1) be viewed as two coupled equations?

Following Schrödinger's observation that $|\Psi|^2$ can be considered as the local density of electronic charge, Max Born soon proposed that this squared magnitude should be considered as the *probability density* of finding the particle with a certain region of space-time [113]. In other words, the probability of finding the particle between x and $x + dx$, y and $y + dy$, z and $z + dz$, and t and $t + dt$ is

$$\int_x^{x+dx} \int_y^{y+dy} \int_z^{z+dz} \int_t^{t+dt} |\Psi(x, y, z, t)|^2 dx dy dz dt , \qquad (B.6)$$

where $|\Psi(x, y, z, t)|^2 = P(x, y, z, t)$ is the probability density in space-time. Although this interpretation of the quantum wave function as a probability amplitude is now universally accepted by physicists, there continue to be differences of opinion about what is meant by the term "probability", as is discussed in Chap. 6.

General Construction of a SE.

An intuitive understanding of the SE can be obtained by comparing it with the classical energy equation

$$\text{total energy} = \text{potential energy} + \text{kinetic energy} ,$$

where the LHS of (B.1) corresponds to the total energy and the RHS terms correspond to the classical potential and kinetic energies, respectively. More generally and compactly, a SE can be written as[2]

$$i\hbar\dot{\Psi} = H\Psi , \qquad (B.7)$$

where the dot indicates a time derivative and H is an *energy operator*. Thus (B.7) is constructed from the classical equation

$$E = H(p_x, x, p_y, y, p_z, z)$$

by replacing the energy E with the operator $+i\hbar\partial/\partial t$, the x-momentum p_x with $-i\hbar\partial/\partial x$, and so on.

As formulated by Schrödinger in 1926, the general aim of a quantum analysis is thus threefold: construct an energy operator from the classical Hamiltonian, find its eigenvalues and eigenfunctions, and write the general solution of the quantum problem as in (B.5). To indicate a sum over discrete eigenvalues (representing bound states) and integration over continuous eigenvalues (representing propagating states or radiation), Leonard Schiff has introduced the symbol [844]

$$\Psi = \mathbf{S}a_n(t)\psi_n \exp(-iE_n t/\hbar) ,$$

for a general solution of (B.7). Although the notation did not catch on with the physics community, the concept is important.

---

[2]This equation appeared on the first-day postmark of the Austrian stamp honoring Schrödinger's 100th anniversary in 1987 [674].

Heisenberg's Uncertainty Principle.

As the structure of the general solution to a quantum problem is a generalized Fourier expansion (also called an eigenfunction expansion), the general solution shares the property that the product of the spreads of a variable and its transform are of order unity. Thus the products of minimum uncertainties in quantum variables are

$$\Delta p_x \times \Delta x \approx \hbar ,$$
$$\Delta p_y \times \Delta y \approx \hbar ,$$
$$\Delta p_z \times \Delta z \approx \hbar ,$$
$$\Delta E \times \Delta t \approx \hbar .$$
(B.8)

Heisenberg and Bohr claimed that these uncertainties establish a fundamental limit to what can be known about physical reality [416], whereas Einstein viewed it as an unsatisfactory aspect of QT [278].

The Klein–Gordon (KG) Equation.

The SE of (B.1) is not invariant to a Lorentz transformation; thus it is not in accord with the SRT, which is to be expected because the separation of total energy into the sum of potential and kinetic components is not relativistically invariant. This is not a problem for the computation of interatomic levels and chemical binding energies because valence electrons are at rest. From (3.15), the general relation between energy and momentum of a moving particle is

$$E^2 = c^2 p^2 + m_0^2 c^4 ,$$

where m_0 is the rest mass ($m_0 = 8$ for a SG kink). Substituting $E \rightarrow i\hbar \partial/\partial t$ and $p \rightarrow i\hbar \partial/\partial x$ leads directly to the wave equation

$$\frac{\partial^2 \Psi}{\partial x^2} - \frac{1}{c^2} \frac{\partial^2 \Psi}{\partial t^2} = \left(\frac{m_0 c}{\hbar} \right)^2 \Psi ,$$
(B.9)

which was derived independently by Oskar Klein and Walter Gordon (among several others, including Schrödinger) in 1926 as a relativistic wave equation with a rest mass [512]. The generalization of this linear PDE to the nonlinear SG equation was discussed in Sect. 3.4.

In the 1920s, and 1930s, KG was of minor interest as an electron model because it fails to represent spin. Currently, the KG equation is used for quantum models of spin-zero particles such as pions, and it can also be used to construct a quantum theory for a SG kink.

References

1. Abbott, E.A.: *Flatland: A Romance of Many Dimensions* (Dover, New York) 1952. (First published in 1884)
2. Ablowitz, M.J. and Clarkson, P.A.: *Solitons, Nonlinear Evolution Equations and Inverse Scattering* (Cambridge University Press, New York) 1991
3. Ablowitz, M.J., Kaup, D.J., Newell, A.C. and Segur, H.: The initial value problem for the sine-Gordon equation, Phys. Rev. Lett. **30** (1973) 1262–1264
4. Ablowitz, M.J., Kaup, D.J., Newell, A.C. and Segur, H.: The inverse scattering transform Fourier analysis for nonlinear problems, Stud. Appl. Math. **53** (1974) 294–315
5. Ablowitz, M.J. and Ladik, J.F.: A nonlinear difference scheme and inverse scattering, Stud. Appl. Math. **55** (1976) 213–229
6. Ablowitz, M.J. and Segur, H.: *Solitons and the Inverse Scattering Transform* (SIAM, Philadelphia) 1981
7. Abraham, T.H.: (Physio)logical circuits: The intellectual origins of the McCulloch–Pitts neural networks, J. Hist. Behavioral Sci. **38**(1) (2002) 3–25
8. Abraham, T.H.: Nicholas Rashevsky's mathematical biophysics, J. Hist. Biology **37** (2004) 333–385
9. Adrian, E.D.: The all-or-none principle in nerve, J. Physiol. (London) **47** (1914) 460–474
10. Agnor, C.B. and Hamilton, D.P.: Neptune's capture of its moon Triton in a binary-planet gravitational encounter, Nature **441** (2006) 192–194
11. Ahlborn, B.K.: *Zoological Physics: Quantitative Models of Body Design, Actions, and Physical Limitations of Animals* (Springer-Verlag, Berlin) 2006
12. Aigner, A.A.: Navier–Stokes equation (in [878]) 2005
13. Aigner, A.A. and Fraedrich, K.: Atmospheric and ocean sciences (in [878]) 2005
14. Airy, G.B.: Tides and waves, *Encyclopedia Metropolitana* **3** (1841)
15. Alexander, D.M. and Krumhansl, J.A.: Localized excitations in hydrogen-bonded molecular crystals, Phys. Rev. B **33** (1986) 7172–7185
16. Alfvén, H.: Existence of electromagnetic-hydrodynamic waves, Nature **150** (1942) 405–406
17. Allee, W.C.: *Animal Aggregations: A Study in General Sociology* (University of Chicago Press, Chicago) 1931
18. Allen, F. et al.: Blue gene: A vision for protein science using a petaflop supercomputer, IBM Systems Journal **40** (2001) 310–327
19. Altenberger, R., Lindsay, K.A., Ogden, J.M. and Rosenberg, J.R.: The interaction between membrane kinetics and membrane geometry in the transmission of action potentials in non-uniform excitable fibres: A finite element approach, J. Neurosci. Meth. **112** (2001) 101–117

20. Amann, O.H., Von Karman, T. and Woodruff, G.B.: *The Failure of the Tacoma Narrows Bridge* (Federal Works Agency, Washington, D.C.) 1941

21. Amit, D.J.: *Modeling Brain Function: The World of Attractor Neural Networks* (Cambridge University Press, Cambridge) 1989

22. Anderson, P.B., Emmeche, C., Finnemann, N.O. and Christiansen, P.V.: *Downward Causation: Minds, Bodies and Matter* (Aarhus University Press, Aarhus, Denmark) 2000

23. Anderson, P.W.: More is different: Broken symmetry and the nature of the hierarchical structure of science, Science **177** (1972) 393–396

24. Anderson, P.W.: Local moments and localized states, Rev. Mod. Phys. **50** (1978) 191–201

25. Anderson, P.W., Arrow, K.J. and Pines, D. (Eds.): *The Economy as a Complex Evolving System* (Addison-Wesley, Redwood City, California) 1988

26. Anderson, R., Storm, J. and Wheal, H.V.: Thresholds of action potentials evoked by synapses on the dendrites of pyramidal cells in the rat hippocampus in vitro, J. Physiol. (London) **383** (1987) 509–526

27. Andronov, A.A., Vitt, A.A. and Khaykin, S.E.: *Theory of Oscillators* (Pergamon Press, New York) 1966. (The original Russian edition was first published in 1937)

28. Arbib, M.A.: Warren McCulloch's search for the logic of the nervous system, Perspectives in Biology and Medicine **43** (2) (2000) 193–216

29. Aristotle: *The Physics* (translated by P.H. Wicksteed and F.M. Cornford) (Harvard University Press, Cambridge, and William Heinemann Ltd, London) 1953

30. Aristotle: *On the Heavens* (Kessinger Publishing, Whitefish, Montana) 2004

31. Aristotle: *Metaphysics* (translated by J. Sachs), (Green Lion Press, Santa Fe, New Mexico) 1999

32. Arnold, M.L.: Natural hybridization and the evolution of domesticated, pest and disease organisms, Molecular Ecology **13** (2004) 997–2007

33. Arnold, M.L.: *Evolution Through Genetic Exchange* (Oxford University Press, Oxford) 2006

34. Arnold, M.L. and Larson, E.J.: Evolution's new look, Wilson Quarterly **28** (4) (2004) 60–73

35. Arnol'd, V.I.: *Ordinary Differential Equations*, 3rd edn. (Springer-Verlag, Berlin) 1992

36. Ascoli, G.A. and Krichmar, J.L.: L-Neuron: A modeling tool for the efficient generation and parsimonious description of dendritic morphology, Neurocomputing **32–33** (2000) 1003–1011

37. Ashby, W.R.: *Design for a Brain* (Chapman & Hall, London) 1906. (First published in 1952)

38. Aspect, A., Dalibard, J. and Roger, G.: Experimental test of Bell's inequalities using time-varying analyzers, Phys. Rev. Lett. **49** (1982) 1804–1807

39. Athorne, A.: Darboux transformation (in [878]) 2005

40. Aydon, C.: *Charles Darwin: The Naturalist Who Started a Scientific Revolution* (Carroll & Graf, New York) 2002

41. Baade, W. and Zwicky, F.: Supernovae and cosmic rays, Phys. Rev. **45** (1934) 138–139

42. Baade, W. and Zwicky, F.: On supernovae, Natl. Acad. Sci. **20** (1934) 254–259

43. Baas, N.A.: Emergence, hierarchies, and hyperstructures. In: *Artificial Life III*, ed. by C.G. Langton, (Addison-Wesley, Reading, Ma.) 1994

44. Bäcklund, A.V.: On ytor med konstant negative krökning, Lunds Universitets Arsskrift **19** (1883) 1–48

45. Baird, M.H.I.: The stability of inverse bubbles, Trans. Faraday Soc. **56** (1960) 213–219

46. Baker, G.L. and Gollub, J.P.: *Chaotic Dynamics* (Cambridge University Press, Cambridge) 1996

47. Baker, J.G., Centrella, J., Choi, D.I., Koppitz, M. and Van Meter, J.: Gravitational wave extraction from an inspiraling configuration of merging black holes, Phys. Rev. Lett. (in press) 2006

48. Ball, R.: Fairy rings of mushrooms (in [878]) 2005

49. Ball, R.: Kolmogorov cascade (in [878]) 2005

50. Ball, R. and Akhmediev, N.A.: *Nonlinear Dynamics: From Lasers to Butterflies* (World Scientific, Singapore) 2003

51. Banerjee, A. and Sobell, H.M.: Presence of nonlinear excitations in DNA structure and their relationship to DNA premelting and to drug intercalation. Journal of Biomolecular Structure and Dynamics **1** (1983) 253–262

52. Bannikov, V.S., Bezruchko, S.M., Grishankova, E.V., Kuz'man, S.V., Mityagin, Yu.A., Orlov, R.Yu., Rozhkov, S.B. and Sokolina, V.A.: Investigation of *Bacillus megaterium* cells by Raman scattering, Dokl. Akad. Nauk SSSR **253** (1980) 479–480 [Dokl. Biophys. **253** (1980) 119–120]

53. Barabási, A.L.: *Linked: How Everything is Connected to Everything Else and What It Means for Business, Science, and Everyday Life* (Plume, Cambridge, Ma.) 2003

54. Barone, A.: Flux–flow effects in Josephson tunnel junctions, J. Appl. Phys. **42** (1971) 2747–2751

55. Barone, A.: The Josephson effect-encounters in macroscopic quantum phenomena, Conference Proceedings Vol. 79 *One hundred years of h,* ed. by E. Beltrametti, G. Giuliani, A. Rimini and N. Robotti (SIF, Bologna) 2002, 95–105

56. Barone, A.: The strong impact of the weak superconductivity, Journal of Superconductivity **17** (5) (2004) 585–592

57. Barone, A., Esposito, F., Magee, C.J. and Scott, A.C.: Theory and applications of the sine-Gordon Equation, Rivista del Nuovo Cimento **1** (1971) 227–267

58. Barone, A. and Paternó, G.: *Physics and Applications of the Josephson Effect* (Wiley, New York) 1982

59. Barrow-Green, J.: *Poincaré and the Three-Body Problem* (Am. Math. Soc., Providence, R.I.) 1996

60. Barthes, M., Almairac, R., Sauvajol, J.L., Currat, R., Moret, J. and Ribet, J.L.: Neutron scattering investigation of deuterated crystalline acetanilide, Europhys. Lett. **7** (1988) 55–60

61. Bartholomay, A.F., Karreman, G. and Landahl, H.D.: Obituary of Nicholas Rashevsky, Bull. Math. Biology **34** (1) (1972) i–iv

62. Barwise, K.J. and Moss, L.S.: *Vicious Circles: On the Mathematics of Non-Wellfounded Phenomena* (CSLI Publications, Stanford) 1996

63. Baxter, R.J.: *Exactly Solvable Models in Statistical Mechanics* (Academic Press, London) 1982

64. Bazin, M.H.: Rapport aux remous et la propagation des ondes, Report to the Academy of Sciences, Paris, 1865

65. Bean, C.P. and Deblois, R.W.: Ferromagnetic domain wall as a pseudorelativistic entity, Bull. Am. Phys. Soc. **4** (1959) 53

66. Beck, C.: *Spatio-temporal Chaos and Vacuum Fluctuations of Quantized Fields* (World Scientific, Singapore) 2002

67. Beck, C.: String theory (in [878]) 2005

68. Belbruno, E.: *Capture Dynamics and Chaotic Motion in Celestial Mechanics* (Princeton University Press, Princeton) 2004

69. Belinski, V. and Verdaguer, E.: *Gravitational Solitons* (Cambridge University Press, Cambridge) 2001

70. Belinski, V.A., Khalatnikov, I.M. and Lifshitz, E.M.: Oscillatory approach to a singular point in the relativistic cosmology, Advances in Physics **19** (1970) 535–573

71. Belinski, V.A., Khalatnikov, I.M. and Lifshitz, E.M.: Solution of the Einstein equations with a time singularity, Advances in Physics **31** (1982) 639–667

72. Bell, J.S.: *Speakable and Unspeakable in Quantum Mechanics* (Cambridge University Press, Cambridge) 1993

73. Bénard, H.: Les tourbillons cellulaires dans une nappe liquide, Rev. Gén. Sci. Pures et Appl. **11** (1900) 1261–1271 and 1309–1328

74. Benedict, R.: *Patterns of Culture* (Houghton Mifflin, Boston) 1989. (First published in 1934)

75. Benjamin, T.B. and Feir, J.E.: The disintegration of wave trains in deep water, J. Fluid Mech. **27** (1967) 417–430

76. Benney, D.J. and Newell, A.C.: The propagation of nonlinear wave envelopes, J. Math. Phys. **46** (1967) 133–139

77. Berg, J.M., Tymoczko, J.L. and Stryer, L.: *Biochemistry*, 5th edn. (W.H. Freeman, New York) 2002

78. Bergson, H.: *Creative Evolution* (Dover, New York) 1998. (A republication of the first English translation in 1911)

79. Bernardo, L.S., Masukawa, L.M. and Prince, D.A.: Electrophysiology of isolated hippocampal dendrites, J. Neurosci. **2** (1982) 1614–1622

80. Bernstein, J.: Über den zeitlichen Verlauf der negativen Schwankung des Nervenstroms, Arch. ges. Physiol. **1** (1968) 173–207

81. Bernstein, J.: Untersuchungen zur Thermodynamik der bioelektrischen Ströme, Arch. ges. Physiol. **92** (1902) 521–562

82. Berry, M.V.: *Principles of Cosmology and Gravitation* (Cambridge University Press, Canbridge) 1989

83. Betyaev, S.K.: Hydrodynamics: problems and paradoxes, Soviet Physics – Uspekhi **38** (3) (1995) 287–316

84. Beurle, R.L.: Properties of a mass of cells capable of regenerating pulses, Phil. Trans. Roy. Soc. London A **240** (1956) 55–94

85. Bianchi, L.: Ricerche sulle superficie a curvatura constante e sulle elicoidi, Annali di Scuola Normale Sup. Pisa **1** (1879) 285–340

86. Bianchi, L.: Sulle transformatione di Bäcklund per le superficie pseudosferiche, Rendiconti Accademia Nazionale dei Lincei **5** (1892) 3–12

87. Bickhard, M.H. and Campbell, D.T.: Emergence. In [22]

88. Bigio, I.J., Gosnell, T.R., Mukeherjee, P. and Saffer, J.D.: Microwave absorption spectroscopy of DNA, Biopolymers **33** (1993) 147–150

89. Binczak, S.: Myelinated nerves (in [878]) 2005
90. Birge, R.T. and Sponer, H.: The heat of dislocation of non-polar molecules, Phys. Rev. **28** (1926) 259–283
91. Birkhoff, G.D.: *Dynamical Systems* (American Mathematical Society, Providence, R.I.) 1926
92. Bissell, C.C.: A.A. Andronov and the development of Soviet control engineering, IEEE Control Systems Magazine **18** (1) (1998) 56–62
93. Bloembergen, N.: *Nonlinear Optics* (World Scientific, Singapore) 1965
94. Boardman, A.: Polaritons (in [878]) 2005
95. Bode, H.W.: *Network Analysis and Feedback Amplifier Design* (Van Nostrand, New York) 1950
96. Bode, M.F. and Evans, A.: *Classical Novae* (Wiley, New York) 1989
97. Boden, M.A.: *The Philosophy of Artificial Life* (Oxford University Press, Oxford) 1996
98. Bogoliubov, N.N. and Mitropolsky, Y.A.: *Asymptotic Methods in the Theory of Nonlinear Oscillations* (Gordon and Breach, New York) 1961
99. Bohm, D.: A suggested interpretation of the quantum theory in terms of "hidden" variables, Phys. Rev. **85** (1953) 166–193
100. Bohm, D.: Proof that probability density approaches $|\psi|^2$ in causal interpretation of quantum theory, Phys. Rev. **89** (1953) 458–466
101. Bohm, D. and Aharonov, Y.: Discussion of experimental proof for the paradox of Einstein, Rosen and Podolski, Phys. Rev. **108** (1957) 1070–1076
102. Bohr, N.: On the constitution of atoms and molecules, Philosophical Magazine **26** (1913) 1–25
103. Bohr, N.: Quantum mechanics and physical reality, Nature **136** (1935) 65
104. Bohr, N.: Can quantum-mechanical description of reality be considered complete? Phys. Rev **48** (1935) 696–702
105. Bollinger, J.J., Heinzen, D.J., Itano, W.M., Gilbert, S.L. and Wineland, D.J.: Test of the linearity of quantum mechanics by rf spectroscopy of the ^{9}Be$^+$ ground state, Phys. Rev. Lett. **63** (1989) 1031–1034
106. Bolton, C.T.: Identification of Cygnus X-1 with HDE 226868, Nature **235** (1972) 271–273
107. Boltzmann, L.: Über die mechanischen Analogien des zweiten Hauptsatzes der Thermodynamik, Journal für die reine und angewandte Mathematik **100** (1887) 201–212
108. Bona, J.L., Pritchard, W.G. and Scott, L.R.: An evaluation of a model equation for water waves, Phil. Trans. Roy. Soc. (London) **302** (1981) 457–510
109. Bonnet, O.: Mémoire sur la théorie des surfaces applicables sur une surface donnée, J. de l'École Imperial Polytechnique **25** (1867) 1–151
110. Borckmans, P. and Dewel, G.: Turing patterns (in [878]) 2005
111. Born, M.: Der Impulse-Energie-Satz in der Elektrodynamik von Gustav Mie, Nach. von der Gesell. der Wissen. (Göttingen), Math. Physikalische Kl. (1914) 23–36
112. Born, M.: *Die Relativitätstheorie Einsteins und ihre physikalischen Grundlagen* (Springer-Verlag, Berlin) 1920
113. Born, M.: Physical aspects of quantum mechanics, Nature **119** (1927) 354–357
114. Born, M.: *The Born–Einstein Letters* (Macmillan, London) 1971
115. Born, M. and Oppenheimer, J.R.: Zur Quantentheorie der Molekeln, Ann. der Physik **84** (1927) 457–484

116. Born, M. and Infeld, L.: Foundations of a new field theory, Proc. Roy. Soc. London A **144** (1934) 425–451

117. Bour, E.: Théorie de la déformation des surfaces, J. de l'École Imperial Polytechnique **19** (1862) 1–48

118. Boussinesq, J.: Théorie des ondes et des remous qui se propagent le long d'un canal rectangular horizontal, en communiquant au liquid contenu dans ce canal des vitesses sensiblement pareilles de la surface eau fond, J. Math. Pures Appl. **17** (1872) 55–108

119. Boyd, P.T., Centrella, J.M. and Klasky, S.A.: Properties of gravitational "solitons", Phys. Rev. D **43** (1991) 379–390

120. Braitenberg, V. and Pulvermüller, F.: Entwurf einer neurologischen Theorie der Sprache, Naturwissenschaften **79** (1992) 102–117

121. Brans, C.H. and Dicke, R.H.: Mach's principle and a relativistic theory of gravitation, Phys. Rev. **124** (1961) 925–935

122. Bray, W.C.: A periodic reaction in homogeneous solution and its relation to catalysis, J. Am. Chem. Soc. **43** (1921) 1262–1267

123. Brazier, M.A.B.: *A History of the Electrical Activity of the Brain* (Pitman, London) 1961

124. Briggs, K.: A precise calculation of the Feigenbaum numbers, Mathematics of Computation **57** (1991) 435–439

125. Brillouin, L.: General relativity theory of gravity and experiments, Proc. Natl. Acad. Sci. USA **57** (1967) 1529–1535

126. Brillouin, L.: *Relativity Reexamined* (Academic Press, New York) 1970

127. Brittain, J.E.: Harold S. Black and the negative feedback amplifier, Proc. IEEE **85** (1997) 1335–1336

128. Brody, T.: *The Philosophy Behind Physics* (Springer-Verlag, Berlin) 1993

129. Brown, J.D., Henneaux, M. and Teitelboim, C.: Black holes in two space-time dimensions, Phys. Rev. D **33** (1986) 319–323

130. Browne, J.: *Charles Darwin: The Power of Place* (Random House, New York) 2002

131. Bruner, L.J.: Stable black soap films, Am. J. Phys. **53** (1985) 177–178

132. Buck, J.B.: Synchronous rhythmic flashing of fireflies, The Quarterly Review of Biology **13** (1938) 301–314

133. Buck, J.B.: Synchronous rhythmic flashing of fireflies. II, The Quarterly Review of Biology **63** (1988) 265–289

134. Bullough, R.: The wave "par excellence", the solitary, progressive great wave of equilibrium of the fluid – an early history of the solitary wave. In: *Solitons*, ed. by M. Lakshmanan (Springer-Verlag, New York) 1988

135. Bullough, R.: Sine-Gordon equation (in [878]) 2005

136. Bunge, M.: *Causality and Modern Science*, 3rd edn. (Dover, New York) 1979

137. Burger, D.: *Sphereland: A Fantasy About Curved Space and an Expanding Universe* (Crowell, New York) 1965

138. Burgers, J.: A mathematical model illustrating the theory of turbulence, Advances in Applied Mechanics **1** (1948) 171–199

139. Burkitt, G.R., Silberstein, R.B., Cadusch, P.J. and Wood, A.W.: Steady-state visual evoked potentials and travelling waves, Clinical Neurophysiology **111** (2000) 246–258

140. Busse, F.H.: Visualizing the dynamics of the onset of turbulence, Science **305** (2004) 1574–1575

141. Busse, F.H.: Fluid dynamics (in [878]) 2005
142. Busse, F.H.: Magnetohydrodynamics (in [878]) 2005
143. Butler, R.P., Marcy, G.W., Fischer, D.A., Brown, T.M., Contos, A.R., Korzennik, S.G., Niesenson, P. and Noyes, R.W.: Evidence for multiple companions to ν Andromedae, Astrophys. J. **526** (1999) 916–927
144. Butz, E.G. and Cowan, J.D.: Transient potentials in neurons with arbitrary geometry, Biophysical J. **14** (1974) 1–22
145. Byrd, P.F. and Friedman, M.D.: *Handbook of Elliptic Integrals for Engineers and Scientists* (Springer-Verlag, New York) 1971
146. Caianiello, E.R.: Outline of a theory of thought processes and thinking machines, J. Theoret. Biol. **1** (1961) 204–235
147. Calini, A.M.: Mel'nikov method (in [878]) 2005
148. Calogero, F. and Degasperis, A.: *The Spectral Transform and Solitons* (North-Holland, Amsterdam) 1982
149. Camp, J.B. and Cornish, N.J.: Gravitational wave astronomy, Annu. Rev. Nucl. Part. Sci. **54** (2004) 525–577
150. Campbell, D.K., Flach, S. and Kivshar, Y.S.: Localizing energy through nonlinearity and discreteness, Physics Today, January 2004, 43–49
151. Campbell, D.K., Schonfeld, J.S. and Wingate, C.A.: Resonance structure in kink–antikink interactions in ϕ^4 theory, Physica D **9** (1983) 1–32
152. Careri, G., Buontempo, U., Galluzzi, F., Scott, A.C., Gratton, E. and Shyamsunder, E.: Spectroscopic evidence for Davydov-like solitons in acetanilide, Phys. Rev. B **30** (1984) 4689–4702
153. Careri, G. and Wyman, J.: Soliton-assisted uniddirectional circulation in a biochemical cycle, Proc. Natl. Acad. Sci. USA **81** (1984) 4386–4388
154. Careri, G. and Wyman, J.: Unidirectional circulation in a prebiotic photochemical cycle, Proc. Natl. Acad. Sci. USA **82** (1985) 4115–4116
155. Carey, B.: Extra second will be added to 2005, *Live Science*, at web address `http://www.livescience.com/technology/050705_leap_second.html` July, 2005
156. Carpenter, G.A.: A geometric approach to singular perturbation problems with applications to nerve impulse equations, J. Diff. Eq. **23** (1977) 335–367
157. Carpenter, G.A.: Periodic solutions of nerve impulse equations, J. Math. Anal. Appl. **58** (1977) 152–173
158. Carroll, S.M.: *Spacetime and Geometry: An Introduction to General Relativity* (Addison-Wesley, San Francisco) 2004
159. Carter, O.L., Presti, D.E., Callisternon, C., Ungerer, Y., Liu, G.B. and Pettigrew, J.D.: Meditation alters perceptual rivalry in Tibetan Buddhist monks, Current Biology **15** (11) (2005) R114–R115
160. Cartwright, J.H.E., García-Ruiz, J.M., Novella, M.L. and Otálora, F.: Formation of chemical gardens, J. Colloid Inerface Sci. **256** (2002) 351–359
161. Cartwright, M.L. and Littlewood, J.E.: On nonlinear differential equations of the second order. I. The equation $\ddot{y} + k(1 - y^2)\dot{y} + y = b\lambda k \cos(\lambda t + a)$, k large, J. London Math. Soc. **20** (1945) 180–189
162. Cartwright, I.L. and Elgin, S.C.R.: Analysis of chromatin structure and DNA sequence organization: Use of the 1,10-phenanthroline-cuprous complex, Nucleic Acids Res. **10** (1982) 5835–5852
163. Casati, G. and Chirikov, B.V.: *Quantum Chaos* (Cambridge University Press, Cambridge) 1995

164. Casten, R.G., Cohen, H. and Lagerstrom, P.A.: Perturbation analysis of an approximation to Hodgkin–Huxley theory, Quart. Appl. Math., **32** (1975) 356–402

165. Castets, V., Dulos, E., Boissonade, J. and De Kepper, P.: Experimental evidence of a sustained standing Turing-type nonequilibrium chemical pattern, Phys. Rev. Lett. **64** (1990) 2953–2956

166. Caudry, P.J., Eilbeck, J.C. and Gibbon, J.D.: The sine-Gordon equation as a model field theory, Nuovo Cimento B **25** (1975) 497–512

167. Chamberlin, M.J.: The selectivity of transcription, Ann. Rev. Biochem. **43** (1974) 721–775

168. Chandrasekhar, S.: The maximum mass of ideal white dwarfs, Astrophysical Journal **74** (1931) 81–82

169. Chandrasekhar, S.: *Hydrodynamic and Hydromagnetic Stability* (Oxford University Press, New York) 1961

170. Chandrasekhar, S.: *The Mathematical Theory of Black Holes* (Oxford University Press, New York) 1983

171. Chase, P.G.: *The Emergence of Culture: The Evolution of a Uniquely Human Way of Life* (Springer-Verlag, New York) 2006

172. Chencier, A.: De la Mécanique céleste à la théorie des systèmes dynamiques, aller et retour: Poincaré et la géométrisation de l'espace des phases (to appear)

173. Chernitskii, A.A.: Born–Infeld equations (in [878]) 2005

174. Chirikov, B.V.: At. Energy **6** (1959) 630. English translation: Resonance processes in magnetic traps, J. Nucl. Energy C **1** (1960) 253–260

175. Chirikov, B.V.: A universal instability of many dimensional oscillator systems, Physics Reports **52** (1979) 263–379

176. Chirikov, B.V. and Vecheslavov, V.V.: Chaotic dynamics of comet Halley, Astronomy and Astrophysics **221** (1989) 146–154

177. Choudury, S.R.: Lorenz equations (in [878]) 2005

178. Christ, N.H. and Lee, T.D.: Quantum expansion of soliton solutions, Phys. Rev. D **12** (1975) 2330–2336

179. Christiansen, P.L., Gaididei, Yu.B., Johansson, M., Rasmussen, K.O., Mezentsev, V.K. and Rasmussen, J.J.: Solitary excitations in discrete two-dimensional nonlinear Schrödinger models with dispersive dipole–dipole interactions, Phys. Rev. B **57** (1998) 11303–11308

180. Christiansen, P.L. and Scott, A.C.: Davydov's Soliton Revisited: Self-Trapping of Vibrational Energy in Protein (Plenum Press, New York) 1990

181. Christensen, T.A., Pawlowski, V.M., Lei, H. and Hildebrand, J.G.: Multi-unit recordings reveal context-dependent modulation of synchrony in odor-specific neural ensembles, Nat. Neurosci. **3** (2000) 927–931

182. Chua, L.O., Desoer, C.A. and Kuh, E.S.: *Linear and Nonlinear Circuits* (McGraw-Hill, New York) 1987

183. Chung, S.H., Raymond, S.A. and Letvin, J.Y.: Multiple meaning in single visual units, Brain Behav. Evol. **3** (1970) 72–101

184. Clemence, G.M.: The relativity effect of planetary motions, Rev. Mod. Phys. **14** (1947) 361–364

185. Clemence, G.M.: Relativity effects in planetary motions, Proc. Am. Philos. Soc. **93** (1949) 532–534

186. Cole, K.S.: *Membranes, Ions and Impulses* (University of California Press, Berkeley) 1968

187. Cole, K.S. and Curtis, H.J.: Electrical impedance of a nerve during activity, Nature **142** (1938) 209

188. Conley, C. and Easton, R.: Isolated invariant sets and isolating blocks, Trans. Am. Math. Soc. **158** (1971) 35–61

189. Conway, F. and Siegelman, J.: *Dark Hero of the Information Age: In Search of Norbert Wiener the Father of Cybernetics* (Basic Books, New York) 2004

190. Cooper, P.J. and Hamilton, L.D.: The A–B conformational change in the sodium salt of DNA, J. Mol. Biol. **16** (2) (1966) 562–563

191. Cornwell, J. (Ed.): *Nature's Imagination: The Frontiers of Scientific Vision* (Oxford University Press, Oxford) 1995

192. Corry, L.: From Mie's electromagnetic theory of matter to Hilbert's unified foundations of physics, Stud. Hist. Phil. Mod. Phys. **30** (1999) 159–183

193. Corry, L., Renn, J. and Stachel, J.: Belated decision in the Hilbert–Einstein priority dispute, Science **278** (1997) 1270–1273

194. Costabile, G., Parmentier, R.D., Savo, B., McLaughlin, D.W. and Scott, A.C.: Exact solutions of the SG equation describing oscillations in a long (but finite) Josephson junction, Appl. Phys. Lett. **32** (1978) 587–589

195. Cottingham, W.N. and Greenwood, D.A.: *An Introduction to the Standard Model of Particle Physics* (Cambridge University Press, Cambridge) 1998

196. Cowan, J.D.: Statistical Mechanics of Nervous Nets. In: *Proceedings of 1967 NATO Conference on Neural Networks* ed. by E.R. Caianiello, Springer-Verlag, Berlin (1969) 181–188

197. Craik, A.D.D.: The origins of water wave theory, Annu. Rev. Fluid Mech. **36** (2004) 1–28

198. Crandall, R.E.: The challenge of large numbers, Scientific American **276** February 1997, 72–78

199. Crane, H.D.: Neuristor – a novel device and system concept, Proc. IRE **50** (1962) 2048–2060

200. Crick, F.H.C.: Linking numbers and nucleosomes, Proc. Natl. Acad. Sci. USA **73** (1976) 2639–2643

201. Cruzeiro-Hansson, L. and Takeno, S.: The Davydov Hamiltonian leads to stochastic energy transfer in proteins, Phys. Lett. A **223** (1996) 383–388

202. Cruzeiro-Hansson, L. and Takeno, S.: Davydov model: The quantum, mixed quantum/classical and full classical systems, Phys. Rev. E **56** (1997) 894–906

203. Curie, P.: Sur la symétrie dans les phénomènes physiques, symétrie d'un champ électrique et d'un champ magnétique, J. Phys. **3** (1894) 393–415

204. Cushing, J.M.: Population dynamics (in [878]) 2005

205. Cvitanović, P.: *Universality in Chaos* (Adam Hilger, Bristol) 1989

206. Dai, H.H. and Huo, Y.: Solitary shock waves and other travelling waves in a general compressible hyperelastic rod, Proc. Roy. Soc. (London) A **456** (2000) 331–363

207. Darboux, G.: Sur une proposition relative aux équations linéaires, C.R. Acad. Sci. Paris **94** (1882) 1343; 1456–1459

208. Darrigol, O.: The spirited horse, the engineer, and the mathematician, Arch. Hist. Exact Sci. **58** (2003) 21–95

209. Darwin, C.R.: *The Origin of Species* (Random House, New York) 1979. (First published in 1859)

210. Darwin, C.R.: *The Autobiography of Charles Darwin* (Barnes & Noble, New York) 2005. (First published in 1887)

211. Dashen, R.F., Hasslacher, B. and Neveu, A.: Particle spectrum in model field theories from semiclassical functional integral techniques, Phys. Rev. D **11** (1975) 3424–3450

212. Dauxois, T. and Peyrard, M.: A nonlinear model for DNA melting. In: *Nonlinear Excitations in Biomolecules*, ed. by M. Peyrard (Springer-Verlag, Berlin) 1995, 127–136

213. Dauxois, T., Peyrard, M. and Bishop, A.R.: Dynamics and thermodynamics of a nonlinear model for DNA denaturation, Phys. Rev. E **47** (1993) 684–695

214. Davis, B.: *Exploring Chaos* (Perseus Books, Reading, Ma.) 1999

215. Davis, W.C.: Explosions (in [878]) 2005

216. Davisson, C. and Germer, L.H.: Diffraction of electrons by a crystal of nickel, Phys. Rev. **30** (1927) 705–740

217. Davydov, A.S.: The theory of contraction of proteins under their excitation, J. Theoret. Biol. **38** (1973) 559–569

218. Davydov, A.S.: *Solitons in Molecular Systems*, 2nd edn. (Reidel, Dordrecht) 1991

219. Day, R.H.: *Complex Economic Systems* (MIT Press, Cambridge, Mass) 1996

220. Deadwyler, S.A., Bunn, T. and Hampson, R.E.: Hippocampal ensemble activity during spatial delayed-nonmatch-to-sample performance in rats, J. Neurosci. **16** (1996) 354–372

221. De Broglie, L.: *Nonlinear Wave Mechanics* (Elsevier, Amsterdam) 1960

222. De Broglie, L.: *Introduction to the Vigier Theory of Elementary Particles* (Elsevier, Amsterdam) 1963

223. De Bruyn, J.R.: Phase transitions (in [878]) 2005

224. Deconinck, B.: Kadomtsev–Petviashvili equation (in [878]) 2005

225. Degallaix, J. and Blair, D.: Gravitational waves (in [878]) 2005

226. de Gennes, P.G.: *Superconductivity of Metals and Alloys* (Perseus Books, New York) 1999

227. Delaney, R.L.: The Mandelbrot set, the Farey tree and the Fibonacci sequence, Am. Math. Monthly **106** (1999) 289–302

228. Derrick, G.H.: Comments on nonlinear equations as models for elementary particles, J. Math. Phys. **5** (1964) 1252–1254

229. Deryabin, M.V. and Hjorth, P.G.: Kolmogorov–Arnol'd–Moser theorem (in [878]) 2005

230. Des Chene, D.: *Spirits and Clocks: Machine and Organism in Descartes* (Cornell University Press, Ithaca, N.Y.) 2001

231. Deutsch, M.: Evidence for the formation of positronium in gases, Phys. Rev. **82** (1951) 455–456

232. Dewitt, B.S.: Quantum mechanics and reality, Physics Today **23** (1970) 30–35

233. Dewitt, B. and Graham, N. (Eds.): *The Many-Worlds Interpretation of the Universe* (Princeton University Press, Princeton) 1973

234. Diacu, F.: The solution of the n-body problem, Math. Intelligencer **18** (3) (1996) 66–70

235. Diacu, F.: Celestial mechanics (in [878]) 2005

236. Diacu, F. and Holmes, P.: *Celestial Encounters: The Origins of Chaos and Stability* (Princeton University Press, Princeton) 1996

237. Dicke, R.H.: New research on old gravitation, Science **129** (1959) 621–624

238. Dicke, R.H.: The Sun's rotation and relativity, Nature **202** (1964) 432–435

239. Dicke, R.H.: The oblateness of the Sun and relativity, Science **184** (1974) 419–429

240. Dicke, R.H. and Goldenberg, H.M.: Solar oblateness and general relativity, Phys. Rev. Lett. **18** (1967) 313–316

241. Dicke, R.H. and Goldenberg, H.M.: The oblateness of the Sun, Astrophys. J. **27** (suppl.) (1974) 131–182

242. Dirac, P.A.M.: Quantum mechanics of many electron systems, Proc. Royal Soc. (London) A **126** (1929) 714–733

243. Dirac, P.A.M.: *The Principles of Quantum Mechanics*, 4th edn. (Oxford University Press, New York) 1982

244. Dittrich M. and Schulten, K.: Zooming in on ATP hydrolysis in F_1, J. of Bioenergetics and Biomembranes **37** (2005) 441–444

245. Dodd, R.K., Eilbeck, J.C., Gibbon, J.D. and Morris, H.C.: *Solitons and Nonlinear Wave Equations* (Academic Press, London) 1982

246. Dorignac, J., Eilbeck, J.C., Salerno, M. and Scott, A.C.: Quantum signatures of breather–breather interactions, Phys. Rev. Lett. **93** (2004) 025504-1-4

247. Döring, W.: Über die Trägheit der Wände zwischen weisschen Bezirken, Zeit. Naturforsch. **31** (1948) 373–379

248. Drazin, P.G. and Johnson, R.S.: *Solitons, An Introduction* (Cambridge University Press, Cambridge) 1989

249. Dunajski, M.: A nonlinear graviton from the sine-Gordon equation, Twistor Newsletter **40** (1996) 43–45

250. Dunajski, M., Mason, L.J. and Woodhouse, N.M.J.: From 2D integrable systems to self-dual gravity, J. Phys. A: Math. Gen. **31** (1998) 6019–6028

251. Dupraw, E.J.: *DNA and Chromosomes* (Holt, Rinehart and Winston, New York) 1970

252. Dupuis, G. and Berland, N.: Belousov–Zhabotinsky reaction (in [878]) 2005

253. Dyachenko, A.I. and Zakharov, V.E.: Modulation instability of Stokes wave – freak wave, JETP Letters **81** (2005) 255–259

254. Dyson, F.W., Eddington, A.S. and Davidson, C.: A determination of the deflection of light by the Sun's gravitational field, from observations made at the total eclipse of May 29, 1919, Philos. Trans. Royal Soc. London **220** A, (1920) 291–333

255. Dysthe, K.B., Trulsen, K. Krogstad, H.E. and Socquet-Juglard, H.: Evolution of a narrow band spectrum of random surface gravity waves, J. Fluid Mech. **478** (2003) 1–10

256. Dysthe, K.B., Socquet-Juglard, H., Trulsen, K. Krogstad, H.E. and Liu, J.: "Freak" waves and large-scale simulations of surface gravity waves (in [685]) 2005

257. Eberhart, R.C., Shi, Y. and Kennedy, J.: *Swarm Intelligence* (Morgan Kaufmann, San Franscisco) 2001

258. Ebraheem, H.K., Shohet, J.L. and Scott, A.C.: Mode locking in reversed field pinch experiments, Phys. Rev. Lett. **88** (2002) 067403-1-4

259. Eckberg, Ö., Wallén, P., Lansner, A., Travén, H., Brodin, L. and Grillner, S.: A computer based model for realistic simulations of neural networks, I: The single neuron and synaptic interaction, Biol. Cybern. **65** (1991) 81–90

260. Eckhardt, B.: Quantum mechanics of classically non-integrable systems, Physics Reports **163** (1988) 205–297

261. Eckhardt, B.: Solar system (in [878]) 2005

262. Eckhardt, B.: Quantum chaos (in [878]) 2005
263. Eckhaus, W.: *Studies in Nonlinear Stability Theory* (Springer-Verlag, New York) 1965
264. Edler, J. and Hamm, P.: Self-trapping of the amide-I band in a peptide model crystal, J. of Chem. Phys. **117** (2002) 2415–2424
265. Edler, J., Hamm, P. and Scott, A.C.: Femtosecond study of self-trapped excitons in crystalline acetanilide, Phys. Rev. Lett. **88** (2002) 067403-1–4
266. Edwards, G.S., Davis, C.C., Saffer, J.D. and Swicord, M.L.: Resonant microwave absorption of selected DNA molecules, Phys. Rev. Lett. **53** (1984) 1284–1287. See also Phys. Rev. Lett. **53** (1984) 2060
267. Edwards, G.S., Davis, C.C., Saffer, J.D. and Swicord, M.L.: Microwave-field-driven acoustic modes in DNA, Biophysical Journal **47** (6) (1985) 799–807
268. Ehresmann, A.C. and Vanbremeersch, J.-P.: Hierarchical evolutive systems: A mathematical model for complex systems, Bull. Math. Biol. **49** (1987) 13–50
269. Eigen, M. and Schuster, P.: *The Hypercycle: A Principle of Natural Self-Organization* (Springer-Verlag, Berlin) 1979
270. Eilbeck, J.C., Lomdahl, P.S. and Newell, A.C.: Chaos in the inhomogeneously driven sine-Gordon equation, Phys. Lett. A **87** (1981) 1–4
271. Eilbeck, J.C., Lomdahl, P.S. and Scott, A.C.: Soliton structure in crystalline acetanilide, Phys. Rev. B **30** (1984) 4703–4712
272. Einstein, A.: Zur elektrodynamik bewegter Körper, Ann. der Physik **17** (1905) 891–921
273. Einstein, A.: Ueber einen die Erzeugung und Verwandlung des Lichtes betreënden heuristischen Gesichtspunkt, **17** (1905) 132–148
274. Einstein, A.: Die Grundlage der allgemeinen Relativitätstheorie, Ann. der Physik **49** (1916) 769–822
275. Einstein, A.: Zum Quantensatz von Sommerfeld und Epstein, Verh. Deutsch. Phys. Ges. **19** (1917) 82–92
276. Einstein, A.: Zur Quantentheorie der Strahlung, Physik. Zeit. **18** (1917) 121–128
277. Einstein, A.: On a stationary system with spherical symmetry consisting of many gravitating masses, Annals of Math. **40** (1905) 922–936
278. Einstein, A.: *Ideas and Opinions* (Crown, New York) 1954
279. Einstein, A., Podolsky, B. and Rosen, N.: Can quantum-mechanical description of physical reality be considered complete? Phys. Rev. **47** (1935) 777–780
280. Ellis, G.F.R., Kirchner, U. and Stoeger, W.R.: Multiverses and physical cosmology, Mon. Not. R. Astron. Soc. **347** (2004) 921–936
281. Ellis, G.F.R. and Stoeger, W.R.: The 'fitting problem' in cosmology, Class. Quantum Grav. **4** (1987) 1697–1729
282. Ellis, J.W.: Molecular absorption spectra of liquids below 3μ, Trans. Faraday Soc. **25** (1929) 888–897
283. Elsaesser, T., Mukamel, S., Murnanae, M.M. and Scherer, N.F. (Eds.): *Ultrafast Phenomena XII* (Springer-Verlag, Berlin) 2000
284. Elsasser, W.M.: *The Physical Foundations of Biology* (Pergamon Press, London) 1958
285. Elsasser, W.M.: *Atom and Organism* (Princeton University Press, Princeton) 1966
286. Elsasser, W.M.: Acausal phenomena in physics and biology, American Scientist **57** (1969) 502–516

287. Elsasser, W.M.: *Reflections on a Theory of Organisms: Holism in Biology* (John Hopkins University Press, Baltimore) 1998. (First published in 1987)
288. Emmeche, C., Koppe, S. and Stjernfelt, F.: Levels, emergence, and three versions of downward causation. In [22]
289. Emmerson, G.S.: *John Scott Russell: A Great Victorian Engineer and Naval Architect* (John Murray, London) 1977
290. Encyclopedia Britannica, **25** (1911) 460
291. Englander, S.W., Kallenbach, N.R., Heeger, A.J., Krumhansl, J.A. and Litwin, S.: Nature of the open state in long polynucleotide double helices: Possibility of soliton excitations, Proc. Natl. Acad. Sci. USA **77** (1980) 7222–7226
292. Enneper, A.: Analytisch-geometrische Untersuchungen, Nach. K. Ges. Wiss. Göttingen (1868) 252
293. Enz, U.: Discrete mass, elementary length, and a topological invariant as a consequence of a relativistic invariant variation principle, Phys. Rev. **131** (1963) 1392–1394
294. Enz, U.: Die Dynamik der blochschen Wand, Helv. Phys. Acta **37** (1964) 245–251
295. Ermentrout, G.B. and Cowan, J.D.: A Mathematical theory of visual hallucination patterns, Biological Cybernetics **34** (1979) 137–150
296. Evans, J.W.: Nerve axon equations, Indiana Univ. Math. J. **21** (1972) 877–885; **22** (1972) 75–90; **22** (1972) 577–593; **24** (1975) 1169–1190
297. Everett, H.: "Relative state" formulation of quantum mechanics, Rev. Mod. Phys. **29** (1957) 454–462
298. Faddeev, L.D.: Quantum completely integrable models in field theory, Soviet Science Rev. of Math. and Phys. C **1** (1981) 107–155
299. Faddeev, L.D. and Takhtajan, L.A.: *Hamiltonian Methods in the Theory of Solitons* (Springer-Verlag, Berlin) 1987
300. Faraday, M.: *Faraday's Chemical History of a Candle* (Chicago Review Press, Chicago) 1988. (Originally published in 1861)
301. Feddersen, H.: Numerical calculations of solitary waves in Davydov's equations, Phys. Scr. **47** (1993) 481–483
302. Feigenbaum, M.: Quantitative universality for a class of nonlinear transformations, J. Statistical Phys. **19** (1978) 25–52
303. Feigenbaum, M.: Private communication, May 2005
304. Fergason, J.L. and Brown, G.H.: Liquid crystals and living systems, J. Am. Oil Chem. Soc. **45** (1968) 120–127
305. Fermi, E., Pasta, J.R. and Ulam, S.M.: Studies of nonlinear problems, Los Alamos Scientific Laboratory Report No. LA1940 (1955). Reprinted in reference [700]
306. Filippov, A.T.: *The Versatile Soliton* (Birkhäuser, Boston) 2000
307. Finkelstein, R., Lelevier, R. and Ruderman, M.: Nonlinear spinor fields, Phys. Rev. **83** (1951) 326–332
308. Fisher, R.A.: The wave of advance of advantageous genes, Ann. Eugen. (now Ann. Hum. Gen.) **7** (1937) 355–369
309. Fisher, S., Windshügel, B., Horak, D., Holmes, K.C. and Smith, J.C.: Structural mechanism of the recovery stroke in the myosin molecular motor, Proc. Natl. Acad. Sci. USA **102** (2005) 6873–6878
310. Fitzhugh, R.: Impulses and physiological states in theoretical models of nerve membrane, Biophys. J. **1** (1961) 445–466

311. Flaschka, H.: Toda lattice (in [878]) 2005
312. Floria, L.M. and Martinez, P.J.: Frenkel–Kontorova model (in [878]) 2005
313. Fodor, J.A.: Special sciences; still autonomous after all these years. In *Philosophical Perspectives, 11, Mind, Causation, and World*, ed. by J.E. Tomberlin (Blackwell, New York) 1997
314. Fontana, W. and Buss, L.W.: The arrival of the fittest: Toward a theory of biological organization, Bull. Math. Biol. **56** (1994) 1–64
315. Forristall, G.Z.: Understanding rogue waves: Are new physics really necessary? (in [685]) 2005
316. Foster, K.R., Epstein, B.R. and Gealt, M.A.: "Resonances" in the dielectric absorption of DNA? Biophys. J. **52** (1987) 421–425
317. Fox, R.E.: *Biological Energy Transduction* (Wiley, New York) 1982
318. Fraga, M.F., Ballestar, E., Paz, M.F., Ropero, S., Setien, F., Ballestar, M.L., Heine-Suñer, D., Cigudosa, J.C., Urioste, M., Benitez, J., Boix-Chornet, M., Sanchez-Aguilera, A., Ling, C., Carlsson, E., Poulsen, P., Vaag, A., Stephan, Z., Spector, T.D., Wu, Y.Z., Plass, C. and Esteller, M.: Epigenetic differences arise during the lifetime of monozygotic twins, Proc. Natl. Acad. Sci. USA **102** (2005) 10604–10609
319. Franken, P.A., Hill, A.E., Peters, C.W. and Weinreich, G.: Generation of optical harmonics, Phys. Rev. Lett. **7** (1961) 118–119
320. Frank-Kamenetskii, D.A.: *Diffusion and Heat Exchange in Chemical Kinetics* (Princeton University Press, Princeton, NJ) 1955
321. Frank-Kamenetskii, M.D.: Physicists retreat again, Nature **328** (1987) 108
322. Frank-Kamenetskii, M.D.: Private communication, April 2006
323. Franklin, W.S.: Book review, Phys. Rev. **6** (1898) 170–175
324. Fransén, E.: Biophysical simulation of cortical associative memory, Doctoral thesis, Royal Institute of Technology, Stockholm, 1996
325. Fransén, E., Lansner, A. and Liljenström, H.: A model of cortical memory based on Hebbian cell assemblies. In: *Computation and Neural Systems*, ed. by F.H. Eeckman and J.M. Bower, (Kluwer, Boston) 1993
326. Fransén, E. and Lansner, A.: Low spiking rates in a population of mutually exciting pyramidal cells, Network **6** (1995) 271–288
327. Fransén, E. and Lansner, A.: A model of cortical associative memory based on a horizontal network of connected columns, Network **9** (1998) 235–264
328. Fraser, J.T.: *The Genesis and Evolution of Time* (Harvester Press, Brighton, England) 1982
329. Fraser, J.T.: *Of Time, Passion, and Knowledge: Reflections on the Strategy of Existence*, 2nd edn. (Princeton University Press, Princeton) 1990
330. Frauenfelder, H., Sligar, S.G. and Wolynes, P.G.: The energy landscapes and motions of protein, Science **254** (1991) 1598–1603
331. Frenkel, Y. and Kontorova, T.: On the theory of plastic deformation and twinning, J. Phys. (USSR) **1** (1939) 137–149
332. Friedberg, R., Lee, T.D. and Sirlin, A.: Class of scalar-field soliton solutions in three space dimensions, Phys. Rev. D **13** (1976) 2739–2761
333. Friedberg, R. and Lee, T.D.: Quantum chromodynamics and soliton models of hadrons, Phys. Rev. D **19** (1978) 2623–2631
334. Friedmann, A.: Über die Krümmung des Raumes, Zeit. für Phys. **10** (1922) 377–386. English translation in: Gen. Rel. Grav. **31** (1999) 1991–2000
335. Friedrich, H. and Wintgen, D.: The hydrogen atom in a uniform magnetic field – an example of chaos, Physics Reports **183** (1989) 37–79

336. Friesecke, G. and Pego, R.L.: Solitary waves on FPU lattices: I. Qualitative properties, renormalization and continuum limit, Nonlinearity **12** (1999) 1601–1627

337. Friesecke, G. and Pego, R.L.: Solitary waves on FPU lattices: II. Linear implies nonlinear stability, Nonlinearity **15** (2002) 1343–1359

338. Friesecke, G. and Pego, R.L.: Solitary waves on Fermi–Pasta–Ulam lattices: III. Howland-type Floquet theory, Nonlinearity **17** (2004) 207–227

339. Friesecke, G. and Pego, R.L.: Solitary waves on Fermi–Pasta–Ulam lattices: IV. Proof of stability at low energy, Nonlinearity **17** (2004) 229–251

340. Friesecke, G. and Wattis, J.A.D.: Existence theorem for travelling waves on lattices, Commun. Math. Phys. **161** (1994) 391–418

341. Frisch, U.: *Turbulence: The Legacy of A.N. Kolmogorov* (Cambridge University Press, Cambridge) 1995

342. Fröhlich, H.: On the theory of superconductivity: The one-dimensional case, Proc. Royal Soc. London A **223** (1954) 296–305

343. Fröhlich, H.: Long range coherence and energy storage in biological sytems, Int. J. Quantum Chem. **2** (1968) 641–649

344. Gabriel, C., Grant, E.H., Tata, R., Brown, P.R., Gestblom, B. and Noreland, E.: Microwave absorption in aqueous solutions of DNA, Nature **328** (1987) 145–146

345. Gabriel, C., Grant, E.H., Tata, R., Brown, P.R., Gestblom, B. and Noreland, E.: Dielectric behavior of aqueous solutions of plasmid DNA at microwave frequencies, Biophys. J. **55** (1989) 29–34

346. Gardner, C.S., Greene, J.M., Kruskal, M.D. and Miura, R.M.: Method for solving the Korteweg–de Vries equation, Phys. Rev. Lett. **19** (1967) 1095–1097

347. Gaspard, P.: Rössler systems (in [878]) 2005

348. Gell-Mann, M.: *The Quark and the Jaguar* (Freeman & Co., New York) 1994

349. Ghazanfar, A.A. and Nicolelis, M.A.L.: Nonlinear processing of tactile information in the thalamocortical loop, J. Neurophysiol. **78** (1997) 506–510

350. Giacometti, G.M., Focesi, A., Giardina, B., Brunori, M. and Wyman, J.: Kinetics of binding of carbon monoxide to *lumbricus* erythrocruorin: A possible model, Proc. Natl. Acad. Sci. USA **72** (1975) 4313–4316

351. Gibbs, J.W.: Graphical methods in the thermodynamics of fluids, Trans. Conn. Acad. Arts Sci. **2** (1873) 309–342

352. Gibbs, J.W.: A method of geometrical representation of the thermodynamic properties of substances by means of surfaces, Trans. Conn. Acad. Arts Sci. **2** (1873) 382–404

353. Gibbs, J.W.: On the equilibrium of heterogeneous substances, Trans. Conn. Acad. Arts Sci. **47** (1876–78) 108–248 and 343–524

354. Gibbs, J.W.: *Elementary Principles in Statistical Mechanics* (Yale University Press, New Haven, Connecticut) 1902

355. Gierer, A. and Meinhardt, H.: A theory of biological pattern formation, Kybernetik **12** (1972) 30–39

356. Gisin, N.: New additions to the Schrödinger cat family, Science **312** (2006) 63–64

357. Glass, L.: Cardiac arrythmias and the electrocardiogram (in [878]) 2005

358. Glass, L. and Mackey, M.C.: *From Clocks to Chaos: The Rhythms of Life* (Princeton University Press, Princeton, NJ) 1988

359. Gleick, J.: *Isaac Newton* (Random House, New York) 2003
360. Glendinning, P.: *Stability, Instability, and Chaos* (Cambridge University Press, Cambridge and New York) 1994
361. Glendinning, P.: Routes to chaos (in [878]) 2005
362. Goel, N.S., Maitra, S.C. and Montroll, E.W.: On the Volterra and other nonlinear models of interacting populations, Rev. Mod. Phys. **43** (1971) 231–276
363. Goenner, H.F.M.: On the history of unified field theories, *Living Reviews in Relativity*, http://www.theorie.physik.uni-goettingen.de/g̃oenner, 26 July 2005
364. Goldstein, H.: *Classical Mechanics* (Addison-Wesley, Cambridge, Ma.) 1951
365. Gollub, J.P, and Swinney, H.L.: Onset of turbulence in a rotating fluid, Phys. Rev. Lett. **35** (1975) 927–930
366. Goodsell, D.S.: *The Machinery of Life* (Springer-Verlag, New York) 1993
367. Goodwin, B.: *How the Leopard Changed its Spots: The Evolution of Complexity* (Scribner's, New York) 1994
368. Gordon, R. and Beloussov, L.: From observations to paradigms; the importance of theories and models, Int. J. Dev. Biol. **50** (2006) 103–111
369. Gould, S.J.: *Ontogeny and Phylogeny* (Belknap Press, Cambridge, Ma.) 1985
370. Gould, S.J.: *Wonderful Life: The Burgess Shale and the Nature of History* (W.W. Norton & Co., New York) 1989
371. Grava, T.: Zero-dispersion limits (in [878]) 2005
372. Green, D.E.: *The Mechanism of Energy Transduction in Biological Systems* (New York Academy of Sciences, New York) 1974
373. Green, D.E. and Ji, S.: Transductional and structural principles of the mitochondrial transducing unit, Proc. Nat. Acad. Sci. USA **70** (1973) 904–908
374. Greene, J.M.: A method for determining a stochastic transition, J. Math. Phys. **20** (1979) 1183–1201
375. Griffith, J.S.: A field theory of neural nets: I. Derivation of field equations, Bull. Math. Biophys. **25** (1963) 187–195
376. Griffith, J.S.: A field theory of neural nets: II. Properties of field equations, Bull. Math. Biophys. **27** (1965) 111–120
377. Griffith, J.S.: *Mathematical Neurobiology: An Introduction to the Mathematics of the Nervous System* (Academic Press, New York) 1971
378. Grimshaw, R.: Water waves (in [878]) 2005
379. Grindrod, P.: *The Theory and Applications of Reaction-Diffusion Equations: Patterns and Waves* (Clarendon Press, Oxford) 1996
380. Grossberg, S.: Adaptive pattern classification and universal recoding I: Parallel development of coding of neural feature detectors, Biol. Cybern. **23** (1976) 121–134
381. Grundler, W. and Keilman, F.: Nonthermal effects of millimeter microwaves on yeast growth, Z. Naturforsch. C **33** (1978) 15–22
382. Grundler, W. and Keilman, F.: Sharp resonances in yeast growth rates prove nonthermal sensitivity to microwaves, Phys. Rev. Lett. **51** (1983) 1214–1216
383. Guckenheimer, J. and Holmes, P.: *Nonlinear Oscillations, Dynamical Systems, and Bifurcations of Vector Fields* (Springer-Verlag, Berlin) 1983
384. Guéron, M., Kochoyan, M. and Leroy, J.L.: A single mode of DNA base-pair opening drives imino proton exchange, Nature **328** (1987) 89–92

385. Guevara, M.R., Glass, L. and Shrier, A.: Phase locking, period doubling bifurcations and irregular dynamics in periodically stimulated cardiac cells, Science **214** (1981) 1350–1353

386. Gustavson, F.G.: On constructing formal integrals of a Hamiltonian system near an equilibrium point, Astron. J. **71** (1966) 670–686

387. Hagen, G.H.L.: Über die Bewegung des Wassers in engen zylindrischen Röhren, Pogg. Ann. **46** (1839) 423–442

388. Haken, H.: Generalized Ginzburg–Landau equations for phase transition-like phenomena in lasers, nonlinear optics, hydrodynamics and chemical reactions, Zeit. für Physik B **21** (1975) 105–114

389. Haken, H.: Analogy between higher instabilities in fluids and lasers, Phys. Lett. A **53** (1975) 77–78

390. Haken, H.: *Quantum Field Theory of Solids* (North-Holland Press, Amsterdam) 1976

391. Haken, H.: *Advanced Synergetics*, 3rd edn. (Springer-Verlag, Berlin) 1993

392. Haken, H.: *Principles of Brain Functioning: A Synergetic Approach to Brain Activity*, 3rd edn. (Springer-Verlag, Berlin) 1996

393. Haken, H.: *Synergetics: Introduction and Advanced Topics* (Springer-Verlag, Berlin) 2004

394. Haken, H.: Synergetics (in [878]) 2005

395. Hakim, M., Lindsay, S.M. and Powell, J.: The speed of sound in DNA, Biopolymers **23** (1984) 1185–1192

396. Halburd, R. and Biondini, G.: Einstein equations (in [878]) 2005

397. Hallinan, J.: Game of life (in [878]) 2005

398. Hamilton, M.P. and Blackstock, D.T.: *Nonlinear Acoustics* (Academic Press, New York) 1998

399. Hamm, P.: Pump–probe measurements (in [878]) 2005

400. Hamm, P.: Private communication

401. Hammack, J.L.: A note on tsunamis: Their generation and propagation in an ocean of uniform depth, J. Fluid Mech. **60** (1973) 769–799

402. Hammack, J.L. and Segur, H.: The Korteweg–de Vries equation and water waves. Part 2. Comparison with experiments, J. Fluid Mech. **65** (1974) 289–314

403. Harris, R.G. (Ed.): *Surface Phenomena* (Cold Spring Harbor, New York) 1933

404. Harth, E.: *The Creative Loop: How the Brain Makes a Mind* (Addison-Wesley, Reading, Ma.) 1993

405. Hasegawa, A. and Tappert, F.: Transmission of stationary nonlinear optical pulses in dispersive dielectric fibers, Appl. Phys. Lett. **23** (1973) 142–144

406. Hasselblatt, B.: Phase space (in [878]) 2005

407. Hawking, S.: *On the Shoulders of Giants* (Running Press, Philadelphia) 2004

408. Hawkins, D.: The spirit of play, a memoir for Stan Ulam, *Los Alamos Science* (special issue) 1987

409. Hayashi, C., Shibayaama, H. and Ueda, Y.: Quasi-periodic oscillations in self-oscillatory systems with external force, (in Japanese) IEEE Technical Report, Nonlinear Theory (December 16, 1961)

410. Hebb, D.O.: Intelligence in man after large removals of cerebral tissue: Report of four left frontal lobe cases, J. Gen. Psychol. **21** (1939) 73–87

411. Hebb, D.O.: On the nature of fear, Physiol. Rev. **53** (1946) 259–276

412. Hebb, D.O.: *Organization of Behavior: A Neuropsychological Theory* (Wiley, New York) 1949

413. Hebb, D.O.: The structure of thought. In: *The Nature of Thought*, ed. by P.W. Jusczyk and R.M. Klein, (Lawrence Erlbaum Associates, Hillsdale, NJ) 1980

414. Hebb, D.O.: *Essay on Mind* (Lawrence Erlbaum Associates, Hillsdale, NJ) 1980

415. Hebling, D.: Traffic flow (in [878]) 2005

416. Heisenberg, W.: *The Physical Principles of the Quantum Theory* (Dover, New York) 1949. (First published in 1930)

417. Heisenberg, W.: *Introduction to a Unified Field Theory of Elementary Particles* (Wiley, New York) 1966

418. Heisenberg, W.: Nonlinear problems in physics, Physics Today **20** (1967) 27–33

419. Heller, E.J.: Bound-state eigenfunctions of classically chaotic Hamiltonian systems: Scars of periodic orbits, Phys. Rev. Lett. **53** (1984) 1515–1518

420. Helmholtz, H.: Messungen über den zeitlichen Verlauf der Zuckung animalischer Muskeln und die Fortpflanzungsgeschwindigkeit der Reizung in den nerven, Arch. Anat. Physiol. (1850) 276–364

421. Helmholtz, H.: On the integrals of hydrodynamic equations that express vortex motion, Phil. Mag. Suppl. **33** (1850) 485–510

422. Henin, F. and Prigogine, I.: Entropy, dynamics, and molecular chaos, Proc. Natl. Acad. Sci. USA **71** (7) (1974) 2618–2622

423. Henn, A., Medalia, O., Steinberg, M., Frandeschi, F. and Sagi, I.: Visualization of unwinding activity of duplex RNA by DbpA, a DEAD box helicase, at single-molecule resolution by atomic force microscopy, Proc. Natl. Acad. Sci. USA **98** (2001) 5007–5012

424. Hénon, M.: A two-dimensional mapping with a strange attractor, Comm. in Math. Phys. **50** (1976) 69–77

425. Hénon, M. and Heiles, C.: The applicability of the third integral of motion, Astron. J. **69** (1964) 73–79

426. Heron, W.: The pathology of boredom, Sci. Am. January 1957

427. Herreras, O.: Propagating dendritic action potential mediates synaptic transmission in CA1 pyramidal cells in situ, J. Neurophysiol. **64** (1990) 1429–1441

428. Henry, B.R.: Personal communication, 1988

429. Henry, B.R.: Local modes in molecules (in [878]) 2005

430. Hensler, G.: Galaxies (in [878]) 2005

431. Hermann, R., Schumacher, G. and Guyard, R.: Scale relativity and quantization of the solar system, Astron. Astrophys. **335** (1998) 281–286

432. Hertz, J., Krogh, A. and Palmer, R.G.: *Introduction to Neural Computation* (Addison-Wesley, Reading, MA) 1991

433. Heyerhoff, M.: The history of the early period of soliton theory, in *Nonlinearity and Geometry*, ed. by D. Wójcik and J. Cieśliński, (Polish Scientific Publishers, Warsaw) 1998

434. Hilbert, D.: Die Grundlagen der physik, Göttingen Nachrichten, Math.-phys. Klasse (1916) 395–407

435. Hill, T.L. and Chen, Y.D.: Stochastics of cycle completions (fluxes) in biochemical kinetic diagrams, Proc. Natl. Acad. Sci. USA **72** (1975) 1291–1295

436. Hille, B.: *Ion Channels of Excitable Membranes*, 3rd edn. (Sinauer Associates, Sunderland, Ma.) 2001

437. Hirota, R.: Exact solutions of the Korteweg–de Vries equation for multiple collisions of solitons, Phys. Rev. Lett. **27** (1971) 1192–1194
438. Hobart, R.H.: On the instability of a class of unitary field models, Proc. Phys. Soc. **82** (1963) 201–203
439. Hodges, A.: *The Enigma* (Simon and Schuster, New York) 1983
440. Hodgkin, A.L. and Huxley, A.F.: A quantitative description of membrane current and its application to conduction and excitation in nerve, J. of Physiology **117** (1952) 500–544
441. Hogan, J.: Planets are round. Will that do? Nature **442** (2006) 724–725
442. Holden, A.V.: Hodgkin–Huxley equations (in [878]) 2005
443. Holden, A.V.: Markin–Chizmadzhev model (in [878]) 2005
444. Holland, P.R.: *The Quantum Theory of Motion: An Account of the de Broglie–Bohm Causal Interpretation of Quantum Mechanics* (Cambridge University Press, Cambridge) 1993
445. Holstein, T.: Studies of polaron motion. I. The molecular crystal model, Ann. Phys. (NY) **8** (1959) 325–342
446. Holstein, T.: Studies of polaron motion. II. The "small" polaron, Ann. Phys. (NY) **8** (1959) 343–389
447. Holstein, T.W., Hobmayer, E. and Technau, U.: Cnidarians: An evolutionarily conserved model system for regeneration? Developmental Dynamics **226** (2003) 257–267
448. Hommes, C.: Economic system dynamics (in [878]) 2005
449. Hopfield, J.J.: Neural networks and physical systems with emergent collective computational abilities, Proc. Natl. Acad. Sci. USA **79** (1982) 2554–2558
450. Hopfield, J.J.: Neurons with graded response have collective computational properties like those of two-state neurons, Proc. Natl. Acad. Sci. USA **81** (1984) 3088–3092
451. Horgan, J.: From complexity to perplexity, Scientific American **272** (June 1995) 104–109
452. Horgan, J.: *The End of Science: Facing the Limits of Knowledge in the Twilight of the Scientific Age* (Broadway Books, New York) 1997
453. Houghton, C.: Instantons (in [878]) 2005
454. Houghton, C.: Yang–Mills theory (in [878]) 2005
455. Hughes, G.E.: *John Buridan on Self-Reference* (Cambridge University Press, Cambridge) 1982
456. Hursh, J.B.: Conduction velocity and diameter of nerve fibres, Am. J. Physiol. **127** (1939) 131–139
457. Huxley, H.E.: The structural basis of muscular contraction, Proc. Royal Soc. London B **178** (1971) 131–149
458. Huxley, A.F.: Muscular contraction, J. Physiol. **243** (1974) 1–43
459. Huygens, C.: *The Pendulum Clock* (Iowa State U. Press, Ames) 1986. (Originally published as *Horologium Oscillatorium* in 1673)
460. Hyman, J.M., McLaughlin, D.W. and Scott, A.C.: On Davydov's alpha-helix solitons, Physica D **3** (1981) 23–44
461. Ikezi, H., Taylor, R.J. and Baker, D.R.: Formation and interaction of ion acoustic solitons, Phys. Rev. Lett. **25** (1970) 11–14
462. INTEL web site: www.intel.com/research/silicon/mooreslaw.htm (2004)
463. Ito, T. and Tanikawa, K.: Long-term integrations and stability of planetary orbits in our Solar system, Mon. Not. R. Astron. Soc. **336** (2002) 483–500

464. Jahnke, W., Henze, C. and Winfree, A.T.: Chemical vortex dynamics in three-dimensional excitable media, Nature **336** (1988) 662–665

465. Jain, M.K.: *The Bimolecular Lipid Membrane: A System* (Van Nostrand Reinhold, New York) 1972

466. James, W.: *Principles of Psychology*, Vol. 1 (Dover, New York) 1950

467. Jessee, B., Gargiulo, G., Razvi, F. and Worcel, A.: Analogous cleavage of DNA by micrococcal nuclease and a 1,10-phenanthroline-cuprous complex, Nucleic Acids Res. **10** (19) (1982) 5823–5834

468. Ji, S.: A general theory of ATP synthesis and utilization, in [372]

469. Johnson, C.T. and Swanson, B.I.: Temperature dependence of the vibrational spectra of acetanilide: Davydov soliton or Fermi coupling? Chem. Phys. Lett. **114** (1985) 547–552

470. Johnson, R.S.: *A Modern Introduction to the Mathematical Theory of Water Waves* (Cambridge University Press, Cambridge and New York) 1997

471. Johnson, S.: *Emergence: The Connected Lives of Ants, Brains, Cities, and Software* (Scribner, New York) 2001

472. Jones, C.K.R.T.: Stability of travelling wave solutions of the FitzHugh–Nagumo system, Trans. Am. Math. Soc. **286** (1984) 431–469

473. Joos, E., Zeh, H.D., Kiefer, C., Giulini, D. Kupsch, J. and Stamatescu, I.O.: *Decoherence and the Appearance of a Classical World in Quantum Theory* 2nd edn. (Springer-Verlag, Berlin) 2003

474. Ju, Y. and Lee, J.: Flame front (in [878]) 2005

475. Juarrero, A.: *Dynamics in Action: Intentional Behavior as a Complex System* (MIT Press, Cambridge, Ma.) 2002

476. Kaluza, T.: Zum Unitätsproblem in der Physik, Sitzungsber. Preuss. Akad. Wiss. (1921) 966–972

477. Kaneko, K. and Tsuda, I.: *Complex Systems: Chaos and Beyond* (Wiley, Chichester and New York) 2000

478. Karpman, V.I. and Kruskal, E.M.: Modulated waves in nonlinear dispersive medium, Sov. Phys. JETP **28** (1969) 277–281

479. Kasner, E. and Newman, J.: *Mathematics and the Imagination* (Dover, New York) 2001. (First published in 1940)

480. Katok, A. and Hasselblatt, B.: *Introduction to the Modern Theory of Dynamical Systems* (Cambridge University Press, Cambridge and New York) 1995

481. Kauffman, S.: *The Origins of Order: Self-Organization and Selection in Evolution* (Oxford University Press, Oxford) 1993

482. Kay, I. and Moses, H.E.: Reflectionless transmission through dielectrics and scattering potentials, J. Appl. Phys. **27** (1956) 1503–1508

483. Keener, J. and Sneyd, J.: *Mathematical Physiology* (Springer-Verlag, New York) 1998

484. Keller, E.F.: *Making Sense of Life: Explaining Biological Development with Models, Metaphors, and Machines*, 2nd printing (Harvard University Press, Cambridge) 2003

485. Keller, E.F. and Lewontin, R.C.: Is biology messy? The New York Review of Books **50** (10) (June 12, 2003)

486. Kelly, P.L.: Self-focusing of optical beams, Phys, Rev. Lett. **15** (1965) 1005–1008

487. Kemeny, G. and Goklany, I.M.: Polarons and conformons, J. Theor. Biol. **40** (1973) 107–123

488. Kemeny, G. and Goklany, I.M.: Quantum mechanical model for conformons, J. Theor. Biol. **48** (1974) 23–38

489. Kennedy, M.P.: Chua's circuit (in [878]) 2005

490. Khodorov, B.I.: *The Problem of Excitability* (Plenum, New York) 1974

491. Kim, J.: *Supervenience and Mind* (Cambridge University Press, Cambridge) 1993

492. Kim, J.: *Mind in a Physical World* (MIT Press, Cambridge) 2000

493. Kim, J.: Making sense of downward causation, in [22]

494. King, A.A.: Phase plane (in [878]) 2005

495. Kivshar, Y.: Optical fiber communications (in [878]) 2005

496. Kivshar, Y.S. and Malomed, B.A.: Dynamics of solitons in nearly integrable systems, Rev. Mod. Phys. **61** (1989) 763–915

497. Klein, A. and Kreis, F.: Nonlinear Schrödinger equation: A testing ground for the quantization of nonlinear waves, Phys. Rev. D **13** (1976) 3282–3294

498. Klein, R.M.: D.O. Hebb: An appreciation. In: *The Nature of Thought*, ed. by P.W. Jusczyk and R.M. Klein (Lawrence Erlbaum Associates, Hillsdale, NJ) 1980

499. Knight, B.W.: The relation between the firing rate of a single neuron and the level of activity in a population of neurons, J. Gen. Physiol. **59** (1972) 767–778

500. Knox, R.S., Maiti, S. and Wu, P.: Search for remote transfer of vibrational energy in proteins, in [180]

501. Knuth, D.E.: Mathematics and computer science: Coping with finiteness, Science **194** (1976) 1235–1242

502. Koch, H.: Sur une courbe continue sans tangente, obtenue par une construction géométrique élémentaire, Archiv för Matemat., Astron. och Fys. 1 **1** (1904) 681–702

503. Koch, P.M. and Van Leeuvan, K.H.: The importance of resonances in microwave "ionization" of excited hydrogen atoms, Physics Reports **255** (1995) 289–403

504. Kohl, G.: Relativität in der Schwebe: Die Rolle von Gustav Mie, Preprint No. 209 (Max Planck Institute for the History of Science, Berlin) 2002

505. Kolmogorov, A., Petrovsky, I. and Piscounov, N.: Étude de l'équation de la diffusion avec croissance de la quantité de matière et son application a un problème biologique, Bull. Univ. Moscow A **1** (1937) 137–149

506. Kolmogorov, A.: Local structure of turbulence in an incompressible fluid for very large Reynolds numbers, Comptes Rendus (Doklady) de l'Academie des Sciences de l'U.R.S.S. **31** (1941) 301–305

507. Kompaneyets, A.S. and Gurovich, V.Ts.: Propagation of an impulse in a nerve fiber, Biophysics **11** (1966) 1049–1052

508. Kornberg, R.D. and Klug, A.: The nucleosome, Scientific American **244** (1981) 52–64

509. Kornmeier, J., Bach, M. and Atmanspacher, H.: Correlates of perceptive instabilities in event-related potentials, Int. J. Bifurcation and Chaos **14** (2) (2004) 727–736

510. Kortweg, D.J. and De Vries, G.: On the change of form of long waves advancing in a rectangular canal, and on a new type of long stationary waves, Philos. Mag. **39** (1895) 422–443

511. Kosevich, A.: Superfluidity (in [878]) 2005

512. Kragh, H.: Equation with many fathers. The Klein–Gordon equation in 1926, Am. J. Phys. **52** (1984) 1024–1033

513. Krinsky, V., Pumir, A. and Efimov, I.: Cardiac muscle models (in [878]) 2005

514. Kuhn, T.: *The Copernican Revolution: Planetary Astronomy in the Development of Western Thought* (Harvard University Press, Cambridge, Ma.) 1957

515. Kuhn, T.: *The Structure of Scientific Revolutions*, 3rd edn. (University of Chicago Press, Chicago) 1996. (First published in 1962)

516. Kuhn, T.: *The Road Since Structure: Philosophical Essays, 1970–1993, with an Autobiographical Interview* (University of Chicago Press, Chicago) 2000

517. Kuramoto, Y.: *Chemical Oscillations, Waves, and Turbulence* (Springer-Verlag, Berlin) 1984

518. Kuvshinov, V.: Black holes (in [878]) 2005

519. Kuvshinov, V. and Kuzmin, A.: Fractals (in [878]) 2005

520. Kuvshinov, V. and Minkevich, A.: Cosmological models (in [878]) 2005

521. Lakshmanan, M. and Rajasekar, S.: *Nonlinear Dynamics: Integrability, Chaos and Patterns* (Springer-Verlag, Berlin) 2003

522. Lamb, H.: *Hydrodynamics*, 6th edn. (Dover, New York) 1932

523. Lamb, G.L., Jr.: Analytical description of ultrashort pulse propagation in a resonant medium, Rev. Mod. Phys. **43** (1971) 99–124

524. Lamb, G.L., Jr.: personal communication, 1974

525. Lamb, G.L., Jr.: Bäcklund transforms at the turn of the century. In: *Bäcklund Transforms*, ed. by R.M. Miura (Springer-Verlag, New York) 1976

526. Lamb, G.L., Jr.: *Elements of Soliton Theory* (Wiley, New York) 1980

527. Landa, P.S.: *Regular and Chaotic Oscillations* (Springer-Verlag, Berlin) 2001

528. Landa, P.S.: Feedback (in [878]) 2005

529. Landa, P.S.: Pendulum (in [878]) 2005

530. Langton, C.G.: *Artificial Life: An Overview* (MIT Press, Cambridge, Ma.) 1995

531. Lansner, A.: Investigations into the pattern processing capabilities of associative nets, Doctoral thesis, Royal Institute of Technology, Stockholm, 1986

532. Lansner, A.: Cell assemblies (in [878]) 2005

533. Lansner, A. and Ekeberg, Ö.: Reliability and recall in an associative network, Trans. IEEE Pattern Anal. Mach. Intell. **PAMI–7** (1985) 490–498

534. Lansner, A. and Fransén, E.: Modelling Hebbian cell assemblies comprised of cortical neurons, Network **3** (1992) 105–119

535. Landau, L.D.: Über die Bewegung der Elektronen im Kristallgitter, Phys. Z. Sowjetunion **3** (1933) 664–665

536. Landau, L.D.: Origin of stellar energy, Nature **141** (1938) 333

537. Laskar, J.: Large scale chaos and the spacing of the inner planets, Astron. Astrophys. **317** (1997) L75–L78

538. Laskar, J.: The limits of Earth orbital calculations for geological time-scale use, Phil. Trans. Roy. Soc. London A **357** (1999) 1735–1759

539. Laskar, J.: On the spacing of planetary systems, Phys. Rev. Lett. **84** (2000) 3240–3243

540. Laurent, G.: Dynamical representation of odors by oscillating and evolving neural assemblies, Trends Neurosci. **19** (1996) 489–496

541. Laurent, G., Wehr, M. and Davidowitz, H.: Temporal representations of odors in an olfactory network, J. Neurosci. **16** (1996) 3837–3847

542. Layne, S.P. and Bigio, I.J.: Raman spectroscopy of *Bacillus megaterium* using an optical multi-channel analyzer, Physica Scripta **33** (1986) 91–96

543. Layne, S.P., Bigio, I.J., Scott, A.C. and Lomdahl, P.S.: Transient fluorescence in synchronously dividing *Escherichia coli*, Proc. Natl. Acad. Sci. USA **82** (1985) 7599–7603

544. Lavrenov, I.: The wave energy concentration at the Agulhas current of South Africa, Natural Hazarda **17** (1998) 117–127

545. Lax, P.D.: Integrals of nonlinear equations of evolution and solitary waves, Commun. Pure Appl. Math. **21** (1968) 467–490

546. Leach, A.R.: *Molecular modelling: Principles and Applications*, 2nd edn. (Prentice-Hall, Harlow) 2001

547. Leduc, S.: *The Mechanism of Life* (Rebman, New York) 1911

548. Lee, P.N., Pang, K., Matus, D.Q. and Martindale, M.Q.: A WNT of things to come: Evolution of Wnt signaling and polarity in cnidarians, Seminars in Cell & Developmental Biology **17** (2006) 157–167

549. Lega, J., Moloney, J.V. and Newell, A.C.: Swift–Hohenberg equation for lasers, Phys. Rev. Lett. **73** (1994) 2978–2981

550. Legéndy, C.R.: On the scheme by which the human brain stores information, Math. Biosci. **1** (1967) 555–597

551. Legéndy, C.R.: The brain and its information trapping device. In: *Progress in Cybernetics* 1, ed. by J. Rose (Gordon and Breach, New York) 1969

552. Legéndy, C.R.: Three principles of brain structure and function, Int. J. Neurosci. **6** (1975) 237–254

553. Lemaître, G.-H.: Un Univers homogene de masse constante et de rayon croissant rendant compte de la vitesse radiale des nebuleuses extragalactiques, Ann. Soc. Sci. Bruxelles A **47** (1927) 49–59

554. Lengyel, I., Kádár, S. and Epstein, I.R.: Quasi-two-dimensional Turing patterns in an imposed gradient, Phys. Rev. Lett. **69** (1992) 2729–2732

555. Lewontin, R.C.: *The Triple Helix: Gene, Organism, and Environment* (Harvard University Press, Cambridge) 2002

556. Lewontin, R.C.: Science and simplicity, The New York Review of Books **50** (7) (May 1, 2003). See also **50** (10) (June 12, 2003)

557. Li, T.Y. and Yorke, J.A.: Period three implies chaos, Amer. Math. Monthly **82** (1975) 985–992

558. Lichtenberg, A.J: Averaging methods (in [878]) 2005

559. Lichtenberg, A.J.: *Phase Space Dynamics of Particles* (Wiley, New York) 1969

560. Lichtenberg, A.J and Lieberman, M.A.: *Regular and Chaotic Dynamics* (Springer-Verlag, New York) 1992

561. Lie, S.: Über Flächen deren Krümmungsradien durch eine Relation verknüpft sind, Arch. f. Math. og Nature **4** (1880) 507–512

562. Likharev, K.K.: *Dynamics of Josephson Junctions and Circuits* (Gordon and Breach, New York) 1984

563. Lillie, R.S.: Factors affecting transmission and recovery in the passive iron wire nerve model, J. Gen. Physiol. **7** (1925) 473–507

564. Lindbergh, C.A.: *The Saturday Evening Post*, June 6, 1953

565. Linde, A.: Chaotic inflation, Phys. Lett. B **129** (1983) 177–181

566. Linde, A.: The self-reproducing inflationary universe, Scientific American November 1994, 48–55

567. Lingrel, J.B.: Na,K-ATPase: Isoform structure, function, and expression, J. Bioenerg. Biomembr. **24** (1992) 263–270

568. Llinás, R.R.: The intrinsic electrophysiological properties of mammalian neurons: Insights into central nervous system function, Science **242** (1988) 1654–1664

569. Llinás, R.R. and Nicholson, C.: Electrophysiological properties of dendrites and somata in alligator Purkinje cells, J. Neurophysiol. **34** (1971) 532–551

570. Lohmann, J.U., Endl, I. and Bosch, T.C.G.: Silencing of developmental genes in *Hydra*, Develop. Biol. **214** (1999) 211–214

571. Lomdahl, P.S., MacNeil, L., Scott, A.C., Stoneham, M.E. and Webb, S.J.: An assignment to internal soliton vibrations of laser-Raman lines from living cells, Phys. Lett. A **92** (1982) 207–210

572. Lonngren, K.E.: Observations of solitons on nonlinear dispersive transmission lines. In: *Solitons in Action*, ed. by K.E. Lonngren and A.C. Scott (Academic Press, New York) 1978

573. Lonngren, K.E.: Ion acoustic soliton experiments in a plasma, Optical Quantum Electronics **30** (1998) 615–630

574. Lonngren, K.E. and Nakamura, Y.: Plasma soliton experiments (in [878]) 2005

575. Lorenz, E.N.: Deterministic nonperiodic flow, J. Atmos. Sci. **20** (1963) 130–141

576. Lorenz, E.N.: *The Essence of Chaos* (University of Washington Press, Seattle) 1993

577. Lotka, A.: Contribution to the theory of periodic reactions, J. Phys. Chem. **14** (1910) 271–271

578. Lotka, A.: Undamped oscillations from the law of mass action, J. Am. Chem. Soc. **42** (1920) 1595–1599

579. Luger, K., Mader, A.W., Richmond, R.K., Sargent, D.F. and Richmond, T.J.: Crystal structure of the nucleosome core particle at 2.8 angstrom resolution, Nature **389** (1997) 251–260

580. Lumley, J.L. and Yaglom, A.M.: Flow, Turbulence and Combustion **66** (2001) 241–286

581. Lund, F. and Regge, T.: Unified approach to strings and vortices with soliton solutions, Phys. Rev. D **14** (1976) 1524–1535

582. Luo, L., Tee, T.J. and Chu, P.I.: Chaotic behavior in erbium-doped fiber-ring lasers, J. Opt. Soc. Am. **15** (1998) 972–978

583. Luther, R.: Räumliche Fortpflanzung chemischer Reactionen, Zeit. Elektrochem. **12** (1906) 596–600. English translation in J. Chem. Ed. **64** (1987) 740–742

584. Lutsenko, S. and Kaplan, J.H.: Organization of P-type ATPases: Significance of structural diversity, Biochemistry **34** (1996) 15607–15613

585. Lüttke, W. and Nonnenmacher, G.A.A.: Reinhard Mecke (1895–1969): Scientific work and personality, J. Mol. Struct. **347** (1995) 1–18

586. Ma, W.-X.: Integrability (in [878]) 2005

587. MacKay, R.S. and Aubry, S.: Proof of the existence of breathers for time-reversible or Hamiltonian networks of weakly coupled oscillators, Nonlinearity **7** (1994) 1623–1643

588. Mackey, M.C. and Glass, L.: Oscillation and chaos in physiological control systems, Science **197** (1977) 287–289

589. MacMillan, M.: *An Odd Kind of Fame: Stories of Phineas Gage* (MIT Press, Cambridge, MA) 2000
590. Maddox, J.: Physicists about to hijack DNA? Nature **324** (1986) 11
591. Maddox, J.: Toward the calculation of DNA, Nature **339** (1989) 577
592. Mainzer, K.: *Thinking in Complexity: The Computational dynamics of Matter, Mind, and Mankind*, 4th edn. (Springer, New York) 2003
593. Mainzer, K.: *Symmetry and Complexity: The Spirit and Beauty of Nonlinear Science* (World Scientific, Singapore) 2005
594. Makhankov, V.G.: *Soliton Phenomenology* (Kluwer Academic, Dordrecht) 1990, 2005
595. Malik, H.S. and Henikoff, S.: Phylogenomics of the nucleosome, Nature Structural Biology **11** (2003) 882–891
596. Malomed, B.: Complex Ginzburg–Landau equation (in [878]) 2005
597. Malomed, B.: Nonlinear Schrödinger equations (in [878]) 2005
598. Mandelbrot, B.B.: How long is the coast of Britain? Statistical self-similarity and fractional dimension, Science **156** (1967) 636–638
599. Mandelbrot, B.B.: *The Fractal Geometry of Nature* (W.H. Freeman, San Francisco) 1983
600. Mandelstam, S.: Soliton operators for quantized sine-Gordon equation, Phys. Rev. D **11** (1975) 3026–3030
601. Manley, J.M. and Rowe, H.E.: Some general properties of nonlinear elements, Proc. IRE **44** (1956) 904–913
602. Mann, R.B., Morsink, S.M., Sikkema, A.E and Steele, T.G.: Semiclassical gravity in $1 + 1$ dimension, Phys. Rev. D **43** (1991) 3948–3957
603. Manneville, P.: Spatiotemporal chaos (in [878]) 2005
604. Manneville, P.: Rayleigh–Bénard convection, thirty years of experimental, theoretical and modeling work. In: *Dynamics of Spatio-Temporal Structures – Henri Bénard Centenary Review*, ed. by I. Mutabazi, E. Guyon, J.E. Wesfreid, Springer Tracts in Modern Physics, 2006
605. Manton, N.S. and Sutcliffe, P.M.: *Topological Solitons* (Cambridge University Press, Cambridge) 2004
606. Marangoni, C.G.M.: Ueber die Ausbreitung der Tropfen einer Flüssigkeit auf der Oberfläche einer anderen, Ann. Phys. Chem. **143** (1871) 337–354
607. Margulis, L. and Sagan, D.: *What Is Life?* (Simon & Schuster, New York) 1995
608. Markin, V.S. and Chizmadzhev, Yu.A.: On the propagation of an excitation for one model of a nerve fiber, Biophysics **12** (1967) 1032–1040
609. Marín, J.L., Eilbeck, J.C. and Russell, F.M.: Localized moving breathers in a 2D hexagonal lattice, Phys. Lett. A **248** (1998) 225–229
610. Marín, J.L., Eilbeck, J.C. and Russell, F.M.: 2D breathers and applications. In: *Nonlinear Science at the Dawn of the 21st Century*, ed. by P.L. Christiansen, M.P. Sørensen and A.C. Scott (Springer-Verlag, Berlin) 2000
611. Marín, J.L., Russell, F.M. and Eilbeck, J.C.: Breathers in cuprite-like lattices, Phys. Lett. A **281** (2001) 21–25
612. Marklof, J.: Energy level statistics, lattice point problems and almost modular functions. In: *Frontiers in Number Theory, Physics and Geometry. Volume 1: On Random Matrices, Zeta Functions and Dynamical Systems* (Les Houches, 9–21 March 2003) (Springer, Berlin) to appear
613. Marshall, A.: *The Principles of Economics*, 8th edn. (Prometheus Books, Amherst, New York) 1997. (Originally published in 1920)

614. Martin, G.M.: Epigenetic drift in aging identical twins, Proc. Natl. Acad. Sci. USA **102** (2005) 10413–10414

615. Maslow, A.H.: *Toward a Psychology of Being*, 2nd edn. (Van Nostrand, New York) 1968

616. Maslow, A.H.: *The Psychology of Science: A Reconnaissance* (Henry Regnery, Chicago) 1969

617. Masmoudi, N.: Rayleigh–Taylor instability (in [878]) 2005

618. Mason, L.: Twistor theory (in [878]) 2005

619. Maturana, H.R. and Varela, F.J.: *Autopoiesis and Cognition* (Dordrecht, Reidel) 1980

620. Maxwell, J.C.: On Boltzmann's theorem on the average distribution of energy in a system of material points, Trans. Camb. Philos. Soc. **12** (1879) 547–570

621. May, R.M.: Simple mathematical models with very complicated dynamics, Nature **261** (1976) 459–469

622. Maynard, E.M., Nordhausen, C.T. and Normann, R.A.: The Utah intracortical electrode array: A recording structure for potential brain–computer interfaces, Electroencephalogr. Clin. Neurophysiol. **102** (1997) 228–239

623. Mayr, E.: *The Growth of Biological Thought* (Harvard University Press, Cambridge) 1982

624. Mayr, E.: The limits of reductionism, Nature **331** (1988) 475

625. Mayr, E.: *What Evolution Is* (Basic Books, New York) 2002

626. Mazur, A.K.: Reversible B↔A transitions in single DNA molecule immersed in a water drop, arXiv:physics/0302012 v1 (5 Feb 2003)

627. McCammon, J.A. and Harvey, S.C.: *Dynamics of Proteins and Nucleic Acids* (Cambridge University Press, Cambridge) 1987

628. McClare, C.W.F.: Chemical machines, Maxwell's demons and living organisms, J. Theoret. Biol. **30** (1971) 1–34

629. McClare, C.W.F.: A "molecular energy" muscle model, J. Theoret. Biol. **35** (1972) 569–595

630. McClare, C.W.F.: Resonance in bioenergetics (in [372]) 1974

631. McCormick, D.A. and Yuste, R.: UP states and cortical dynamics. In: *Microcircuits: The Interface Between Neurons and Global Brain Function*, ed. by S. Grillner and A.M. Graybiel (MIT Press, Cambridge) 2006

632. McCowan, J.: On the solitary wave, Philos. Mag. **32** (1891) 45–58; 553–555

633. McCulloch, W.S. and Pitts, W.H.: A logical calculus of the ideas immanent in nervous activity, Bull. Math. Biophys. **5** (1943) 115–133

634. McCulloch, W.S., Carnap, R., Brunswik, E., Bishop, G.H., Meyers, R., Von Bonin, G., Menger, K. and Szent-Gyorgyi, A.: Committee on mathematical biology, Science **123** (1956) 725

635. McCulloh, K.A., Sperry, J.S. and Adler, F.R.: Water transport in plants obeys Murray's law, Nature **421** (2003) 939–942

636. McCulloh, K.A., Sperry, J.S.: The evaluation of Murray's law in *Psilotum nudum* (Psilotaceae), an analogue of ancestral vascular plants, American Journal of Botany **92** (2005) 985–989

637. McGhee, J.D. and Felsenfeld, G.: Nucleosome Structure, Ann. Rev. Biochem. **49** (1980) 1115–1156

638. McHugh, T.J., Blum, K.I., Tsien, J.Z., Tonegawa, S. and Wilson, M.A.: Impaired hippocampal representation of space in CA1-specific NMDAR1 knockout mice, Cell **87** (1996) 1339–1349

639. McKean, H.: Nagumo's equation, Adv. Math. **4** (1970) 209–223
640. McKenna, J: Tacoma Narrows Bridge collapse (in [878]) 2005
641. McLaughlin, D.W. and Scott, A.C.: Perturbation analysis of fluxon dynamics, Phys. Rev. A **18** (1978) 1652–1680
642. McLaughlin, J.B. and Martin, P.C.: Transition to turbulence of a statically stressed fluid, Phys. Rev. Lett. **33** (1974) 1189–1192
643. McLaughlin, J.B. and Martin, P.C.: Transition to turbulence in a statically stressed fluid system, Phys. Rev. A **12** (1975) 186–203
644. McMahon, B.H. and LaButte, M.X.: Protein structure (in [878]) 2005
645. Mecke, R.: Valency and deformation vibrations of multi-atomic molecules. III. Methane, acetylene, ethylene and halogen derivatives, Zeit. für Phys. Chem. B **17** (1932) 1–20
646. Mecke, R.: Dipolmoment und chemische Bindung, Zeit. für Elektrochem. **54** (1950) 38–42
647. Mecke, R., Baumann, W. and Freudenberg, K.: Das Rotationsschwingungsspectrum des Wasserdampfes, Zeit. für Physik **81** (1933) 313–465
648. Mecke, R. and Ziegler, R.: Das Rotationsschwingungsspektrum des Acetylens (C_2H_2), Zeit. für Physik **101** (1936) 405–417
649. Mel, B.W.: Synaptic integration in an excitable dendritic tree, J. Neurophysiol. **70** (1993) 1086–1101
650. Meinhardt, H.: *Models of Biological Pattern Formation* (Academic Press, London) 1982
651. Meinhardt, H. and Koch, A.: Biological pattern formation – from basic mechanisms to complex structures, Rev. Mod. Phys. **66** (1994) 1481–1507
652. Meiss, J.D.: Standard map (in [878]) 2005
653. Mel'nikov, V.K.: On the stability of the center for time periodic perturbations, Trans. Moscow. Math. Soc. **12** (1963) 1–57
654. Melzak, R. and Scott, T.H.: The effects of early experience on the response to pain, J. Comp. Phys. Psychol. **50** (1957) 155–161
655. Mie, G.: Grundlagen einer Theorie der Materie, Ann. der Physik **37** (1912) 511–534; **39** (1912) 1–40; **40** (1913) 1–66
656. Miljkovic-Licina, M., Gauchat, D. and Galliot, B.: Neuronal evolution: Analysis of regulatory genes in a first-evolved nervous system, the hydra nervous system, BioSystems **76** (2004) 75–87
657. Milner, P.M.: The cell assembly: Mark II, Psychol. Rev. **64** (1957) 242–252
658. Milner, P.M.: The mind and Donald O. Hebb, Sci. Am. (January 1993) 124–129
659. Minding, E.F.A.: Wie sich entscheiden lässt, ob zwei gegebene krumme Flächen aufeinander abwickelbar sind oder nicht, J. für reine und angewandte Math. **19** (1839) 370–387
660. Minkowski, H.: Die Grundleichungen für die elektromagnetischen Vorgäge in bewegten Körpern, Göttingen Nachrichten **8** (1908) 59–111. Translated in *The Principle of Relativity*, Dover, New York, 1923
661. Minorsky, N.: *Nonlinear Oscillations* (Van Nostrand, New York) 1962
662. Minsky, M. and Papert, S.: *Perceptrons* (MIT Press, Cambridge) 1969
663. Misner, C.W.: Mixmaster universe, Phys. Rev. Lett. **22** (1969) 1071–1074
664. Mitchell, P.: Coupling of phosphorylation to electron and proton transfer by a chemiosmotic type of mechanism, Nature **191** (1961) 144–148
665. Mollenauer, L.F.: *Solitons in Optical Fibers: Fundamentals and Applications to Telecommunications* (Academic Press, New York) 2005

666. Møller, J.V., Juul, B. and Le Maire, M.: Structural organisation, ion transport, and energy transduction of P-type ATPases, Biochim. Biophys. Acta **1286** (1996) 1–51

667. Moloney, J.V.: Nonlinear optics (in [878]) 2005

668. Moloney, J.V. and Newell, A.C.: *Nonlinear Optics*, 2nd edn. (Westview Press, Boulder Colorado) 2004

669. Monod, J.: *Chance and Necessity: An Essay on the Natural Philosophy of Modern Biology* (Vintage Press, New York) 1972

670. Monod, J., Wyman, J. and Changeux, J.P.: On the nature of allosteric transitions: A plausible model, J. Mol. Biol. **12** (1965) 88–118

671. Monod, J.: On chance and necessity. In: *Studies in the Philosophy of Biology*, ed. by F.J. Ayala and T. Dobzhansky (MacMillan, London) 1974, 357–375

672. Moore, G.E.: Cramming more components onto integrated circuits, Electronics **38** (1965) 114–117

673. Moore, R.O. and Biondini, G.: Harmonic generation (in [878]) 2005

674. Moore, W.: *Schrödinger: Life and Thought* (Cambridge University Press, Cambridge) 1989

675. Morbidelli, A.: Interplanetary kidnap, Nature **441** (2006) 162–163

676. Mori, M. and Umemura, M.: The evolution of galaxies from primeval irregulars to present-day ellipticals, Nature **440** (2006) 644–647

677. Mornev, O.: Elements of the "optics" of autowaves. In: *Self-Organization of Autowaves and Structures Far from Equilibrium*, ed. by V.I. Krinsky (Springer-Verlag, Berlin) 1984

678. Mornev, O.: Geometrical optics, nonlinear (in [878]) 2005

679. Mornev, O.: Gradient system (in [878]) 2005

680. Mornev, O.: Zeldovich–Frank-Kamenetsky equation (in [878]) 2005

681. Morris, C. and Lecar, H.: Voltage oscillations in the barnacle giant muscle. Biophys. J. **71** (1981) 193–213

682. Mosekilde, E.: *Topics in Nonlinear Biology* (World Scientific, Singapore) 1996

683. Mueller, P., Rudin, D.O., Tien, H.T. and Westcott, W.C.: Reconstitution of cell membrane structure *in vitro* and its transformation into an excitable system, Nature **194** (1962) 979–980

684. Mueller, P., Rudin, D.O.: Action potentials induced in bimolecular lipid membranes, Nature **217** (1968) 713–719

685. Müller, P. and Henderson, D.: Rogue waves, *Proceedings of the Aha Huliko'a Winter Workshop, University of Hawaii at Manoa*, January 24–28, 2005

686. Murray, C.D.: The physiological principle of minimum work. I. The vascular system and the cost of blood volume, Proc. Natl. Acad. Sci. USA **12** (1926) 207–214

687. Murray, C.D.: The physiological principle of minimum work. II. Oxygen exchange in capillaries, Proc. Natl. Acad. Sci. USA **12** (1926) 299–304

688. Murray, C.D.: The physiological principle of minimum work applied to the angle of branching of arteries, J. Gen. Physiol. **9** (1926) 835–841

689. Murray, C.D.: A relationship between circumference and weight and its bearing on branching angles, J. Gen. Physiol. **10** (1927) 725–7291

690. Murray, J.D.: *Mathematical Biology II* (Springer-Verlag, New York) 2004

691. Murray, N. and Holman, M.: The origin of chaos in the outer solar system, Science **283** (1999) 1877–1881

692. Muto, V., Holding, J. Christiansen, P. and Scott, A.C.: Solitons in DNA, J. Biomol. Struct. Dyn. **5** (1988) 873–894

693. Muto, V., Scott, A.C. and Christiansen, P.: Thermally generated solitons in a Toda lattice model of DNA, Phys. Lett. A **136** (1989) 33–36

694. Muto, V., Lomdahl., P. and Christiansen, P.: Two-dimensional discrete model for DNA dynamics: Longitudinal wave propagation and denaturation, Phys. Rev. A **42** (1990) 7452–7458

695. Mygind, J.: Josephson junctions (in [878]) 2005

696. Nagel, E., Newman, J.R. and Hofstadter, D.R.: *Gödel's Proof* (New York University Press, New York) 2002

697. Nagumo, J., Arimoto, S. and Yoshizawa, S.: An active impulse transmission line simulating nerve axon, Proc. IRE **50** (1962) 2061–2070

698. Nappi, C., Lisitskiy, M.P., Rotoli, G., Cristiano, R. and Barone, A.: New fluxon resonant mechanism in annular Josephson tunnel structures, Phys. Rev. Lett. **93** (2004) 187001-1–4

699. Naugolnykh, K. and Ostrovsky, L.: *Nonlinear Wave Processes in Acoustics* (Cambridge University Press, Cambridge) 1998

700. Newell, A.C. (Ed.): *Nonlinear Wave Motion*, AMS Lectures in Applied Mathematics **15** (American Mathematical Society, Providence, R.I.) 1974

701. Newell, A.C.: *Solitons in Mathematics and Physics* (SIAM, Philadelphia) 1984

702. Newell, A.C.: Inverse scattering method or transform (in [878]) 2005

703. Newman, D.V.: Emergence and strange Attractors, Philosophy of Science **63** (1996) 245–260

704. Newman, D.V.: Chaos, emergence, and the mind–body problem, Australasian Journal of Philosophy **79** (2001) 180–196

705. Newman, W.R.: *Atoms and Alchemy: Chemistry and the Experimental Origins of the Scientific Revolution* (University of Chicago Press, Chicago) 2006

706. Newton, I.: *Optiks*, based on the 4th edn. of 1730 (Dover, New York) 1952

707. Newton, P.K.A. and Aref, H.: Chaos vs. turbulence (in [878]) 2005

708. Nicolelis, M.A.L., Baccala, L.A., Lin, R.C.S. and Chapin, J.K.: Sensorimotor encoding by synchronous neural ensemble activity at multiple levels of the somatosensory system, Science **268** (1995) 1353–1358

709. Nicolelis, M.A.L., Fanselow, E.E. and Ghazanfar, A.A.: Hebb's dream: The resurgence of cell assemblies, Neuron **19** (1997) 219–221

710. Nicolelis, M.A.L., Ghazanfar, A.A., Faggin, B.M., Votaw, S. and Oliveira, L.M.O.: Reconstructing the engram: Simultaneous, multisite, many single neuron recordings, Neuron **18** (1997) 529–537

711. Nicolis, G.: *Introduction to Nonlinear Science* (Cambridge University Press, Cambridge) 1995

712. Nicolis, G.: Brusselator (in [878]) 2005

713. Nicolis, G.: Chemical kinetics (in [878]) 2005

714. Nicolis, G.: Nonequilibrium statistical mechanics (in [878]) 2005

715. Nicolis, G. and Prigogine, I.: *Self-organization in Nonequilibrium Systems* (Wiley, New York) 1977

716. Nicolis, G. and Prigogine, I.: *Exploring Complexity,* (W. Freeman, New York) 1989.

717. Nicolis, J.S.: *Dynamics of Hierarchical Systems: An Evolutionary Approach* (Springer-Verlag, Berlin) 1986

718. Neyts, K. and Beeckman, J.: Liquid crystals (in [878]) 2005

719. Nilsson, N.J.: *Learning Machines*, 2nd edn. (Morgan Kaufmann, San Mateo) 1990

720. Nordhausen, C.T., Maynard, E.M. and Normann, R.A.: Single unit recording capabilities of a 100 microelectrode array, Brain Res. **726** (1996) 129–140

721. Nordsletten, D.A., Blackett, S., Bentley, M.D., Ritman, E.L. and Smith, N.P.: Structural morphology of renal vasculature, Am. J. Physiol., Heart Circ. Physiol. **291** (2006) H296–H309

722. Nottale, L., Schumacher, G., and Gay, J.: Scale relativity and quantization of the solar system, Astron. Astrophys. **322** (1997) 1018–1025

723. Olsen, M., Smith, H. and Scott, A.C.: Solitons in a wave tank, Am. J. Phys. **52** (1984) 826–830

724. Onorato, M., Osborne, A.R., Serio, M. and Bertone, S.: Freak waves in random oceanic sea states, Phys. Rev. Lett. **86** (2001) 5831–5834

725. Oppenheimer, J.R. and Snyder, H.: On continued gravitational contraction, Phys. Rev. **54** (1939) 455–459

726. Oppenheimer, J.R. and Volkoff, G.M.: On massive neutron cores, Phys. Rev. **55** (1939) 374–381

727. Osborne, A.R. and Burch, T.L.: Internal solitons in the Andaman Sea, Science **208** (1980) 451–460

728. Osborne, A.R., Onorato, M. and Serio, M.: The nonlinear dynamics of rogue waves and holes in deep water gravity wave trains, Phys. Lett. A **275** (2000) 386–393

729. Ostrovsky, L.: Modulated waves (in [878]) 2005

730. Ostrovsky, L.: Shock waves (in [878]) 2005

731. Ostrovsky, L.: Propagation of wave packets and space-time self-focusing in a nonlinear medium, Sov. Phys. JETP **24** (1967) 797–800

732. Ostrovsky, L. and Hamilton, M.: Nonlinear acoustics, (in [878]) 2005.

733. Ostrovsky, L.A. and Potapov, A.I.: *Modulated Waves: Theory and Applications* (John Hopkins University Press, Baltimore) 1999

734. Ostrovsky, L.A. and Stepanyants, Yu.A.: Do internal solitons exist in the ocean? Rev. Geophys. **27** (1989) 293–310

735. Ostrovsky, L. and Sverdlov, M.: Hurricanes and tornadoes (in [878]) 2005

736. Ott, E.: *Chaos in Dynamical Systems* (Cambridge University Press, Cambridge and New York) 1993

737. Ourjoumtsev, A., Tualle-Brouri, R., Laurat, J. and Grangier, P.: Generating optical Schrödinger kittens for quantum information processing, Science **312** (2006) 83–85

738. Overbye, D.: Vote makes it official: Pluto isn't what it used to be, The New York Times, 25 August 2006

739. Painter, P.R., Edén, P. and Bengtsson, H.U.: Pulsatile blood flow, shear force, energy dissipation and Murray's law, Theor. Biol. and Medical Modelling **3** (2006) 31

740. Pais, A.: *Niels Bohr's Times: In Physics, Philosophy, and Polity* (Oxford University Press, Oxford) 1991

741. Parnas, I. and Segev, I.: A mathematical model for conduction of action potentials along bifurcating axons, J. Physiol. (London) **295** (1979) 323–343

742. Parry, W.E.: *Three Voyages for the Discovery of a Northwest Passage From the Atlantic to the Pacific and Narrative of an Attempt to Reach the North Pole*, Vol. 1 (Harper & Brothers, New York) 1840

743. Pastushenko, V.F. and Markin, V.S.: Propagation of excitation in a model of an inhomogeneous nerve fiber. II. Attenuation of pulse in the inhomogeneity, Biophysics **14** (1969) 548–552

744. Pastushenko, V.F., Markin, V.S. and Yu.A. Chizmadzhev: Propagation of excitation in a model of an inhomogeneous nerve fiber. III. Interaction of pulses in the region of the branching node of a nerve fibre, Biophysics **14** (1969) 929–937

745. Pastushenko, V.F., Markin, V.S. and Yu.A. Chizmadzhev: Propagation of excitation in a model of an inhomogeneous nerve fiber. IV. Branching as a summator of nerve pulses, Biophysics **14** (1969) 1130–1138

746. Pechkin, A.A.: Operationalism as the philosophy of Soviet physics: The philosophical bachgrounds of L.I. Mandelstam and his school, Synthese **124** (2000) 407–432

747. Pedersen, N.F.: Superconductivity (in [878]) 2005

748. Pekar, I.: *Untersuchungen über die Elektronentheorie der Kristalle* (Akademie-Verlag, Berlin) 1954

749. Pelinovsky, D.: Manley–Rowe relations (in [878]) 2005

750. Pelloni, B.: Burgers equation (in [878]) 2005

751. Penfield, P.: *Frequency–Power Formulas* (Wiley, New York) 1960

752. Penfield, W. and Perot, P.: The brain's record of auditory and visual experience – A final summary and discussion, Brain **86** (1963) 595–696

753. Penrose, R.: Gravitational collapse and spacetime singularities, Phys. Rev. Lett. **14** (1976) 57–59

754. Penrose, R.: Nonlinear gravitons and curved twistor theory, Gen. Rel. and Grav. **7** (1976) 31–52

755. Penrose, R.: *The Emperor's New Mind* (Penguin Books, New York) 1991

756. Penrose, R.: *Shadows of the Mind* (Oxford University Press, New York) 1994

757. Peregrine, D.H.: The fascination of fluid mechanics, J. Fluid Mech. **106** (1995) 59–80

758. Perring, J.K. and Skyrme, T.H.R.: A model unified field equation, Nucl. Phys. **31** (1962) 550–555

759. Petty, M.: Langmuir–Blodgett films (in [878]) 2005

760. Peyrard, M. and Bishop, A: Statistical mechanics for a nonlinear model for DNA denaturation, Phys. Rev. Lett. **62** (1989) 2755—2758

761. Peyrard, M.: Biomolecular solitons (in [878]) 2005

762. Pikovsky, A. and Rosenlum, M.: Van der Pol equation (in [878]) 2005

763. Planck, M.: Ueber das Gesetz der Energieverteilung im Normalspektrum, Ann. der Physik **4** (3) (1901) 553–563

764. Poincaré, H.: *Science and Method* (St. Augustine's Press, Chicago) 2001. (First published in 1903)

765. Poincaré, H.: Sur la dynamique de l'électron, Comptes Rendus Acad. Sci. **140** (1905) 1504–1508

766. Poiseuille,, J.L.M.: Recherches expérimentelles sur le mouvement des liquides dans les tubes de très petits diamètres, C.R. Acad. Sci. **11** (1840) 961–967 and 1041–1049; **12** (1841) 112–115

767. Pojman, J.A.: Polymerization (in [878]) 2005

768. Pojman, J.A., Ilyashenko, V.M. and Khan, A.M.: Free-radical frontal polymerization: Self-propagating thermal reaction waves, J. Chem. Soc. Faraday Trans. **92** (1996) 2825–2837

769. Polidori, J.: *The Vampyre: And Other Tales of the Macabre* (Oxford University Press, New York) 2001. (First published in 1819)

770. Popper, K.R.: *Quantum Theory and the Schism in Physics* (Rowman & Littlefield, Lanham, Maryland) 1985

771. Pribram, K.H.: The neurophysiology of remembering, Sci. Am. January 1969, 73–85

772. Price, H.: *Time's Arrow and Archimedes' Point: New Directions for the Physics of Time* (Oxford University Press, New York) 1997

773. Prigogine, I.: *Nonequilibrium Statistical Mechanics* (Wiley, New York) 1962

774. Pritchard, R.M., Heron, W. and Hebb, D.O.: Visual perception approached by the method of stabilized images, Can. J. Psychol. **14** (1960) 67–77

775. Pritchard, R.M.: Stabilized images on the retina, Sci. Am. June 1961, 72–79

776. Prohofsky, E.W.: Solitons hiding in DNA and their possible significance in RNA transcription, Phys. Rev. A **38** (1988) 1538–1541

777. Pushkin, D.O. and Aref, H.: Cluster coagulation (in [878]) 2005

778. Quetelet, A. and Verhulst, P.F.: Annuaire de l'Académie Royale des Sciences de Belgique **16** (1850) 97–124

779. Rabinovich, M.I. and Rulkov, N.F.: Chaotic dynamics (in [878]) 2005

780. Rabinovich, M..S. and Rosvold, H.E.: A closed field intelligence test for rats, Can. J. Psychol. **4** (1951) 122–128

781. Rabon, E.C. and Reuben, M.A.: The mechanism and structure of the gastric H,K-ATPase, Annu. Rev. Physiol. **52** (1990) 321–344

782. Raby, O.: *Radio's First Voice: The Story of Reginald Fessenden* (MacMillan of Canada, Toronto) 1970

783. Rall, W.: Branching dendrite trees and motoneuron resistivity, Expl, Neurol. **1** (1959) 491–527

784. Rall, W.: Theory of properties of dendrites, Ann. New York Acad. Sci. **96** (1959) 1017–1091

785. Rall, W.: Electrophysiology of a dendrite neuron model, Biophys. J. **2** (1962) 145–167

786. Rañada, A.F., Trueba, J.L. and Donoso, J.M.: Ball lightning (in [878]) 2005

787. Rashevsky, N.: Some physico-mathematical aspects of nerve conduction, Physics **4** (1933) 341–349

788. Rashevsky, N.: Some physico-mathematical aspects of nerve conduction. II, Physics **6** (1936) 308–314

789. Rashevsky, N.: Topology and life: In search of general mathematical principles in biology and sociology, Bulletin of Math. Biophys. **16** (1954) 317–348

790. Rashevsky, N.: *Mathematical Biophysics: Physico-Mathematical Foundations of Biology*, Volumes I and II, 3rd edn. (Dover, New York) 1960. (First edition published in 1938, second edition in 1948)

791. Rayleigh, Lord: On waves, Philos. Mag. **1** (1876) 257–279

792. Rayleigh, Lord: Investigation of the character of the equilibrium of an incompressible heavy fluid of variable density, Proc. London Math. Soc. **14** (1883) 170–177

793. Rayleigh, Lord: On the convective currents in a horizontal layer of fluid when the higher temperature is on the under side, Phil. Mag. **32** (1916) 529–546

794. Recami, E. and Scott, A.C.: Tachyons and superluminal motion (in [878]) 2005

795. Reid, E.W.: Report on experiments upon "absorption without osmosis", Brit. Med. J. **1** (1892) 323–326

796. Reid, E.W.: Transport of fluid by certain epithelia, J. Physiol. **26** (1901) 436–444

797. Reinisch, G.: Macroscopic Schroedinger quantization of the early chaotic solar system, Astron. Astrophys. **337** (1998) 299–310

798. Remoissenet, M.: *Waves Called Solitons*, 3rd edn. (Springer-Verlag, Berlin and New York) 1999

799. Reynolds, C.W.: Flocks, herds, and schools: A distributed behavioral model, Computer Graphics **21** (4) (1987) 25–34

800. Reynolds, O.: On the experimental investigation of the circumstances which determine whether the motion of water shall be direct or sinuous, and the law of resistance in parallel channels, Phil. Trans. Roy. Soc. London A **35** (1883) 935—982

801. Rich, A.: The double helix: a tale of two puckers, Nature Structural Biology **10** (2003) 247–249

802. Richardson, L.F.: *Weather Prediction by Numerical Process* (Cambridge University Press, Cambridge) 1922

803. Richmond, T.J., Finch, J.T., Rushton, B., Rhodes, D. and Klug, A.: Structure of the nucleosome core particle at 7 angstrom resolution, Nature **311** (1984) 532–537

804. Richter, J.P.: *The Notebooks of Leonardo da Vinci* (Dover, New York) 1970. (First published in 1883)

805. Riesen, A.H.: The development of visual perception in man and chimpanzee, Science **106** (1947) 107–108

806. Rikitake, T.: Oscillations of a system of disk dynamos, Proc. Cambridge Phil. Soc. **174** (1958) 89–105

807. Rinzel, J. and Keller, J.: Traveling-wave solutions of a nerve conduction equation, Biophys. J. **13** (1973) 1313–1337

808. Robertson, H.P.: Relativistic cosmology, Rev. Mod. Phys. **5** (1933) 62–90

809. Robley, A., Reddiex, B. Arthur, T., Pech, R. and Forsyth, D.: Interactions between feral cats, foxes, native carnivores, and rabbits in Australia, Arthur Rylah Institute for Environmental Research, Department of Sustainability and Environment, Melbourne, Australia

810. Rochester, N., Holland, J.H., Haibt, L.H. and Duda, W.L.: Tests on a cell assembly theory of the action of a brain using a large digital computer, Trans. IRE Inf. Theory **2** (1956) 80–93

811. Rogers, C.: Bäcklund transformations (in [878]) 2005

812. Rosen, G.: Particle-like solutions to nonlinear scalar wave theories, J. Math. Phys. **6** (1965) 1269–1272

813. Rosen, N. and Rostenstock, H.B.: The force between particles in a nonlinear field theory, Phys. Rev. **85** (1952) 257–259

814. Rosen, R.: A relational theory of biological systems, Bull. Math. Biophys. **20** (1958) 245–260

815. Rosen, R.: A relational theory of biological systems, II, Bull. Math. Biophys. **21** (1959) 109–128

816. Rosen, R.: Nicholas Rashevsky 1899–1972, Progress in Theoretical Biology **2** (1972) 11–14

817. Rosen, R.: *Life Itself: A Comprehensive Inquiry Into the Nature, Origin, and Fabrication of Life* (Columbia University Press, New York) 1991

818. Rosen, R.: *Essays On Life Itself* (Columbia University Press, New York) 2000

819. Rosen, R.: *Autobiographical Reminiscences* (unpublished) mid-1990s (ed. by Judith Rosen)

820. Rosenblatt, F.: *Principles of Neurodynamics* (Spartan Books, New York) 1959

821. Rosenblum, M. and Pikovsky, A.: Synchronization (in [878]) 2005
822. Rosenthal, W.: Results of the MAXWAVE project (in [685]) 2005
823. Rosser, J.B. Jr.,: On the complexities of complex economic dynamics, J. Econ. Perspectives **13** (1999) 169–192
824. Rössler, O.E.: An equation for continuous chaos, Phys. Lett. A **57** (1976) 397–398
825. Rössler, U.K., Bogdanov, A.N. and Pfleiderer, C.: Spontaneous skyrmion ground states in magnetic metals, Nature **442** (2006) 797–801
826. Rubinstein, J.: Sine–Gordon equation, J. Math. Phys. **11** (1970) 258–266
827. Ruelle, D.: *Chaotic Evolution and Strange Attractors* (Cambridge University Press, Cambridge) 1989
828. Ruelle, D. and Takens, F.: On the nature of turbulence, Commun. Math. Phys. **20** (1971) 167–192
829. Russell, F.M.: The observation in mica of charged particles from neutrino interactions, Phys. Lett. B **25** (1967) 298–300
830. Russell, F.M.: Identification and selection criteria for charged lepton tracks in mica, Nucl. Tracks Radiat. Meas. **15** (1988) 41–44
831. Russell, F.M. and Collins, D.R.: Lattice-solitons and non-linear phenomena in track formation, Radiation Meas. **25** (1995) 67–70
832. Russell, F.M. and Collins, D.R.: Lattice-solitons in radiation damage, Nucl. Instrum. Meth. in Phys. Res. B **105** (1995) 30–34
833. Russell, J.S.: Report on Waves, British Association for the Advancement of Science (1845)
834. Russell, J.S.: *The Wave of Translation in the Oceans of Water, Air and Ether* (Trübner, London) 1885
835. Sachs, G., Shin, J.M., Briving, C., Wallmark, B. and Hersey, S.: The pharmacology of the gastric acid pump: The H+,K+ ATPase, Annu. Rev. Toxicol. **35** (1995) 277–305
836. Saint-Venant, A.B.: Mouvements des molécules de l'onde solitaire, Comptes Rendu **101** (1885) 1101–1105, 1215–1218, and 1445–1447
837. Sakaguchi, H. and Brand, H.R.: Localized patterns for the quintic complex Swift–Hohenberg equation, Physica D **117** (1998) 95–105
838. Salerno, M.: Discrete model for DNA promoter dynamics, Phys. Rev. A **44** (1991) 5292–5297
839. Salerno, M.: Dynamical properties of DNA promoters, Phys. Lett. A **167** (1992) 49–53
840. Salerno, M. and Kivshar, Y.: DNA promoters and nonlinear dynamics, Phys. Lett. A **193** (1994) 263–266
841. Schalch, T., Duda, S., Sargent, D.F. and Richmond, T.J.: X-ray structure of a tetranucleosome implications for the chromatin fibre, Nature **436** (2005) 138–141
842. Schechter, B.: Taming the fourth dimension, New Scientist **183** (2456) (2004) 26–29
843. Schemer, L., Jiao, H.Y., Kit, E. and Agnon, Y.: Evolution of a nonlinear wave field along a tank: Experiments and numerical simulations based on the Zakharov equation, J. Fluid Mech. **427** (2000) 107–129
844. Schiff, L.I.: *Quantum Mechanics* (McGraw–Hill, New York) 1949
845. Schlipp, P.A. (Ed.): *Albert Einstein: Philosopher Scientist*, The Library of Living Philosophers, Vol. 7, 1949

846. Schliwa, M. and Woehlke, G.: Molecular motors, Nature **422** (2003) 759–765
847. Schluenzen, F., Tocilj, A., Zarivach, R., Harms, J., Glueman, M., Janell, D., Bashan, A, Bartels, H., Agmon, I., Franceschi, F. and Yonath, A.: Structure of functionally activated small ribosomal subunit at 3.3 Å resolution, Cell **102** (2000) 615–623
848. Schöll, E.: Diodes (in [878]) 2005
849. Schrieffer, J.R.: *Theory of Superconductivity* (Perseus Books, New York) 1971
850. Schrödinger, E.: Quantisierung als Eigenwertproblem, Ann. der Physik **79** (1926) 361–376; 489–527; 734–756; **80** (1926) 437–490; **81** (1926) 108–139
851. Schrödinger, E.: Contributions to Born's new theory of the electromagnetic field, Proc. Royal Soc. London A **150** (1935) 465–477
852. Schrödinger, E.: Die gegenwärtige Situation der Quantenmechanik, Naturwissenschaften **23** (1935) 807–812, 823–828, 844–849
853. Schrödinger, E.: *What Is Life?* (Cambridge University Press, Cambridge) 1944. (Republished in 1967)
854. Schultz, S.G.: A century of (epithelial) transport physiology: From vitalism to molecular cloning, Am. J. Physiol. – Cell Physiol. **274** (1998) 13–23
855. Schwarzschild, K.: Über das Gravitationsfeld eines Kugel aus inkompressibler Flüssigkeit nach der Einsteinschen Theorie, Sitz. der k. Preussischen Akademie der Wiss. **1** (1916) 424–434
856. Sciama, D.W.: On the origin of inertia, Notices of the Royal Astronomical Society **113** (1953) 34–42
857. Scott, A.C.: Neuristor propagation on a tunnel diode loaded transmission line, Proc. IEEE **51** (1963) 240
858. Scott, A.C.: Distributed device applications of the superconducting tunnel junction, Solid State Electronics **7** (1964) 137–146
859. Scott, A.C.: A nonlinear Klein–Gordon equation, Am. J. Physics **37** 1969 52–61
860. Scott, A.C.: Distributed multimode oscillators of one and two spatial dimensions, Trans. IEEE **CT–17** (1970) 55–60
861. Scott, A.C.: *Active and Nonlinear Wave Propagation in Electronics* (Wiley, New York) 1970
862. Scott, A.C.: Tunnel diode arrays for information processing and storage, Trans. IEEE **SM–1** (1971) 267–275
863. Scott, A.C.: Information processing in dendritic trees, Math. Biosci. **18** (1973) 153–160
864. Scott, A.C.: The electrophysics of a nerve fiber, Rev. Mod. Phys. **11** (1975) 487–553
865. Scott, A.C.: *Neurophysics* (John Wiley & Sons, New York) 1977
866. Scott, A.C.: Neurodynamics (a critical survey), J. Math. Psychol. **15** (1977) 1–45
867. Scott, A.C.: Citation classic – Soliton – New concept in applied science, Current Contents **34** (1979) 162
868. Scott, A.C.: The laser-Raman spectrum of a Davydov soliton, Phys. Lett. A **86** (1981) 60–62
869. Scott, A.C.: Dynamics of Davydov solitons, Phys. Rev. A **26** (1982) 578–595; ibid. **27** 2767
870. Scott, A.C.: Soliton oscillations in DNA, Phys. Rev. A **31** (1985) 3518–3519
871. Scott, A.C.: Soliton in biological molecules, Comments Mol. Cell. Biol. **3** (1985) 5–57

872. Scott, A.C.: Davydov's soliton, Physics Reports **217** (1992) 1–67

873. Scott, A.C.: *Stairway to the Mind* (Springer-Verlag, New York) 1995

874. Scott, A.C.: *Neuroscience: A Mathematical Primer* (Springer-Verlag, New York) 2002

875. Scott, A.C.: *Nonlinear Science: Emergence and Dynamics of Coherent Structures*, 2nd edn. (Oxford University Press, Oxford) 2003

876. Scott, A.C.: Reductionism revisited, Journal of Consciousness Studies **11** (2) (2004) 51–68

877. Scott, A.C.: The development of nonlinear science, La Rivista del Nuovo Cimento **27** (10–11) (2004) 1–115

878. Scott, A.C. (Ed.): *Encyclopedia of Nonlinear Science* (Taylor & Francis, New York) 2005

879. Scott, A.C.: Distributed oscillators (in [878]) 2005

880. Scott, A.C.: Emergence (in [878]) 2005

881. Scott, A.C.: Matter, nonlinear theory of (in [878]) 2005

882. Scott, A.C.: Multiplex neuron (in [878]) 2005

883. Scott, A.C.: Nerve impulses (in [878]) 2005

884. Scott, A.C.: Neuristor (in [878]) 2005

885. Scott, A.C.: Rotating-wave approximation (in [878]) 2005

886. Scott, A.C.: Threshold phenomena (in [878]) 2005

887. Scott, A.C., Bigio, I.J. and Johnson, C.T.: Polarons in acetanilide, Phys. Rev. B **39** (1989) 517–521

888. Scott, A.C., Chu, F.Y.F. and McLaughlin, D.W.: The soliton: A new concept in applied science, Proc. IEEE **61** (1973) 1443–1483

889. Scott, A.C., Chu, F.Y.F. and Reible, S.A.: Magnetic flux propagation on a Josephson transmission line, J. Appl. Phys. (1976) 3272–3286

890. Scott, A.C., Eilbeck, J.C. and Gilhøj, H.: Quantum lattice solitons, Physica D **78** (1994) 194–213

891. Scott, A.C., Gratton, E., Shyansunder, E. and Careri, G.: IR overtone spectrum of the vibrational soliton in crystalline acetanilide, Phys. Rev. B **32** (1985) 5551–5553

892. Scott, A.C. and Johnson, W.J.: Internal flux motion in large Josephson junctions, Appl. Phys. Lett. **14** (1960) 316–318

893. Scott, A.C. and Vota-Pinardi, U.: Velocity variations on unmyelinated axons, J. Theoret. Neurobiol. **1** (1982) 150–172

894. Scott, A.C. and Vota-Pinardi, U.: Pulse code transformations on axonal trees, J. Theoret. Neurobiol. **1** (1982) 173–195

895. Searle, J.R.: *Expression and Meaning: Studies in the Theory of Speech Acts* (Cambridge University Press, Cambridge) 1985

896. Seeger, A., Donth, H. and Kochendörfer, A.: Theorie der Versetzungen in eindimensionalen Atomreihen, Zeit. für Physik **134** (1953) 173–193

897. Shapiro, I.I., Ash, M.E. and Smith, W.B.: Icarus: Further confirmation of the relativistic perihelion precession, Phys. Rev. Lett. **20** (1968) 1517–1518

898. Shapiro, I.I.: A century of relativity, Rev. Mod. Phys. **71** (1999) S41–S53

899. Shelley, M.W.: *Frankenstein, or The Modern Prometheus* (Penguin Classics, New York) 2003. (First published in 1818)

900. Sherman, T.F.: On connecting large vessels to small: The meaning of Murray's law, J. Gen. Physiol. **78** (1981) 431–453

901. Shinozaki, K., Yoda, K, Hozumi, K. and Kira, T.: A quantitative analysis of plant form – The pipe model theory. I. Basic analyses, Nihon Seitai Gakkai shi **14** (1964) 97–105

902. Shinozaki, K., Yoda, K, Hozumi, K. and Kira, T.: A quantitative analysis of plant form – The pipe model theory. II. Further evidence of the theory and its application in forest ecology, Nihon Seitai Gakkai shi **14** (1964) 133–139

903. Shohet, J.L.: Private communication, 2004

904. Shohet, J.L.: Nonlinear plasma waves (in [878]) 2005

905. Shohet, J.L., Barmish, B.R., Ebraheem, H.K. and Scott, A.C.: The sine-Gordon equation in reversed-field pinch experiments, Physics of Plasmas **11** (2004) 3877–3887

906. Sievers, A.J. and Takeno, S.: Intrinsic localized modes in anharmonic crystals, Phys. Rev. Lett. **39** (1988) 3374–3379

907. Skou, J.C.: The influence of some cations on an adenosine triphosphatease from pripheral nerves, Biochim. Biophys. Acta **23** (1957) 394–401

908. Skou, J.C. and Esmann, M.: The Na,K-ATPase, J. Bioenerg. and Biomembr. **24** (1992) 249–261

909. Slater, J.C.: *Quantum Theory of Matter* (McGraw–Hill, New York) 1951

910. Smale, S.: Differentiable dynamical systems, Bull. Amer. Math. Soc. **73** (1967) 747–817

911. Smalheiser, N.R.: Walter Pitts, Perspectives in Biology and Medicine **43** (2) (2000) 217–226

912. Smil, V.: Global warming (in [878]) 2005

913. Smith, D.O.: Morphological aspects of the safety factor for action potential propagation at axon branch points in the crayfish, J. Physiol. (London) **301** (1980) 261–269

914. Smith, P.J.C. and Arnott, A.: LALS: A linked-atom least squares reciprocal space refinement system incorporating stereochemical restraints to supplement sparse diffraction data, Acta Cryst. A **34** (1978) 3–11

915. Smolin, L.: *The Trouble with Physics: The Rise of String Theory, the Fall of Science, and What Comes Next* (Houghton Mifflin, Boston) 2006

916. Snow, C.P.: *The Two Cultures* (Cambridge University Press, Cambridge) 1993. (First published in 1959)

917. Sobell, H.M.: Actinomycin and DNA transcription, Proc. Natl. Acad. Sci. USA **82** (1985) 5328–5331

918. Sobell, H.M.: Kink–antikink bound states in DNA structure. In: *Biological Macromolecules and Assemblies*, Vol. 2: *Nucleic Acids and Interactive Proteins*, ed. by F.A. Jurnak and A. McPherson (John Wiley & Sons, New York 1985) 171–232

919. Sobell, H.M.: DNA premelting (in [878]) 2005

920. Sobell, H.M.: personal communication, 2006

921. Socquet-Juglard, H., Dysthe, K., Trulsen, K., Krogstad, H.E. and Liu, J.: Probability distributions of surface gravity waves during spectral changes, J. Fluid Mech. **542** (2005) 195–216

922. Solari, H.G. and Natiello, M.: Lasers (in [878]) 2005

923. Sørensen, M.P.: Perturbation theory (in [878]) 2005

924. Stapp, H.P.: *Mind, Matter, and Quantum Mechanics* (Springer-Verlag, Berlin) 1993

925. Stephenson, S.L. and Stempen, H.: *Myxomycetes: A Handbook of Slime Molds* (Timber Press, Portland, Oregon) 1994

926. Stern, M.D.: Theory of excitation–contraction coupling in cardiac muscle, Biophys. J. **63** (1992) 497–517

927. Steuerwald, R.: *Über Enneper'sche Flächen und Bäcklund'sche Transformation* (Bayerischen Akad. Wiss., München) 1936

928. Stewart, I.: Catastrophe theory in physics, Rep. Prog. Phys. **45** (1982) 185–221

929. Stoeger, W.R.: Private communication, 4 May 2006

930. Stokes, D.L. and Green, N.M.: Structure and funcntion of the calcium pump, Annu. Rev. Biophys. Biomol. Struct. **32** (2003) 445–468

931. Stokes, G.G.: *Mathematical and Physical Papers* (Cambridge University, Cambridge) 1880

932. Strandl, S.: *The History of the Machine* (Dorset Press, New York) 1979

933. Strogatz, S.H.: *Nonlinear Dynamics and Chaos* (Addison-Wesley, Reading, Ma.) 1994

934. Strogatz, S.H.: *Synch: How Order Emerges from Chaos in the Universe, Nature, and Daily Life* (Hyperion, New York) 2003

935. Stryer, L.: *Biochemistry*, 4th edn. (W.H. Freeman, New York) 1995

936. Stuart, G. and Sakmann, B.: Active propagation of somatic action potentials into neocortical pyramidal cell dendrites, Nature **367** (1994) 69–72

937. Stuart, G., Spruston, N. and Häusser, M.: *Dendrites* (Oxford University Press, Oxford) 1999

938. Süli, Á., Dvorak, R. and Freistetter, F: The stability of the terrestrial planets with a more massive 'Earth', Mon. Not. R. Astron. Soc. **363** (2005) 241–250

939. Sutcliffe, P.: Skyrmions (in [878]) 2005

940. Svendsen, I.A. and Buhr Hansen, J.: On the deformation of periodic long waves over a gently sloping bottom, J. Fluid Mech. **87** (1978) 433–448

941. Swain, J.D.: Tensors (in [878]) 2005

942. Takeno, S. and Homma, S.: Topological solitons and modulated structure of bases in DNA double helices – A dynamic plane base-rotator model, Prog. Theor. Phys. **70** (1987) 308–311

943. Takhtajan, L.A. and Faddeev, L.D.: Essentially nonlinear one-dimensional model of classical field theory, Theor. Math. Phys. **21** (1974) 1046–1057

944. Tanuiti, T. and Washimi, H.: Self trapping and instability of hydrodynamic waves along the magnetic field in a cold plasma, Phys. Rev. Lett. **21** (1968) 209–212

945. Tappert, F. and Varma, C.M.: Asymptotic theory of self trapping of heat pulses in solids, Phys. Rev. Lett. **25** (1970) 1108–1111

946. Tasaki, I., Ishii, K. and Ito, H.: On the relation between the conduction-rate, the fibre-diameter and the internodal distance of the medullated nerve fibre, Jpn. J. Med. Sci. **9** (1944) 189–199

947. Tayfun, M.A.: Narrow-band nonlinear sea waves, J. Geophys. Res. **85** (1980) 1548–1552

948. Taylor, G.I.: The formation of a blast wave from a very intense explosion. I. theoretical discussion. II. The atomic explosion of 1945, Proc. Roy. Soc. (London) A **201** (1950) 159–186

949. Taylor, G.I.: The instability of liquid surfaces when accelerated in a direction perpendicular to their planes, Proc. Roy. Soc. (London) A **201** (1950) 192–196

950. Taylor, J.B.: Unpublished 1968 report CLM-PR-12 at Culham Laboratories (England), referenced in [374] and noted by J.D. Meiss (personal communication)

951. Taylor, R.P.: Lévy flights (in [878]) 2005

952. Taylor, R.P., Newbury, R., Micolich, A.P., Fromhold, T.M., Linke, H., Davies, A.G. and Martin, T.P.: A review of fractal conductance fluctuations in ballistic semiconductor devices. In: *Electron Transport in Quantum Dots*, ed. by J.P. Bird (Kluwer Academic/Plenum, New York) 2003

953. Taylor, R.P., Spehar, B., Wise, J.A., Clifford, C.W.G., Newell, B.R., Hägerhäll, C.M., Purcell, T. and Martin, T.P.: Perceptual and physiological response to the visual complexity of fractals, Journal of Nonlinear Dynamics, Psychology, and Life Sciences **9** (2005) 89–114

954. Tegmark, M.: The importance of quantum decoherence in brain processes, Phys. Rev. E **61** (2000) 4194–4206

955. Tegmark, M.: Measuring spacetime: From the Big Bang to black holes, Science **296** (2002) 1427–1433

956. Tegmark, M.: Parallel universes, Scientific American (May 2003) 41–51

957. Ter Haar, D.: Talk at the Technical University of Denmark, Lyngby, 1990

958. Thacker, H.B.: Exact integrability in quantum field theory and statistical systems, Rev. Mod. Phys. **53** (1981) 253–285

959. Thom, R.: *Structural Stability and Morphogenesis an Outline of a General Theory of Models* (Addison-Wesley, Reading, Ma.) 1989. (First published in 1972)

960. Thoma, F., Koller, T. and Klug, A.: Involvement of histone H1 in the organization of the nucleosome and of the salt-dependent superstructures of chromatin, J. Cell. Biol. **83** (1979) 403–427

961. Thompson, D.W.: *On Growth and Form*, abridged edn. (Cambridge University Press, Cambridge) 1961. (First published in 1917 with an enlarged edition in 1942)

962. Thompson, J.M.T. and Stewarty, H.B.: *Nonlinear Dynamics and Chaos: Geometrical Methods for Engineers and Scientists*, 2nd edn. (Wiley, New York) 2002

963. Thompson, W.R. and Heron, W.: The effects of restricting experience on the problem-solving capacity of dogs, Can. J. Psychol. **8** (1954) 17–31

964. Thorne, K.S.: *Black Holes and Time Warps: Einstein's Outrageous Legacy* (W.W. Norton, New York) 1994

965. Tien, H.T. and Ottova, A.L.: The lipid bilayer concept and its experimental realization: From soap bubbles, the kitchen sink, to bilayer lipid membranes, J. Membr. Sci. **189** (2001) 83–117

966. Tinkham, M.: *Introduction to Superconductivity* (McGraw-Hill, New York) 1996

967. Tobias, D.J. and Freites, J.A.: Molecular dynamics (in [878]) 2005

968. Toda, M.: Vibration of a chain with nonlinear interactions, J. Phys. Soc. Japan **22** (1967) 431–436

969. Toda, M.: *Theory of Nonlinear Lattices* (Springer-Verlag, Berlin) 1981

970. Tropp, H.S.: The Origins and History of the Fields Medal, Historia Mathematica **3** (1976) 167–181

971. Turin, L.: Personal communication, December 2004

972. Turing, A.: Can a machine think? Mind **59** (1950) 433–460

973. Turing, A.: The chemical basis of morphogenesis, Philos. Trans. Roy. Soc. B **237** (1952) 37–72

974. Tuszyński, J.A., Paul, R., Chatterjee, R., and Sreenivasan, S.R.: Relationship between Fröhlich and Davydov models of biological order, Phys. Rev. A **30** (1984) 2666–2675

975. Tuszyński, J.A.: Critical phenomena (in [878]) 2005

976. Tuszyński, J.A.: Ferromagnetism and ferroelectricity (in [878]) 2005

977. Tuszyński, J.A.: Fröhlich theory (in [878]) 2005

978. Tuszyński, J.A.: Renormalization groups (in [878]) 2005

979. Tuszyński, J.A.: Scheibe aggregates (in [878]) 2005

980. Ueda, Y.: *The Road to Chaos* (Aerial Press, Santa Cruz) 1992

981. Vale, R.D. and Milligan, R.A.: The way things move: Looking under the hood of molecular motor proteins, Science **288** (2000) 88–95

982. Van der Pol, B.: On relaxation oscillations, Philos. Mag. **2** (1926) 978–992

983. Van der Pol, B.: The nonlinear theory of electric oscillations, Proc. IRE **22** (1934) 1051–1086

984. Van der Pol, B.: On a generalization of the non-linear differential equation $d^2u/dt^2 - \varepsilon(1 - u^2)du/dt + u = 0$, Proc. Acad. Sci. Amsterdam A **60** (1957) 477–480

985. Van der Pol, B. and Van der Mark, J.: Frequency demultiplication, Nature **120** (1927) 363–364

986. Van der Pol, B. and Van der Mark, J.: The heartbeat considered as a relaxation oscillation, and an electric model of the heart, Philos. Mag. **6** (1928) 763–765

987. Van der Heijden, G.: Butterfly effect (in [878]) 2005

988. Van Gulick, V.: Reduction, emergence and other recent options on the mind/body problem, Journal of Consciousness Studies **8** (9–10) (2001) 1–34

989. Van Zandt, L.L.: Resonant misrowave absorption by dissolved DNA, Phys. Rev. Lett. **57** (1986) 2085–2087

990. Varela, F.J., Maturana, H.R. and Uribe, R.: Autopoiesis: The organization of living systems, its characterization and a model, Biosystems **5** (1974) 187–196

991. Vázquez, L., Pascual, P. and Jiménez, S.: Charge density waves (in [878]) 2005

992. Vázquez, L. and Zorano, M.P.: FitzHugh–Nagumo equation (in [878]) 2005

993. Verhulst, P.F.: Recherches mathématiques sur la loi d'accroissement de la population, Nouv. Mém. de l'Academie Royale des Sci. et Belles-Lettres de Bruxelles **18** (1845) 1–41

994. Veselov, A.P.: Huygens principle (in [878]) 2005

995. Vessot, R.F.C., Levine, M.W., Mattison, E.M., Blomberg, E.L., Hoffman, T.E., Nystrom, G.U., Farrel, B.F., Decher, R., Eby, P.B., Baugher, C.R., Watts, J.W., Teuber, D.L. and Wills, F.D.: Test of relativistic gravitation with a space-borne hydrogen maser, Phys. Rev. Lett. **45** (1980) 2081–2084

996. Vol'kenshtein, M.V.: The conformon, J. Theor. Biol. **34** (1972) 193–195

997. Volterra, V.: Principes de biologie mathématique, Acta Biotheoretica **3** (1937) 1–36

998. Von Bertalanffy, L.: Untersuchungen über die Gesetzlichkeit des Wachstums. I. Allgemeine Grundlagen der Theorie; Mathematische und physiologische

Gesetzlichkeiten des Wachstums bei Wassertieren. Arch. Entwicklungsmech. **131** (1934) 613–652

999. Von Bertalanffy, L.: Zu einer allgemeinen Systemlehre, Biologia Generalis **195** (1949) 114–129

1000. Von Bertalanffy, L.: The concepts of systems in physics and biology, Bulletin of the British Society for the History of Science **1** (1949) 44–45

1001. Von Bertalanffy, L.: The theory of open systems in physics and biology, Science **111** (1950) 23–29

1002. Von Bertalanffy, L.: *General System Theory: Foundations, Development* (George Braziller, New York) 1968

1003. Von Bertalanffy, L.: *Perspectives on General Systems Theory* (George Braziller, New York) 1975

1004. Von Hippel, A.R.: *Dielectrics and Waves* (Wiley, New York) 1954

1005. Von Senden, M.: *Space and Sight: The Perception of Space and Shape in the Congenitally Blind Before and After Operation* (Methuen & Co., London) 1960 (a republication of *Raum-und Gestaltauffassung bei Operierten vor und nach der Operation*, Barth, Leipzig, 1932)

1006. Von Steinbüchel, Wittman, M. and Szelag, E.: Temporal constraints of perceiving, generating, and integrating information: Clinical indications, Restorative Neurology and Neuroscience **14** (1999) 167–182

1007. Voorhees, B.H.: Axiomatic theory of hierarchical systems, Behav. Sci. **28** (1983) 24–34

1008. Wadati, M.: Quantum inverse scattering method (in [878]) 2005

1009. Waddington, C.H.: *Organizers and Genes* (Cambridge University Press, Cambridge) 1940

1010. Walker, A.G.: On Milne's theory of world-structure, Proc. Lond. Math. Soc. **42** (1936) 90–127

1011. Walker, G.H. and Ford, J.: Amplitude instability and ergodic behaviour for conservative nonlinear oscillator systems, Phys. Rev. **188** (1969) 416–432

1012. Washimi, H. and Tanuiti, T.: Propagation of ion acoustic solitary waves of small amplitude, Phys. Rev. Lett. **17** (1966) 996–998

1013. Watson, J.D.: *The Double Helix: A Personal Account of the Discovery of the Structure of DNA* (Touchstone, New York) 1968

1014. Waxman, S.G.: Regional differentiation of the axon, a review with special reference to the concept of the multiplex neuron, Brain Res. **47** (1972) 269–288

1015. Webb, S.J.: Laser-Raman spectroscopy of living cells, Physics Reports **60** (1980) 201–224

1016. Weber, J.: Detection and generation of gravitational waves, Phys. Rev. **117** (1960) 306–313

1017. Weber, J.: Observation of the thermal fluctuations of a gravitational-wave detector, Phys. Rev. Lett. **17** (1966) 1228–1230

1018. Weber, J.: Evidence for discovery of gravitational radiation, Phys. Rev. Lett. **22** (1966) 1320–1324

1019. Weber, M.J.: *Handbook of Laser Wavelengths* (CRC Press, New York) 1999

1020. Wehr, M. and Laurent, G.: Odor encoding by temporal sequences of firing in oscillating neural assemblies, Nature **384** (1996) 162–166

1021. Weibel, E.R. and Gomez, D.M.: Architecture of the human lung, Science **137** (1962) 577–585

1022. Weinberg, S.: *Gravitation and Cosmology: Principles and Applications of the General Theory of Relativity* (John Wiley & Sons, New York) 1972

1023. Weinberg, S.: Newtonianism, reductionism and the art of congressional testimony, Nature **330** (1987) 433–437

1024. Weinberg, S.: The limits of reductionism, Nature **331** (1988) 475–476

1025. Weinberg, S.: Precision test of quantum mechanics, Phys. Rev. Lett. **62** (1989) 485–488

1026. Weinberg, S.: *Dreams of a Final Theory: The Search for the Fundamental Laws of Nature* (Pantheon Books, New York) 1992

1027. Weinberg, S.: Reductionism redux, The New York Review of Books **42** (15) 5 October 1995

1028. Weiss, M.T.: Quantum derivation of energy relations analogous to those for nonlinear reactances, Proc. IRE **45** (1957) 1012–1013

1029. West, B.J.: Branching laws (in [878]) 2005

1030. West, B.J. and Deering, W.: Fractal physiology for physicists: Lévy statistics, Phys. Rep. **246** (1994) 1–100

1031. Weyl, H.: Feld und Materie, Ann. der Physik **65** (1921) 541–563

1032. Wheeler, J.A.: Assessment of Everett's "relative state" formulation of quantum theory, Rev. Mod. Phys. **29** (1957) 463–465

1033. Whitham, G.B.: *Linear and Nonlinear Waves* (Wiley, New York) 1974

1034. Whittaker, E.T.: *A Treatise on the Analytical Dynamics of Particles and Rigid Bodies*, 4th edn. (Cambridge University Press, Cambridge) 1937

1035. Widrow, B. and Angell, J.B.: Reliable, trainable networks for computing and control, Aerospace Eng. (September, 1962) 78–123

1036. Wiener, N.: *Cybernetics: Control and Communication in the Animal and the Machine*, 2nd edn. (Wiley, New York) 1961. (First published in 1948)

1037. Wiener, N.: Nonlinear prediction and dynamics, Proc. Third Berkeley Symp. **3** (1956) 247–252

1038. Wiener, N.: *Nonlinear Problems in Random Theory* (Wiley, New York) 1958

1039. Wiggins, S.: *Global Bifurcations and Chaos* (Springer-Verlag, New York) 1988

1040. Wilhelmsson, H.: Alfvén waves (in [878]) 2005

1041. Will, C.M.: *Was Einstein Right? Putting General Relativity to the Test* (Basic Books, New York) 1986

1042. Williams, R.J.: *Biochemical Individuality* (U. of Texas Press, Austin) 1956

1043. Wilson, H.R.: *Spikes, Decisions, and Actions: The Dynamical Foundations of Neuroscience* (Oxford University Press, Oxford) 1999

1044. Wilson, H.R. and Cowan, J.D.: Excitatory and inhibitory interactions in localized populations of model neurons, Biophys. J. **12** (1972) 1–24

1045. Wilson, H.R. and Cowan, J.D.: A mathematical theory of the functional dynamics of cortical and thalamic nervous tissue, Kybernetik **13** (1973) 55–80

1046. Wilson, M.A. and McNaughton, B.L.: Dynamics of the hippocampal ensemble code for space, Science **261** (1993) 1055–1058

1047. Wilson, T.A.: Design of the bronchial tree, Nature **213** (1967) 668–669

1048. Winfree, A.T.: The prehistory of the Belousov–Zhabotinsky oscillator, Journal of Chemical Education **61** (1984) 661–663

1049. Winfree, A.T.: *When Time Breaks Down: The Three-Dimensional Dynamics of Electrochemical Waves and Cardiac Arrhythmias* (Princeton University Press, Princeton) 1987

1050. Winfree, A.T.: *The Geometry of Biological Time* (Springer-Verlag, New York) 2001
1051. Winfree, A.T.: Dimensional analysis (in [878]) 2005
1052. Wisdom, J.: Chaotic motion in the solar system, Icarus **72** (1987) 241–275
1053. Woit, P.: *Not Even Wrong: The Failure of String Theory and the Search for Unity in Natural Law* (Basic Books, New York) 2006
1054. Wojtkowkski, M.P.: Lyapunov exponents (in [878]) 2005
1055. Woodcock, C.L.: A milestone in the odyssey of higher-order chromatin structure, Nature Structural & Molecular Biology **12** (2005) 639–640
1056. Worthington, A.M.: *A Study of Splashes* (MacMillan, New York) 1963
1057. Worthington, A.M. and Cole, R.S.: Impact with a liquid surface, studied by the aid of instantaneous photography, Phil. Trans. Roy. Soc. A **189** (1897) 137–148
1058. Wu, J.Y., Cohen, L.B. and Falk, C.X.: Neuronal activity during different behaviors. In: *Aplysia*: A distributed organization? Science **263** (1994) 820–823
1059. Wyman, J.: The turning wheel: A study in steady states, Proc. Natl. Acad. Sci. USA **72** (1975) 3983–3987
1060. Xie, A., Van Der Meer, L. Hoff, W. and Austin, R.H.: Long-lived Amide-I vibrational modes in myoglobin, Phys. Rev. Lett. **84** (2000) 5435–5438
1061. Yakushevich, L.V.: *Nonlinear Physics of DNA*, 2nd edn. (Wiley-VCH, Weinheim, Germany) 2004. (The first edition was published in 1998)
1062. Yakushevich, L.V.: DNA solitons (in [878]) 2005
1063. Yarin, A.L.: Drop impact dynamics: Splashing, spreading, receding, bouncing ..., Annu. Rev. Fluid Mech. **38** (2006) 159–192
1064. Yomosa, S.: Soliton excitations in deoxyribonucleic acid (DNA) double helices, Phys. Rev. A **27** (1903) 2120–2125
1065. Yomosa, S.: Solitary excitations in deoxyribonucleic acid (DNA) double helices, Phys. Rev. A **30** (1903) 474–480
1066. Young, J.Z.: Structure of nerve fibers and synapses in some invertebrates, Cold Spring Harbor Symp. Quant. Biol. **4** (1936) 1–6
1067. Young, L.S.: Horseshoes and hyperbolicity in dynamical systems (in [878]) 2005
1068. Yukalov, V.I.: Bose–Einstein condensation (in [878]) 2005
1069. Zabusky, N.J. (Ed.): *Topics in Nonlinear Physics: Proceedings of the Physics Session* (Springer-Verlag, New York) 1967
1070. Zabusky, N.J.: Solitons and bound states of the time independent Schrödinger equation, Phys. Rev. **168** (1968) 124–128
1071. Zabusky, N.J.: Personal communication, January 2005
1072. Zabusky, N.J.: Fermi–Pasta–Ulam, solitons and the fabric of nonlinear and computational science: History, synergetics, and visiometrics, Chaos **15** (2005) (in press)
1073. Zabusky, N.J. and Galvin, C.J.: Shallow-water waves, the Korteweg–de Vries equation and solitons, J. Fluid Mech. **47** (1971) 811–824
1074. Zabusky, N.J. and Kruskal, M.D.: Interaction of solitons in a collisionless plasma and the recurrence of initial states, Phys. Rev. Lett. **15** (1965) 240–243
1075. Zaikin, A.N. and Zhabotinsky, A.M.: Concentration wave propagation in two-dimensional liquid-phase self-oscillating systems, Nature **225** (1970) 535–537

1076. Zakharov, V.E.: Collapse of Langmuir waves, Soviet Phys. JETP **35** (1972) 908–914

1077. Zakharov, V.E., Dyachenko, A.I. and Vasilyev, O.A.: New method for numerical simulation of a nonstationary potential flow of incompressible fluid with a free surface, Eur. J. Mech. B (Fluids) **21** (2002) 283–291

1078. Zakharov, V.E., Manakov, S.P., Novikov, S.P. and Pitaevskii, L.P.: *Theory of Solitons* (Consultants Bureau, New York) 1984

1079. Zakharov, V.E. and Shabat, A.B.: Exact theory of two-dimensional self-modulation of waves in nonlinear media, Soviet Phys. JETP **34** (1972) 62–69

1080. Zalta, E.N.: *Stanford Encyclopedia of Philosophy*, available at the website `http://plato.stanford.edu/`

1081. Zeeman, E.C.: Differential equations for the heartbeat and nerve impulse, *Dynamical Systems*, (Proc. Sympos. Univ. Bahia, Salvador, 1971) (Academic Press, New York) 1977 683–741

1082. Zeeman, E.C.: *Catastrophe Theory: Selected Papers (1972–1977)* (Addison-Wesley, Reading, Ma.) 1977

1083. Zeldovich, Y.B. and Frank-kamenetskii, D.A.: On the theory of uniform flame propagation, Doklady Akademii Nauk SSSR **19** (9) (1938) 693–697

1084. Zhabotinsky, A.M.: Periodic movement in the oxidation of malonic acid in solution (Investigation of the kinetics of Belousov's reaction), in Russian, Biofizika **9** (1964) 306–311

1085. Zhang, C.T.: Soliton excitations in deoxyribonucleic acid (DNA) double helices, Phys. Rev. A **35** (1987) 886–891

1086. Zhang, Z., Devarajan, P., Dorfman, A.L. and Morrow, J.S.: Structure of the ankyrin-binding domain of alpha-Na,K-ATPase, J. Biol. Chem. **273** (1998) 18681–18684

1087. Zolotaryuk, A.V.: Polarons (in [878]) 2005

THE FRONTIERS COLLECTION

Series Editors:
A.C. Elitzur M.P. Silverman J. Tuszynski R. Vaas H.D. Zeh